George Henry Elliott

European Light-House Systems

Being a Report of a Tour of Inspection Made in 1873

George Henry Elliott

European Light-House Systems
Being a Report of a Tour of Inspection Made in 1873

ISBN/EAN: 9783337152178

Printed in Europe, USA, Canada, Australia, Japan

Cover: Foto ©berggeist007 / pixelio.de

More available books at **www.hansebooks.com**

EUROPEAN LIGHT-HOUSE SYSTEMS;

BEING A

REPORT OF A TOUR OF INSPECTION

MADE IN 1873,

BY

MAJOR GEORGE H. ELLIOT,
Corps of Engineers, U.S.A., Member and Engineer-Secretary of the Light-House Board,

Under the Authority of
Hon. WILLIAM A. RICHARDSON,
Secretary of the Treasury.

LONDON:
LOCKWOOD & CO., 7 STATIONERS'-HALL COURT,
LUDGATE HILL.
1875.

TABLE OF CONTENTS.

	Page.
Extract from Light-House Board Report for 1873	9
Outward voyage, and deficiency in fog-signals on transatlantic steamers	15
The Trinity House, London	18
Fog-signal experiments near Dover, under direction of Professor Tyndall	22
Extracts from Professor Tyndall's report	25
Sir Frederick Arrow's remarks on Professor Tyndall's report	59
South Foreland electric lights	66
Roman pharos in Dover Castle	70
Examination of the Doty lamp	74
Experiments with lights in Westminster clock-tower	75
Trinity-House depot at Blackwall	76
Improvements in lamps for light-houses	80
Inspection-tour of the North Sea lights of England in the Trinity House steam-yacht Vestal	90
Iron light-houses off the mouth of the Thames	99
Orfordness lights	101
Depot at Yarmouth	101
Haisborough lights	104
Experiments at night with Haisborough lights	109
Newarp light-ship	113
Cockle light-ship	115
Spurn Point lights	115
Flamborough Head light	117
Whitby lights	118
Souter Point light	125
Coquet light	127
Inner Farne Island lights	130
Longstone light	130
Return to London	132
Second tour of inspection.—the southwest coast of England	132
Light-house depot at Isle of Wight	132
Saint Catherine light	133
The Needles light	134
Lights of the Bill of Portland	135
The Start light	135
The Eddystone light	136
Saint Anthony light	137
Plymouth Breakwater light	137
The Lizard lights	139
The Wolf light	140
Runnlestone bell-buoy	141
Seven Stones light-ship	145
Longships light	145
Godrevy light	149
"The Stones" buoy, off Godrevy	149

CONTENTS.

	Page.
Holyhead light	149
North Stack fog-signal station	155
South Stack light	156
Visit to Commissioners of Irish Lights, Dublin	158
Howth Baily light	159
Wicklow Head light	160
Gas-apparatus for fixed lights	161
Gas-apparatus for intermittent lights	162
Gas-apparatus for revolving lights	165
Gas-apparatus for group flashing lights	165
Gas-apparatus for triform fixed lights	166
Gas-apparatus for triform intermittent lights	167
Gas-apparatus for triform revolving lights	167
Gas-apparatus for triform group flashing lights	167
Experiments with the triform gas-light	167
Illuminating powers of gas-lights	169
Cost of gas-light apparatus	170
Cost of gas-light apparatus, (triform)	171
Cost of maintenance of gas-lights	171
Illumination of beacons by gas	172
Wigham's gas-gun	174
Visit to Commissioners of Northern Lights, Edinburgh	175
Stevenson's holophone	179
Visit to lens manufactory of Chance, Brothers & Co	183
Visit to Commission des Phares, Paris	184
Dépôt des Phares of France	185
M. Reynaud's observations concerning mineral oil	193
Lens-makers of Paris	203
Lepaute's observations concerning light-house burners	208
Light-houses at the mouth of the Seine	215
Phare de l'Hôpital	217
Feu de port at Honfleur	218
Phare de Fatouville	219
Phares de la Hève	220
Description of the electric lights at La Hève	226
International exhibition at Vienna	249
Submarine foundations for harbor light-houses	250
Models of light-houses	250
Light-house apparatus	251
Iron tower for fourth-order light	254
Swedish light-houses	257
Austrian fog-horn	261
Osnaghi's reflectors	262
Return voyage	266

LIST OF ILLUSTRATIONS.

PLATES.

	Page.
Frontispiece: The Wolf Rock light-house	1
I. South Foreland, general plan of light-houses	66
II. South Foreland, ground-plan of engine-house	68
III. South Foreland, lantern and lens at	70
IV. Six-wick Trinity House burner	76
V. Maplin Sand light-house	100
VI. Haisborough, general plan	105
VII. Upper part of English light-house tower	108
VIII. Ventilating window for light-house tower	110
IX. Souter Point, chart of vicinity of	120
X. Souter Point, plan and details of low light-room at	122
XI. Souter Point, plan of machine-room	122
XII. Souter Point, general plan	124
XIII. Souter point, ground-plan	124
XIV. Souter Point, section of lantern, lens, and low light-room	126
XV. Souter Point, east elevation	126
XVI. Coquet Island, chart of vicinity of	128
XVII. The Longstone light-house	130
XVIII. The Eddystone light-house	136
XIX. The Irish gas-light for light-houses—108 jets	138
XX. Land's End, chart of vicinity of	142
XXI. Comparative sections of rock light-houses	144
XXII. Holmes's fog-horn apparatus	146
XXIII. South Stack fog or "occasional" light	156
XXIV. Howth Baily gas light-house	158
XXV. Howth Baily gas light-house	160
XXVI. Wicklow Head gas light-house	162
XXVII. One-hundred-and-eight-jet burner	164
XXVIII. Twenty-eight-jet burner	168
XXIX. Triform burners	172
XXX. Triform lenses	174
XXXI. Swedish light-ship	204
XXXII. Revolving catadioptric apparatus for light-ships	206
XXXIII. Burners of French light-house lamps	208
XXXIV. Interior adjustments of level in mineral-oil light-house lamps	214
XXXV. Exterior adjustments of level in mineral-oil light-house lamps	212
XXXVI. Mineral-oil lamp of 1845, Lepaute's gas-burner, Doty lamp	214
XXXVII. La Hève, elevation and general plan	220
XXXVIII. La Hève, ground plan of machine-room	226
XXXIX. La Hève, section of machine-room	230
XL. La Hève, front elevation of magneto-electric machine	232
XLI. La Hève, side elevation of magneto-electric machine	234
XLII. La Hève, plan of magneto-electric machine	238
XLIII. La Hève, details of magneto-electric machine	240

		Page.
XLIV.	La Hève, switches	242
XLV.	La Hève, regulators	246
XLVI.	Lanterns and lenses for electric and oil lights at Cape Grisnez	248
XLVII.	Submarine foundation for harbor light-houses	250
XLVIII.	Lens for electric light	254
XLIX.	Iron tower for fourth-order light	256
L.	Austrian fog-trumpet	260

FIGURES.

No.		
1.	Roman pharos in Dover Castle	73
2.	Douglass gas-burner	79
3.	Five-gallon oil-can	91
4.	Chimney-gauge	92
5.	Buoy-finder at Yarmouth	102
6.	Bilge-keels of light-ships	104
7.	Red cut, Spurn Point	116
8.	Fastening for red panes, Whitby	119
9.	Lamp-guard	119
10.	General view of establishment at Souter Point	120
11.	Lens and lanterns, Souter Point	124
12.	Low light-room, Coquet Island	128
13.	Filling oil-butts at Saint Catherine's	134
14.	Red light, Plymouth Breakwater	138
15.	Rundlestone bell-buoy, off Land's End	145
16.	Three-wick lamp at Holyhead	150
17.	English and American lanterns	154
18.	Faraday's wind-guard, North Stack	155
19.	Diagram illustrating revolving intermittent gas-lights	163
20.	Stevenson's holophone, vertical section	180
21.	Stevenson's holophone, front elevation	180
22.	Mineral-oil test, (areometer)	184
23.	Mineral-oil test, (flashing point)	184
24.	Sautter's lantern for electric light	207
25.	La Hève, plan of electric light-room	222
26.	La Hève, vertical section of electric light-room	223
27.	La Hève, lens for electric light	224
28.	Swedish light-house, details	257
29.	Swedish light-house, details	257
30.	Osnaghi's reflector for flashing light	262
31.	Osnaghi's reflector for fixed light	263

PUBLISHER'S PREFACE.

THE very important and interesting information in regard to the present condition of European seacoast illumination, and its cognate aids to commerce, which is found in the following pages, was collected by Major Elliot, of the United States Engineers, in 1873, and appearing soon afterwards in the form of a report to the Lighthouse Board, it attracted much attention and commendation.

It will be observed that Major Elliot had extraordinary and excellent facilities, and that he made the most of them in obtaining a knowledge of the immense improvements which, in later years, have been made in one of the most important of Governmental services.

His report, which may be considered to be a complete exhibit of the state of the lighthouse service at the present time, has not been accessible to the public, and in order to satisfy the demand both in this country and in Europe, from engineers, scientists, persons interested in commerce, and general readers, the Publisher issues the present edition.

NEW YORK, JUNE, 1875.

EXTRACT FROM THE REPORT OF THE LIGHT-HOUSE BOARD TO THE SECRETARY OF THE TREASURY FOR THE FISCAL YEAR ENDING 30TH JUNE, 1873.

The Light-House Board, during the past year, desirous of acquainting itself minutely with any improvements which of late years may have been introduced into the light-house service in Europe, obtained the sanction of the honorable the Secretary of the Treasury to commission Major Elliot, of the Corps of Engineers of the Army, and engineer-secretary of the Board, to visit Europe and report upon everything which he might observe relative to light-house apparatus and the management of light-house systems. He has lately returned, after having gathered information which will prove of importance in its application in our country, as will be evident from his preliminary report.

Major Elliot was everywhere received with marked cordiality, and every facility was given him to inspect the various coasts and systems of administration, of which full information was furnished him, together with the drawings and models necessary for a perfect acquaintance with the latest improvements which have been adopted in Great Britain and on the continent.

The special thanks of the Board are due to His Royal Highness the Duke of Edinburgh, the master; to Sir Frederick Arrow, the Deputy Master, and the Elder Brethren of Trinity House, for the warmth of their reception and the marked distinction they conferred upon him as the representative of the Board; and to M. Léonce Reynaud, Inspector-general of the *Corps des Ponts et Chaussées* and Director of the French light-house service, for his efforts to make the visit of Major Elliot profitable to his country and agreeable to himself.

"TREASURY DEPARTMENT,
"OFFICE OF THE LIGHT-HOUSE BOARD,
"*Washington, September 17, 1873.*

"Professor JOSEPH HENRY, L. L. D., *Chairman:*

"I have the honor to make a preliminary report of my journey of inspection of the light-house establishments of Europe, which I have made by direction of the Board, with the approval of the honorable the Secretary of the Treasury, and from which I returned a few days since.

"I sailed on the 30th of April, and, after a pleasant voyage, reached Liverpool on the 10th of May, observing *en route* the light-houses on the Irish coast, and the light-ships and buoys on the approaches to Liverpool.

"On the 13th of May I arrived at London, and was cordially received by Sir Frederick Arrow, the Deputy Master, and the Elder Brethren of the Trinity House, which has charge of the lights of England and a supervisory control over those of Scotland and Ireland. The Trinity House kindly offered me an opportunity of making a tour of inspection of the light-houses, &c., on the coasts of England in the steamers which were about to take the annual supplies, and at the first session after my arrival a delegation of the elder brethren was appointed to accompany me.

"I remained in London some weeks to take advantage of this opportunity, and in the mean while my time was occupied in inspecting the depots, lamp-shops, photometric test-rooms, &c., belonging to the Trinity House; also plans of light-houses, lenses, and other optical apparatus used on the coasts of Great Britain.

"In company with Professor Tyndall, the scientific adviser, some of the Elder Brethren and the engineer of the Trinity House, I visited Dover to attend the inauguration of fog-signal experiments, which, under Professor Tyndall's direction, are now being carried on at an experimental station on the cliffs near the great electric lights of South Foreland. The light-house authorities of Great Britain are fully alive to the necessity for powerful fog-signals, and are anxiously seeking to find the best machine, not only to inform the mariner enveloped in fog of his approach to the coast, but, by distinguishing characteristics of sound, to indicate to him on what part of the coast he is. The Board will remember that Sir Frederick Arrow and Captain Webb, of the Elder Brethren, visited the United States during the summer of last year to be present at some of our experiments with the steam-whistle, the horn, and the siren at Portland Harbor. I think the Board has been impressed that on coasts where fog is habitual, as those of New England, California, and some of the great lakes, fog-signals are fully as important as lights, and the English seem to be approaching a similar conclusion.

"Professor Tyndall told me that he intends to make an exhaustive series of experiments with all fog-signals now in use, to determine the best. Both he and the Elder Brethren are especially pleased at the action of our Board in sending an American siren for use in the Dover experiments. In these experiments the signals are observed from several vessels cruising in the Straits of Dover, at different distances from the fog-signal station and under varying conditions of wind and weather. The signals tested when I was at Dover were a steam-whistle, an air-whistle, and a trumpet, very much like the American invention of Daboll, but patented by Professor Holmes, and in use at several English light-stations. The experimenters have since included a cannon and our own siren. The experiments are not yet concluded, and Sir Frederick kindly promised to inform me of the results.

"The delay in London gave me an opportunity of examining the lamps invented by Mr. Douglass, the distinguished engineer of Trinity House, which present improvements of the greatest importance as regards both the British lights and those of other countries. Not only is the brilliancy of the flame very much increased by ingenious methods of promoting combustion, but the consumption of oil is actually decreased. In British light-houses and on the continent colza-oil is generally used, though, for the sake of economy, mineral oil is being rapidly substituted for it, and *the French government has made an order for a general change to mineral oil in all the light-houses of the republic.*

"When the Trinity House tender had been made ready, I embarked with two of the Elder Brethren (Admiral Collinson, C. B., and Captain Weller) to inspect the British lights on the shore of the North Sea, and visited nearly every one on the coast from the mouth of the Thames to the Tweed, (the boundary of Scotland,) including the gas-light at Haisborough and a new electric light at Souter Point below the river Tyne. At Haisborough the gas-light was established for experimental comparison with an oil-light a few hundred yards from it, there being two towers, as at Cape Ann on the coast of Massachusetts. At Souter Point the electric light is necessary, because the coast near the Tyne is enveloped by a dense volume of smoke, produced by the immense number of manufactories on the river between Shields and Newcastle.

"I had excellent opportunities for testing the different varieties of lights in all kinds of weather, and particularly the gas and electric lights in fog. I was especially shown the system of marking the positions of rocks and shoals by means of what Trinity House calls "red cuts," i. e. by covering proper sections of the dioptric apparatus with red glass screens; and at different places on the northeast coast of England I made several boat-excursions at night to test the utility of the system.

"After my return from the inspection of the northeast coast I embarked with Captain Webb, of the Elder Brethren, at Portsmouth, and inspected the light-houses on the Isle of Wight and the southern coast, and passed around Land's End as far as St. Ives, on the west coast of Cornwall, visiting the celebrated light-house on the Wolf Rock, off Land's End, which is the most recent and difficult of all the English examples of light-house engineering. I regretted that I could not land at the Eddystone light-house, but the sea, although usually not so dangerous as at the Wolf, was too heavy when I passed it to make a landing practicable.

"Besides the light-houses on the coast, I particularly observed the light-ships and the system of buoyage; and I will here mention that the English use revolving apparatus in their light-ships in many cases, and they are found much more useful than fixed lights. I would recommend to the immediate consideration of the Board the propriety of distinguishing in this way some of our numerous light-ships off the coast of Massachusetts and in Long Island Sound.

"The English also find no difficulty in using fog-signals operated by hot-air engines in their light-ships, and I saw several, in one case hearing the signal distinctly at a distance of eight miles.

"From England I went to France, and had conference with M. Reynaud, *l'Inspecteur Général des Ponts et Chaussées*, and director of the French light-house service, and M. Allard, the chief engineer, who is in charge of the office of the *Commission des Phares;* also with the three lens-manufacturers of Paris.

"I was much interested in seeing our own optical apparatus in all stages of its manufacture; in learning the modes adopted by French engineers of testing the lenses, burners, and mineral oil; and in examining the most complete *dépôt des phares* in the world, where are shown examples of all stages in the progress of the science of light-house illumination, from the first efforts of FRESNEL, inventor of the system which bears his name, to the latest improvement of the present time.

"I visited the lights at the mouth of the Seine, and the double electric lights of La Hève at Sainte Adresse, near Havre.

"I afterward proceeded to Vienna and examined the light-house apparatus at the industrial exhibition, consisting of models, drawings, and photographs of light-houses from different countries, including our own. A package of these, which I made up a short time before I went to Europe, I was glad to learn, on my return, obtained a diploma of honor.

"After returning from Vienna I visited several light-houses on the coast of Wales, including two very interesting ones, that at Holyhead and one at the "South Stack."

"The first-named, a new one, though quite ready, was not lighted until some days after my inspection; it combined all the latest improvements of the English in regard to lens, lamps, and lantern.

"At South Stack is a light which is lowered down the cliff in foggy weather when the light is obscured by fog and it is clear below, a plan which I had before thought of as applicable to our lights on the elevated cliffs of the Pacific coast.

"I also visited Ireland and Scotland, the former by special invitation from the Board of Commissioners of Irish Lights, and I had an excellent opportunity of seeing two of the light-houses (Howth Baily and Wicklow Head) where the illuminant used is gas, of which Professor Tyndall when in the United States, expressed so favorable an opinion and which has been applied only by the Irish Board, except in the case I have mentioned, viz, the experimental light at Haisborough, on the east coast of England.

"These gas-lamps can be increased in an instant, when the weather becomes thick or foggy, from twenty-eight to forty-eight, sixty-eight, or eighty-eight, even to one hundred and eight jets for dense fog, and the inventor, Mr. Wigham, of Dublin, exhibited to me apparatus for producing a light from three hundred and twenty-four jets in the same lens apparatus.

"At Edinburgh I visited the Board of Commissioners of Scottish Lights, and had an interesting and instructive interview with Mr. Thomas Stevenson, engineer of the Board and a member of the family of celebrated Scottish engineers.

"I also visited the very extensive manufactory of light-house lenses of Chance Brothers & Company, near Birmingham, who are the furnishers of light-house apparatus to the Trinity House, and who also supply in a large degree the Irish and Scottish Boards, as well as India, China, and South America. Chance Brothers claim that their optical apparatus is superior to the French, and they certainly have a great advantage in having for the constant supervision of their work a gentleman of high scientific attainments.

"I carried with me a special letter from the honorable the Secretary of State to the ministers and consuls of the United States in Europe, and I received every facility and courtesy from them and from the officials of the countries which I visited.

"I have full notes of my inspection, and at an early day shall have the honor to present to you a detailed report of what I saw differing from our own system.

"In closing this preliminary report, I will say that the great questions which are occupying the attention of the light-house authorities of Europe, and in which the different establishments are in competition with each other, are:

"*What is the best illuminant?* and

"*What is the best means for producing the most perfect combustion?*

"I will only add that while the British and French systems are necessarily very much like our own, I saw many details of construction and administration which we can adopt to advantage, (and which I shall exhibit in my detailed report,) while there are many in which we excel. Our shore fog-signals, particularly, are vastly superior both in number and power. They are in advance of us in using both the gas and electric lights in positions of special importance, in the use of azimuthal condensing prisms for certain localities; in the character of their lamps; in the use of fog-signals in light-ships; in their light-ships with revolving lights, and, more than all, in the character of their keepers, who are in service during good behavior until death or superannuation, who are promoted for merit, and whose lives are insured by the government for the benefit of their families.

"I am much indebted to Mr. Paul J. Pelz, chief draughtsman to the

Board, who accompanied me by its permission and with the approval of the honorable the Secretary of the Treasury, as my secretary, and who has made many sketches for the illustration of my report, and who has, in other ways, been of much assistance to me in the execution of the duty assigned me.

"Very respectfully,

"GEORGE H. ELLIOT,
"*Major of Engineers U. S. A., Engineer-Secretary.*"

REPORT.

OUTWARD VOYAGE.

As stated in the preliminary report of my journey of inspection of the European light-house establishments, I sailed from New York in the steamship Cuba, of the Cunard line, on the 30th of April, 1873. *Date of sailing*

During the voyage I had many interesting conversations with the commander, Captain McCauley, in regard to the lights of the United States, France, and England. In reference to our own lights he stated that they were in general satisfactory to mariners, and had been of great service to him on many occasions, during his long service in the Cunard Company, especially in running between Boston and New York and between Halifax and Boston. *Conversation with Captain McCauley concerning American lights.*

With respect to brilliancy, the English and American lights are, in his opinion, about equal, but those of the French he considers superior to either. *Brilliancy of American, French, and English lights.*

He said of the electric lights, of which the English and French have several, that they penetrated fog much more successfully than the common oil-light, and aids to navigation in fog are, in his opinion, vastly more important than for fair weather. *Electric lights penetrate fog.*

Our fog-signals he praised highly, saying that the steam-whistle at Cape Ann and the siren at Sandy Hook had often been of great service to him, and he confidently relied on hearing them at distances of from six to eight miles. He thought it would be much to the interest of commerce if the British government would place similar signals at important points, as the channel approaches to Great Britain were nearly as much subject to fogs as is our eastern coast. *Fog-signals of great advantage to mariners. Fog-signals needed on English coasts.*

Our Nantucket Shoals he considered to be badly lighted, and called my attention to the fact that on the coasts of Great Britain, in similar localities, light-ships are placed at distances of ten to twelve miles apart; such I found afterward to be the case. He thought a light-ship should be placed off the Rose and Crown Shoal, (which is due east from Sankaty Head, on the Island of Nantucket,) so that a vessel could take a course to it from the Highlands (Cape Cod) light, and thence to the Nantucket New South Shoals light-ship, which he thought should be moved farther out. *Insufficient lighting of Nantucket Shoals. Light-ship needed off Rose and Crown Shoal.*

Meeting with steamer on Newfoundland Banks in a thick fog.

On the fifth night out from New York (the night of the 4th of May) we met a steamship on the Banks off Newfoundland, but there being at the time one of those dense fogs prevalent at some seasons of the year in that part of the Atlantic, we did not see her, and *only knew of her proximity by the sound of her whistle*, a fact which impressed

Importance of fog signals on transatlantic steamers.

me with the importance of powerful fog-signals on the steamships plying between America and Europe on this much-frequented track.*

Danger of collisions with vessels during fogs.

In addition to the large number of steamships the number of sailing-vessels is very great, and the tales of narrow escapes from collision, especially with fishermen anchored on the Banks, which one hears while (enveloped in dense fog) he is steaming along at a high rate of speed, very much impair the confidence which is naturally inspired by vessels like those of the Cunard line and commanders like Captain McCauley; for it is evident that want of efficient fog-signals cannot be compensated for by strength of ship or skill of

Insufficiency of whistles in use on steamers.

officers. The whistles in use are, I am told, frequently insufficient in power, and, being placed abaft the foresails and in front of the great smoke-pipes, are in such positions that the *sound-shadows* often cover the precise directions in which it is most essential the warning should be conveyed.

Position and kind of fog-signals should be determined by an international commission.

I am of the opinion that not only the *position* but the *kind* of fog-signals to be used in transatlantic steamers should be regulated by a joint commission of the governments interested, and that, before deciding these questions, not only the whistle, but the Daboll trumpet and the siren which we use at our fog-signal stations on shore, as well as the Austrian fog-trumpet, (shown in Plate L,) should be considered.

As the power necessary to operate these signals is on these steamships always at hand, it is not, as in the light-house service, a question of cost of maintenance, but the questions to be decided on are:

Questions to be determined.

First. What is the most efficient instrument for the purpose?

Second. What is the most advantageous position practicable for it?

Question of best position for steamer's fog-signal.

This position, it is evident, must be one in no way interfering with the management of the sails and rigging, and where no danger exists of the signal being carried away by the sea.

---

* From information derived from my friend, Mr. George W. Blunt, of New York, and from other sources, it appears that on an average from eight to ten steamers cross the Banks *every day* going from America to Europe and *vice versa*.

The question of ship-lights should also be determined by the same commission, and I have no doubt the magneto-electric light, which I believe has been spoken of before in this connection, and which is fully described in this report, will have favorable consideration, since the steam-power necessary for operating the magnetic machines is constantly available. *Ship-lights.*

On the morning of Friday, the 9th of May, we made the southwesterly point of Ireland, and had a good opportunity of seeing the important light-house on Fastnet Rock, off Cape Clear. This tower, having nearly vertical sides, which spread with a curve near the base, is 92 feet high; and, together with the appendages, presents a very picturesque appearance, being surrounded by a high retaining-wall, necessary, apparently, for the formation of a platform large enough for the establishment. *Fastnet Rock light.*

The lantern has the vertical sash-bars introduced into our service from the French. A broad band of red contrasts strongly with the color of the main body of the tower, which is built of iron.

As the Cuba steamed along the south coast of Ireland, and from two to four miles distant from the shore, a good view of the neat light-stations was afforded.

A very interesting one was the Old Head of Kinsale with its tall tower, on which two red bands distinguished it as a day-mark. *Old Head of Kinsale light.*

Each of the stations appears to have capacious grounds walled in with stone, and all are neatly whitewashed. The buildings connected with light-houses are generally of one story, covering a large area. We passed Ballycottin light, which stands 195 feet above the sea, and when we stopped at Queenstown to deliver the mails, we saw on the eastern head of the harbor the handsome light at Point Roche. *Ballycottin light.* *Point Roche light.*

Off the mouth of this harbor is an extensive shoal, the upper end marked by a bell-boat, and the lower by a can-buoy, on which the word "Danger" was painted in white letters. *Bell-boat.*

A few hours after leaving Queenstown we passed the light-house on Tuskar Rock, at the entrance of St. George's Channel, evidently an important station. The tower, to which is attached the double dwelling for the keepers, is 100 feet high. Tuskar Rock is several miles from shore, in the great highway to Liverpool, so that vessels entering or leaving St. George's Channel pass quite near it. *Tuskar Rock light.*

It was on this rock that the Cunard steamship Tripoli struck a short time ago, and it is evident that the powerful fog-signal that the English government proposes to place at this point is much needed.

Arrival at the Mersey. On the morning of the 10th we arrived at the mouth of the Mersey, and, after waiting an hour or more for sufficient tide to take us over the bar, we proceeded up the river to Liverpool.

Bell-buoy on the bar. On the bar we passed a large bell-buoy, shaped like our nun-buoys, above the water-line, except that it rested on a large bearing-surface, projecting a foot or more beyond its sides.

The sea being quite smooth, the bell was silent, as is too often the case with this very unreliable kind of signal.

Buoyage of channel. The channel of the river is marked by frequent buoys; on the starboard hand red "can," and on the port black "nun."

Light-ships with revolving lights. We passed several light-ships, some of which, as Captain McCauley informed me, have *revolving lights*, an important fact to be noted, since a revolving is seen much farther than a fixed light, and, when light-ships are numerous, as off the southeastern Irish coast, in the approaches to Liverpool, or on the shoals off the coast of Massachusetts, distinguishing characteristics are as necessary as for shore-lights.

Docks, &c., at Liverpool. I spent a day examining the great docks at Liverpool, and at Birkenhead, on the opposite side of the river, and became much interested in the immense walls, the gates and bridges, swung by hydraulic power, and many other objects which this is not the place to describe.

TRINITY HOUSE, LONDON.

Visit to the Trinity House. Soon after reaching London I called at the Trinity House, where I was received with great politeness by Sir Frederick Arrow, the Deputy Master, who, with Captain Webb, of the Elder Brethren, visited America during the summer of 1872 for the purpose of attending our fog-signal experiments made in the harbor of Portland, in the State of Maine, after which we had the pleasure of meeting them at Washington.

Improvements in lamps. Sir Frederick expressed his gratification at the attentions he received in the United States, and after an interesting conversation regarding our respective establishments, particularly relating to our fog-signals and to the very great improvements in light-house lamps made by the Trinity House, (whereby the quantity of light from the "four-wick" *Advantage gained.* lamp for large sea-coast light-houses had been increased 22 per cent., while the consumption of oil had actually been decreased more than 19 per cent., an advantage of over 41 per cent. in favor of the new lamp for large towers, and a still greater one for smaller sea-coast and harbor lights,) he said that since I informed him of my intended visit to Europe he had made several engagements for me, including a dinner at the Lord Mayor's on the 21st of May, in honor of the return of the Master of Trinity House, the Duke of Edinburgh; several cruises around the coast of England in the steam-yachts of the corporation, which were shortly to start on their annual supply-voyages to the light-stations, and a visit to Dover, to be present at some fog-signal experiments to be undertaken by the Elder Brethren under direction of Professor Tyndall.

It was during this visit that I had the pleasure of meeting Mr. Robin Allen, for many years the Secretary of the Trinity House, and Mr. Edwards, private secretary of the Deputy Master, who accompanied him on his visit to the United States.

During my stay in London I made frequent visits to the Trinity House, and was very soon after my arrival introduced to Mr. Douglass, the talented Engineer of the establishment, and to most of the Elder Brethren; the pleasure was also afforded me of meeting my friend Captain Webb, and I was glad to hear that it was with him that one of my cruises among the British light-houses was to be made.

Mr. Douglass showed me his plans of some of the more important English light-houses, particularly that of the Wolf Rock, off Land's End, as well as his drawings of lanterns and lamps. *Plans of light-houses. Wolf Rock. Drawings of lamps.*

It is noticeable that the English, in their lanterns, use diagonal sash-bars and low parapets, (or unglazed parts,) differing in this respect from the French and ourselves. *Sash-bars in English lanterns.*

Mr. Douglass was, as I afterward found the French and other light-house engineers of Europe to be, especially interested in the subject of lamps as well as that of material for illumination, these subjects being considered of most importance at the present time in light-house administration. *Interest of European engineers in lamps and oils for light-houses.*

Within the last five or six years improvements have been made from time to time in lenticular apparatus, but they are of trifling importance when contrasted with *the great increase of power and concurrent decrease of expense* of sea-coast lights as compared with the system in use in Europe a few years ago, and with ours of the present time. *Increase in power of light and decrease in expense.*

These vast ameliorations have been produced by— *Causes producing these changes.*
1st. *The introduction of mineral oil for light-house illumination.*

2d. *The improvements in "burners" for lamps, resulting from experiments made to determine the best form of lamps for burning mineral oil in light-houses, which improvements apply equally to lamps burning oil of a mineral, animal, or vegetable origin.*

Photometric experiments. This matter will be more fully noted when I come to describe the depot at Blackwall. I was shown the room devoted to photometric experiments, where a six-wick lamp burning colza-oil, a four-wick lamp for mineral oil, and small one-wick lamps of both the new and old styles were burning for my inspection.

There was a very remarkable difference of color and brilliancy between the flames of the improved and the other lamps, that of the latter being of a dirty yellowish hue, while that of the former, being more plentifully supplied with air, appeared perfectly white, surpassing even the excellent gas-light of London.

Corporation of Trinity House. The Corporation of Trinity House, or, according to the original charter, "The Master, Wardens, and Assistants of the Guild, Fraternity, or Brotherhood of the Most Glorious and Undivided Trinity, and of St. Clement, in the parish of Deptford Strond, in the county of Kent," existed as early *Date of charter.* as the reign of Henry VII, and was incorporated by royal charter during the reign of Henry VIII.

In the year 1565, in the reign of Queen Elizabeth, the corporation was empowered, by act of Parliament, "to preserve ancient sea-marks and to erect beacons, marks, and signs of the sea," but it was more than a century, *i. e.,* *Date of first owning light-houses.* not until 1680, before the corporation constructed or owned any light-houses. After that date it from time to time purchased the lights which were owned by individuals or by *Entire control of English lights vested in Trinity House.* the Crown, and also erected new ones. In 1836 an act of Parliament vested in the Trinity House the entire control of the light-houses of England and Wales, and gave it certain powers over the Irish and Scotch lights.

Light-dues. Prior to the act of 1836 the charge was from one-sixth of a penny to one penny per ton on all ships at each time of passing a light-house, but by this act uniform light-dues of a halfpenny per ton were established.

At Bell Rock. The charge of one penny per ton at Bell Rock light-house is the only exception to this uniform rate. By further pro- *What shipping is free.* visions of the act, national ships, fishing-vessels, and vessels in ballast are exempt from light-dues.

Light-houses not owned by the Trinity House. It should be mentioned that only the light-houses for general use are owned by the Trinity House, harbor and other local lights being constructed and maintained at the expense of the cities or localities which they especially benefit; but the Trinity House not only has over them a supervisory control in regard to their sites and plans, but inspects them from time to time, thus securing their efficiency.

The Elder Brethren, twenty-nine in number, comprise sixteen active members, including two officers of the Navy, and thirteen honorary members, all of whom are elected by the body as vacancies occur. *The Elder Brethren.*

The honorary members include his royal highness the Prince of Wales, some of the ministers to the Crown, several members of the nobility and of Parliament. *Honorary members.*

The Duke of Edinburgh is the present Master, but the Deputy Master, who is elected by the Elder Brethren from their active list, is the executive officer. *Master and deputy master.*

Out of the annual revenues £350 are paid to each of the active members; these members are organized into committees, which meet twice a week except when absent on duty. *Salaries. Committees.*

The entire board holds weekly sessions, at which the matters before considered in committee are disposed of. *Weekly sessions.*

The corporation of the Trinity House includes also the Junior Brethren, who are elected by the Elder Brethren, and simply form a reserve from which the Elder Brethren add to their own number when vacancies occur. *Junior Brethren.*

The Junior Brethren have no duties.

Since 1854 the Trinity House has been subordinate to the Board of Trade, whose president is one of the Queen's Ministers. *The Trinity House subordinate to Board of Trade.*

All light-dues collected by the corporation of Trinity House go into a general fund called "the mercantile-marine fund," from which is paid the cost of the maintenance of the light-house establishment and of the erection of new lights. This fund is under the control of the Board of Trade, whose authority must be obtained for the erection of any new light-house or for any important change in administration. *The mercantile-marine fund.*

This subordination to the Board of Trade extends to the light-house boards of Scotland and Ireland, causing, I was told, much inconvenience and embarrassment. *Light-house boards of Scotland and Ireland subordinate to Board of Trade.*

Modifications in the light-house administration of Great Britain have been from time to time suggested and changes may be desirable; but, judging by my own observations, the English lights, many more of which I saw than of the Irish and Scotch, are certainly managed in a most efficient manner, and, since the great improvements introduced by the eminent engineer, Mr. Douglass, the English may fairly be said *Observations upon the character of the British system.*

to have placed themselves on an equality with the French, who have so long led the world in the matter of light-house illumination.

Trinity House. Trinity House is an ancient structure on Tower Hill, opposite the old Tower of London, in "The City." It has a handsome freestone front in classic style. The main entrance is on the ground-floor through a capacious hall, where are exhibited models of many of the most celebrated light-houses of England, and also of beacons and buoys.

Office accomodations. Ample accommodations are afforded for the officers, for the Board and committees, for the Engineer's Department, and for the photometric experiments, and, in addition, there is a grand banqueting hall and *salon*.

I am under obligations to General Schenck, the American minister; to Mr. Moran, the secretary; and to Captain Ramsay, United States Navy, attaché of the legation, for offers of assistance, of which my previous acquaintance with Sir Frederick Arrow and Captain Webb and the kindness of the Elder Brethren made it unnecessary for me to avail myself.

FOG-SIGNAL EXPERIMENTS NEAR DOVER.

Experiments with fog-signals. On the 19th of May I proceeded to Dover, to be present at the commencement of an extensive series of fog-signal experiments, to be undertaken by the Trinity House, under the supervision of their scientific adviser, Professor Tyndall.

There were present on this occasion, besides Professor Tyndall, several of the Elder Brethren, the Engineer of the Trinity House, a representative from the Board of Trade, and the Inspector of Irish lights.

The experimenters divided themselves into two parties, and embarked on two steam-yachts, for the purpose of practically testing the sounds while afloat.

The limited time at my disposal did not allow me the pleasure of accepting an invitation to join them, and I had only an opportunity of observing the machines themselves at the experimental station at South Foreland. This I did before inspecting the South Foreland light-houses, to which I at once proceeded.

Disposition of the signals. There were two sets of signals used; one placed on the summit of the cliff near the engine-house belonging to the light-station, at an elevation of 275 feet above the sea; the other near the foot of the cliff, at an elevation of 40 feet, and near the bottom of an old shaft, by which it was reached from the upper station.

The machines used were steam-whistles, air-whistles, and fog-trumpets. The trumpets and the air-whistles were connected with two air-chambers, supplied with air by pumps driven by the engine used for the electric lights. {Manner of operating.}

An Ericsson engine was at the station, but was not used.

The steam-whistles were supplied by a twenty horse-power upright engine.

The trumpets were shaped like those used in our own service. The steam-whistles, with a diameter of 12 inches, had a height of 14 inches, the space between the lip and the disk being $1\frac{1}{4}$ inches. The air-whistles, with a diameter of 6 inches, were $9\frac{1}{2}$ inches high, the lips being placed $1\frac{1}{2}$ inches from the disks. {Shape and dimensions of signals.}

The steam-whistles were blown under a pressure of 64 pounds; the air-trumpets and whistles under a pressure of 18 pounds. Behind these latter were reflectors about 12 by 15 feet, slightly curved toward the land. {Pressure used. Sound-reflectors.}

The day was stormy, there being a high wind from the eastward accompanied by rain, and the Strait of Dover was pretty rough, but on the whole the weather was favorable for the purpose. The two parties continued afloat all day, and the signals were sounded until dark.

On shore, so near the signals, and while inspecting the light-houses, I could not determine in regard to the qualities of the different sounds as well as could those on board the yachts, but, so far as I could judge, the air-whistles and trumpets were decidedly superior to the steam (12-inch) whistles. The note of the latter was much more shrill than that found by us to best serve the purpose for which this instrument is designed, and the condensation of the steam, and consequent drip of water* were so great, I was convinced, as to greatly impair the vibration. {Air-whistles and trumpets superior to steam-whistles.}

The results of the trial of the 12-inch steam-whistle were from some cause much less satisfactory than in our experiments at Portland, and at our light-stations, the sound produced being certainly no louder than that of the 6-inch air-whistle. {Results less satisfactory than at Portland, Me.}

On the return of the experimental parties in the evening, there was a general expression of disappointment in regard to all the signals. {Opinions of experimenters.}

It was stated that the yachts ran outside the limits of sound at comparatively short distances; that the air trumpets and whistles were heard much farther than the steam-whistles, while a gun fired at Dover Castle was heard at much greater distance than any of the signals at the station. {Short distance at which the sound failed to reach the parties.}

---

* It may have been water thrown out with the steam in consequence of the insufficient height of the whistle above the boiler.

The yachts were out in the channel occupying various positions in relation to the wind, and the signals were regularly sounded according to a programme previously arranged.

The experiments, in which were afterward included one of our own whistles and sirens, were to be continued for some months. Following will be found a list of the printed questions to be considered and answered by the experimenters.

Questions to be decided.

"SOUTH FORELAND FOG-SIGNAL EXPERIMENTS.—QUESTIONS PROPOSED TO BE DETERMINED.

"First. What is the most efficient height above the sea-surface for the signals?

"Second. What are the comparative values of air and steam for sounding whistles and horns?

"Third. Which is the more efficient instrument—whistle or horn?

"Fourth. What is the proper pressure, having regard to efficiency and economy, at which air or steam should be employed for whistles or horns?

"Fifth. What is the relative range of the same whistle or horn with various pressures of steam or air?

"Sixth. What is the relative range of long and short blasts from the same instrument, and what is the minimum duration of the blasts of maximum efficiency?

"Seventh. What is the most efficient note for a fog-signal?

"Eighth. What is the relative range of the highest and lowest notes of the same instrument?

"Ninth. What is the relative range of one and two whistles or horns of the same power?

"Tenth. What is the relative range of the horn in the direction of its axis, and at 45° and 90° respectively from the direction of its axis?

"Eleventh. Is the horn used with maximum efficiency by always keeping it pointed to windward, by using more than one horn and distributing the sound over the phonic arcs or by rotating one horn?

"Twelfth. Is any appreciable advantage gained by using reflectors in conjunction with whistles or horns; and, if so, what shape is preferable?

"Thirteenth. What horse-power is required to sound the most efficient signal (whistle or trumpet) for giving an effective range (of two miles) in fog and against wind at force 9 of the Beaufort scale?

"Fourteenth. How is the propagation and distribution of sound affected by different atmospheric conditions?"

It has been found by General Duane, of the Corps of Engineers United States Army, and light-house engineer of the New England coast, in his experiments made to determine the best form of boilers for steam fog-signals, that, as the steam used is at a high pressure, and is drawn off at intervals, and there is a consequent tendency to foam and to throw out water with the steam, a horizontal tubular boiler (locomotive) with rather more than one-half of the interior space allowed for steam-room, is best adapted for the purpose.

Experiments of General Duane.

Boilers best to be used for fog-signals.

The steam-dome must be very large and be surmounted by a steam-pipe 12 inches in diameter. Both dome and pipe were first made small, and were gradually enlarged until no difficulty with regard to foaming remained.

Steam-dome and steam-pipe.

The steam should be drawn off at a point 10 feet above the water-level in the boiler.

Drawing-off point for the steam.

The main points, therefore, to be observed in regard to the *boiler*, are to have plenty of steam-room and to draw the steam from a point high above the water-level.

In regard to the *bell of the whistle*, the best results have been obtained by making the diameter two-thirds of its length, and the "set" of the bell, *i. e.*, the vertical distance of the lower edge above the cup, from one-fourth to one-third of the diameter for a pressure of steam of from 50 to 60 pounds.

Form of whistle.

These conditions were not fulfilled in the Dover experiments at the time of my visit, and I have no doubt that this accounts in some measure for the disappointing results of the trials with the steam-whistle.

Just as this report is going to press I have received from Sir Frederick Arrow, Professor Tyndall's report of the fog-signal experiments at South Foreland, which the former has been kind enough to send me at the moment of its publication in order that I might make use of it here, and I take pleasure in interpolating at this place some extracts of much interest.

Report of Professor Tyndall.

EXTRACTS FROM PROFESSOR TYNDALL'S REPORT ON FOG-SIGNALS.

* * * * * * *

"*May* 20, 1873.— * * * There was nothing, as far as I am aware of, in our knowledge of the transmission of sound through the atmosphere, to invalidate the founding upon these experiments of the general conclusion that, as a fog-signal, the gun possessed a clear mastery over the horns. No observation, to my knowledge, had ever been made to show that a sound once predominant would not be always predominant, or that the atmosphere on different days

Guns not always superior to horns as signals.

would show preferences to different sounds. A complete reversal of the foregoing conclusion was, therefore, not to be anticipated; still, on many subsequent occasions, it was completely reversed.

* * * * * *

Signals on top of cliff superior to those at bottom.
"The observations of May 19 and 20 proved that no advantage was gained by placing the horns at the bottom of the cliff. With scarcely an exception the higher horns proved in all cases slightly superior to the lower ones. In subsequent experiments, therefore, the higher horns alone were for the most part invoked.

* * * * * *

"*June* 3.—At seven miles we halted; the sound of the horns was here very distinct, the steam-whistle being also well heard.

"While in this position an exceedingly heavy rain-shower approached at a galloping speed. It could hardly have been borne forward with such velocity by the wind, which had only a force of 2. Its advance was probably due to the rapid successive condensation of different parts of the same continuous cloud. The sounds were not sensibly impaired during the continuance of the rain. Not till subsequently was the influence of such a shower in clearing the air understood. At eight miles the whistles were still heard and the horns better heard. At nine miles the whistles ceased to be heard, while the horns continued to be fairly audible.

Use of two horns.
"In no case did any sensible inequality show itself between the sound of the single horn and that of the pair of horns. The beats of the two horns were, however, very characteristic at the longer distances. The blasts of the horns were not of uniform strength; even in the same blast sudden swellings out and fallings off of the sounds were observed.

* * * * * *

Effect of the acoustic shadow.
"*June* 11.—* * * The fall of the sound is not caused directly by the acoustic shadow, for it occurs when the instruments are in view; but the limit of the acoustic shadow is close at hand. A little within the line joining the Foreland and the pier end the instruments are cut off by a projection of the cliff near the station. * * * All the sea-space between this 'boundary' and the cliff under Dover Castle is in the shadow. Into this, however, the direct waves diverge, and lose intensity by their divergence, the portion of the wave nearest the shadow suffering most. Hence, I doubt not, one cause of the decay of the sound in the position here referred to. The interference of sound reflected from the cliff with the direct sound doubtless also contrib-

"July 1.— * * * Here a word of reflection on our observations may be fitly introduced. It is an opinion entertained in high quarters that the waves of sound are reflected at the limiting surfaces of the minute particles which constitute haze and fog, the alleged waste of sound in fog being thus explained. Dr. Robinson, for example, defines fog to be 'a mixture of air and globules of water,' and states 'that at each of the innumerable surfaces where these two touch a portion of the vibration is reflected and lost.' Theoretically it may be so; but if this were an efficient practical cause of the stoppage of sound, it would be difficult to understand how to-day, in a thick haze, the sound reached a distance of twelve and three-quarters miles; and that on May 20, in a calm and hazeless atmosphere, the maximum reach of the sounds was only from five to six miles. Such facts foreshadow a revolution in our notions regarding the action of haze and fogs upon sound.'

_{Accepted theory of loss of sound in fog.}

_{Dr. Robinson's definition of fog.}

* * * * * * *

"July 2.—In the foregoing observations we have had very remarkable fluctuations in the range of the sound, that range varying from three or four miles on May 19 to ten and one-half or twelve and three-quarters on the 1st of July. The direction and force of the wind, known to exercise a potent influence upon sound, entirely fail to account for these fluctuations, nor could any other observed meteorological element be held responsible for them. Prior to July 3 surmises more or less vague had passed through my mind regarding them; but all remained uncertain until on the 3d surmise and perplexity were, to a great extent, displaced by clear physical demonstration.

_{Fluctuations in range of sound.}

"On July 3 we first steamed to a point 2.9 miles southwest by west of the signal-station. No sounds, not even the guns, were heard at this distance. At two miles they were equally inaudible. But this being the position in which the sounds, though strong in the axis, invariably subsided, we steamed to the exact bearing from which our observations had been made upon July 1. At 2.15 p. m., and at a distance of three and three-quarters miles from the station, with calm air and a smooth sea, the horns and whistle (American) were sounded, but they were inaudible. Surprised at this result, I signaled for the guns. They were all fired, but, though the smoke seemed at hand, no sound whatever reached us. On July 1, in this bearing, the range

_{Acoustic opacity of air July 3.}

of both horns and guns was ten and a half miles. We steamed in to three miles, paused, and listened with all attention, but neither horn nor whistle was heard. The guns were again signaled for; five of them were fired in succession, but not one of them was heard. We steamed in on the same bearing to two miles, and had the guns fired point-blank at us. The howitzer and the mortar, with 3-pound charges, yielded a feeble thud, while the 18-pounder was wholly unheard. Applying the law of inverse squares, it follows that, with air and sea in an apparently worse condition, the sound at two miles distance on July 1 must have had at least five-and-twenty times the intensity which it possessed at the same distance on the 3d.

"With the Foreland so close to us, the sea so calm, and the air so transparent, it was difficult, indeed, to realize that the guns had been fired or the trumpets sounded at all. What could have caused this extraordinary stifling of the sound? Had it been converted by internal friction into heat? Or had it been wasted in partial reflections at the limiting surfaces of non-homogeneous masses of air? A few words will render this question intelligible to the general reader. Sulphur in homogeneous crystals is exceedingly transparent to radiant heat, whereas the ordinary brimstone of commerce is highly impervious to it. Why? Because the brimstone of commerce does not possess the molecular continuity of the crystal, but is a mere aggregate of minute grains not in perfect optical contact with each other. Where this is the case, a portion of the heat is always reflected on entering and on quitting a grain. Hence when the grains are minute and numerous this reflection is so often repeated that the heat is entirely wasted before it can plunge to any depth into the substance. A snow-ball is opaque to light for the same reason. It is not optically continuous ice, but an aggregate of grains of ice, and the light which falls upon the snow being reflected at the limiting surfaces of the snow-granules, fails to penetrate the snow to any depth. Thus, by the mixture of air and ice, two transparent substances, we produce a substance as impervious to light as a really opaque one. The same remark applies to foam, to clouds, to common salt, indeed to all transparent substances in powder. They are all impervious to light, not through the real absorption or extinction of the light, but through internal reflection.

Humboldt's observations at the Falls of the Orinoco. "Humboldt, in his observations at the Falls of the Orinoco, is known to have applied these principles. He found the noise of the falls three times louder by night than by day, though in that region the night is far noisier than the day. The plain between him and the falls consisted of spaces of grass and rock intermingled. In the heat of the day he found the temperature of the rock to be 30° higher than that of the grass. Over every heated rock, he concluded, rose a column of air rarefied by the heat, and he ascribed the deadening of the sound to the reflections which it endured at the limiting surfaces of the rarer and the denser air. This philosophical explanation made it generally known that a non-homogeneous atmosphere is unfavorable to the transmission of sound.

"But what, on July 3, with a calm sea as a basis for the atmosphere, could so destroy its homogeneity as to enable it to quench in so short a distance so vast a body of sound? As I stood upon the deck of the Irené pondering this question I became conscious of the exceeding power of the sun beating against my back and heating the objects near me. Beams of equal power were falling on the sea, and must have produced copious evaporation. That the vapor generated should so rise and mingle with the air as to form an absolutely homogeneous mixture I considered in the highest degree improbable. It would be sure, I thought, to streak and mottle the atmosphere with spaces in which the air would be in different degrees saturated, or it might be displaced by the vapor. At the limiting surfaces of these spaces, though invisible, we should have the conditions necessary to the production of partial echoes, and the consequent waste of the sound.

Reasons for the non-homogeneity of the atmosphere.

"Curiously enough, the conditions necessary for the testing of this explanation immediately set in. At 3.15 p. m. a cloud threw itself athwart the sun and shaded the entire space between us and the South Foreland. The production of vapor was checked by the interposition of this screen, that already in the air being at the same time allowed to mix with it more perfectly; hence the probability of improved transmission. To test this inference I had the steamer turned and urged back to our last position of inaudibility. The sounds, as I expected, were distinctly though faintly heard. This was at three miles' distance. At three and three-quarters miles we had the guns fired, both point-blank and elevated. The faintest pop was all that we heard; but we did hear a pop, whereas we had previously heard nothing, either here or three-quarters of a mile nearer. We steamed out to four and a quarter miles, where the sounds were for a moment faintly heard; but they fell away as we waited, and though the greatest quietness reigned on board,

Clouds causing the air to transmit sound more readily.

and though the sea was without a ripple, we could hear nothing. We could plainly see the steam-puffs which announced the beginning and end of a series of trumpet-blasts, but the blasts themselves were quite inaudible.

"It was now 4 p. m., and my intention at first was to halt at this distance, which was beyond the sound-range, but not far beyond it, and see whether the lowering of the sun would not restore the power of the atmosphere to transmit the sound. But, after waiting a little, the anchoring of a boat was suggested, so as to liberate the steamer for other work; and though loath to lose the anticipated revival of the sound myself, I agreed to this arrangement. Two men were placed in the boat and requested to give all attention so as to hear the sound if possible. With perfect stillness around them they heard nothing. They were then instructed to hoist a signal if they should hear the sounds, and to keep it hoisted as long as the sound continued.

"At 4.45 we quitted them and steamed toward the South Sand Head light-ship. Precisely 15 minutes after we had separated from them the flag was hoisted. The sound had at length succeeded in piercing the body of air between the boat and the shore.

"We continued our journey to the light-ship, went on board, and heard the report of the lightsmen. Returning toward the Foreland, in answer to a signal expressing a wish to communicate with us, we manned a boat and pulled to the shore. The exhaustion of the ammunition was reported, but the horns and whistle continued to sound. We steamed out to our anchored boat, and then learned that when the flag was hoisted the horn-sounds were heard; that they were succeeded after a little time by the whistle-sounds, and that both increased in intensity as the evening advanced. On our arrival, of course we heard the sounds ourselves.

"The explanation given above of the stoppage of the sound is in perfect harmony with these observations. But we pushed the test further by steaming farther out. At five and three-quarters miles we halted and heard the sounds. At six miles we heard them distinctly, but so feebly that we thought we had reached the limit of the sound-range. But while we waited the sound rose in power. We steamed to the Varne buoy, which is seven and three-quarters miles from the signal-station, and heard the sounds there better than at six miles distance. We continued our course outward to ten miles, halted there, but heard nothing.

"At eight miles' distance the sound in the evening was at least as well heard as at two miles in the morning. That this could occur it was necessary, in accordance with the law of inverse squares, that the sound at two miles' distance should have risen in the evening to an intensity at least sixteen times that which it possessed in the morning.

Signals heard at eight miles in the evening.

"Steaming on to the Varne light-ship, which is situated at the other end of the Varne Shoal, we hailed the master, and were informed by him that up to 5 p. m., nothing had been heard. At that hour the sounds began to be audible. He described one of them as 'very gross, resembling the bellowing of a bull,' which very accurately characterizes the sound of the large American steam-whistle. At the Varne light-ship, therefore, the sounds had been heard toward the close of the day, though it is twelve and three-quarters miles from the signal-station. On our return to Dover Bay at 10 p. m. we heard the sounds, not only distinct but loud, where nothing could be heard in the morning.

"I have already referred to the winds and currents which establish themselves round the South Foreland. Mr. Holmes was, as usual, there on July 3, and he informed me that, from the motion of the smoke of some passing steamers and from the sails of sailing-vessels, he could recognize a curious circulation of the air from land to sea. The wind would sometimes hug the cliff to the northeast of the Foreland; then bend around and move toward the South Sand Head light-ship. And, in point of fact, the wind at the light-vessel had been southwest, with a force of 3 nearly the whole of the day; whereas with us it had passed from southwest by west to a dead calm, and afterward to southeast. On shore also it had shifted from southwest to southeast. The atmospheric conditions between the light-vessels and the Foreland were, therefore, different from those between us and the Foreland; and the consequence was that at the time when we were becalmed and heard nothing the light-keepers, with the larger component of a wind of 3 acting against the sounds, heard them plainly all day.

Winds and currents about the Foreland.

"But both the argument and the phenomena have a complementary side, which we have now to consider. A stratum of air three miles thick on a perfectly calm day has been proved competent to stifle both the cannonade and the horn sounds employed at the South Foreland; while the observations just recorded seem to point distinctly to the mixture of air and aqueous vapor as the cause of this extraordinary phenomenon. Such a mixture could fill the atmosphere with

Reflection of sound.

an impervious *acoustic cloud* on a day of perfect *optical* transparency. But, granting this, it is incredible that so great a body of sound could utterly disappear in so short a distance without rendering any account of itself. Supposing, then, instead of placing ourselves behind the acoustic cloud, we were to place ourselves in front of it, might we not, in accordance with the law of conservation, expect to receive by reflection the sound that had failed to reach us by transmission? The case would then be strictly analogous to the reflection of light from an ordinary cloud to an observer placed between it and the sun.

<small>Echoes observed.</small>
"My first care in the early part of the day in question was to assure myself that our inability to hear the sound did not arise from any derangement of the instruments on shore. Accompanied by Mr. Edwards, who was good enough on this and some other days to act as my amanuensis, at 1 p. m. I was rowed to the shore, and landed at the base of the South Foreland cliff. The body of air which had already shown such extraordinary power to intercept the sound, and which manifested this power still more impressively later in the day, was now in front of us. On it the sonorous waves impinged, and from it they were sent back to us with astonishing intensity. The instruments, hidden from view, were on the summit of a cliff 235 feet above us, the sea was smooth and clear of ships, the atmosphere was without a cloud, and there was no object in sight which could possibly produce the observed effect. From the perfectly transparent air the echoes came, at first with a strength apparently but little less than that of the direct sound, and then dying gradually and continuously away. A remark made by my talented companion in his note-book at the time shows how the phenomenon affected him. 'Beyond saying that the echoes seemed to come from the expanse of ocean, it did not appear possible to indicate any more definite point of reflection.' Indeed, no such point was to be seen; the echoes reached us, as if by magic, from absolutely invisible walls.

"Here, I doubt not, we have the key to many of the mysteries and discrepancies of evidence which beset this question. The foregoing observations show that there is no need to doubt either the veracity or capability of the conflicting witnesses, for the variations of the atmosphere are more than sufficient to account for theirs. The mistake, indeed, hitherto has been, not in reporting incorrectly, but in neglecting the monotonous operation of repeating the observations during a sufficient time. I shall have occasion to remark subsequently on the mischief likely to arise from giving instructions to mariners founded on observations of this incomplete character.

"The more accurate comprehension of various historic occurrences will be rendered possible by these observations. In his lecture entitled 'Wirkungen aus der Ferne,' the eminent Berlin philosopher, Dove, has collected some striking cases of this kind. During the battle of Cassano, on the Adda, between the Duc de Vendôme and the Prince Eugene, an army-corps stationed under the duke's brother, five miles up the river, failed to join the battle through not hearing the cannonade. In a river-valley, particularly on a warm day, it would, in my opinion, be very perilous to place much dependence upon sound. Near Montereau, on the Seine, during the battle between Napoleon I and the King of Würtemberg, which lasted seven hours, no sound of the conflict was heard by Prince Schwartzenberg, thirteen miles up the river. A Prussian officer sent thither at noon first heard the cannonade at a distance of four and a half miles from the field of battle. This happened on a day apparently resembling in point of mildness and serenity our 3d of July. In the battle of Liegnitz, where Frederick the Great overthrew Laudon, the sound of the battle was unheard by Field-Marshal Daun, who was posted on a height four and a half miles from the battle-field. Dove himself recounts the fact of his having failed to catch a single shot of the battle of Katzbach, at four and a half miles distance, while he plainly heard the cannonade of Bautzen, eighty miles away.

Striking cases of non-transmission of sound.

"The stoppage of the sound in the foregoing cases Dove referred, and doubtless correctly, to the non-homogeneous character of the air. He also notes the exceedingly interesting observation that in certain clear winter days, when the sun has already attained some power, the semaphore is difficult to decipher, the reason being that by the solar warmth upward currents of warm and downward currents of cold air (similar to those of Humboldt on the plain of Antures) are established, and that such days are also unfavorable to the transmission of sound. In another passage, however, he seems to endorse the prevalent notion that the transparency of the air and its power to transmit sound go hand in hand; whereas in our experiments days of the highest optical transparency proved themselves acoustically most opaque.

Other instances of non-homogeneous atmosphere.

"'Over water,' says Sir John Herschel, 'or a surface of ice, sound is propagated with remarkable clearness and

Opinion of Sir John Herschel.

strength;' and he refers to the well-known case of Lieutenant Foster, who, in the polar expedition of Captain Parry, carried on a conversation across the frozen harbor of Port Bowen, which is a mile and a quarter wide. But as regards smoothness, water could hardly be in a better condition than the sea between the Irené and the South Foreland on the 3d of July. Still, though aided by reflection from the sea's surface, the sound was powerless to penetrate the air. And in regard to Lieutenant Foster's observation, there cannot, I think, be a doubt that the extraordinary acoustic transparency of the polar atmosphere is mainly due to the absence of that flocculence which in our observations proved so hostile to the transmission of the sound. To the same cause is, I believe, to be ascribed the hearing of cannonades at the extraordinary distances of eighty, one hundred and eighty, and two hundred miles, mentioned by Sir John Herschel in his essay on sound. Had Humboldt himself been aware of the observations here recorded, might not his classical observation also have been connected with the vapor raised from the Orinoco by a tropical sun?

Experiments conducted by the French Bureau des Longitudes. "In the celebrated experiments conducted by the commission of the French Bureau des Longitudes in 1822, two stations were chosen, 11.6 miles apart, the one at Montlhery, and the other at Villejuif, near Paris. Two remarkable phenomena, which have a special interest in relation to our observations, presented themselves to the observers; the one was that while the report of every gun fired at Montlhery was exceedingly well heard at Villejuif, by far the greater number of those fired at Villejuif failed to be heard at Montlhery. In reference to this point Arago, the writer of the report, with that philosophic reserve which he showed in other matters, expressed himself thus: 'Quant aux différences si remarquables d'intensité que le bruit du canon a toujours présentées suivant qu'il se propageait du nord au sud entre Villejuif et Montlhery, ou du sud au nord entre cette seconde station et la première, nous ne chercherons pas aujourd'hui à les expliquer, parceque nous ne pourrions offrir au lecteur que des conjectures dénuées des preuves.' To another phenomenon he also directs attention, offering not only a description, but an explanation: 'Avant de terminer cette note, nous ajouterons seulement que tous les coups tirés à Montlhery y étaient accompagnés d'un roulement semblable a celui du tonnerre, et qui durait 20″ à 25″. Rien de pareil n'avait lieu à Villejuif; il nous est arrivé seulement d'entendre, à moins d'une seconde d'intervalle, deux coups distincts du canon de Montlhery. Dans *deux* autres circonstances le bruit de ce canon a été accompagné d'un roulement prolongé. Ces phénomènes n'ont jamais eu lieu qu'au moment d'apparition de quelques nuages ; par un ciel complètement serein le bruit était unique et instantané. Ne serait-il pas permis de conclure de là qu'à Villejuif les coups multiples du canon de Montlhery résultaient d'échos formés dans les nuages, et de tirer de ce fait un argument favorable à l'explication qu'ont donnée quelque physiciens du roulement du tonnerre ?'

"It is not here stated that at Montlhery the clouds were seen when the echoes were heard. The explanation of the Montlhery echoes is in fact an inference from observations made at Villejuif. I think that inference requires qualification. Some hundreds of cannon-shots have been fired at the South Foreland, many of them when the heavens were completely free from clouds, and never in a single case has a '*roulement*' similar to that noticed at Montlhery been absent. It follows, moreover, so hot upon the direct sound as to present scarcely a sensible breach of continuity between the sound and the echo. This could not be the case if the clouds were its origin. A reflecting cloud, even at the short distance of 1,000 yards, would leave a silent interval of five seconds between the sound and the echo. Had such an interval been observed at Montlhery it could hardly have escaped record by the philosophers stationed there.

"But, to fall back from reasoning upon facts, it is certain that air of perfect visual transparency is competent to produce echoes of great intensity and long duration. I shall have further occasion to refer to such echoes; for it was not with whistles, nor trumpets, nor guns, that these echoes in our observations reached their greatest development, but with the steam-siren, to be described farther on. The blasts sounded by this instrument number, I believe, about twenty thousand; but whatever might be the state of the weather, cloudy or serene, stormy or calm, no single blast of the siren failed to be accompanied by echoes of astonishing strength.

Echoes produced by transparent air.

"The other point referred to, which Arago declined to discuss, presents a grave difficulty. No attempt, as far as I am aware of, has since been made to solve it, or even to show that a solution is conceivable. I think the foregoing observations might be shown to have some bearing upon the point. Arago makes incidently the significant remark that, on June 22, when only one out of twelve of the shots fired at Villejuif was heard, and that feebly, at Montlhery, 'l'hygromètre avait marché beaucoup à l'humidité ;' and farther on he speaks of the air as 'tout près du terme de l'humidité extrême.' I believe myself safe in saying that air thus moving rapidly toward its point of saturation is sure to yield echoes; and the fact that echoes were heard at Montlhery and not at Villejuif is a proof of the different hygrometric condition of the air at the two stations. With the light wind recorded in the report, Montlhery would probably be swathed by vapor from the valley of the Seine. It seems to me by no means impossible to imagine a distribution of vapor sufficient to produce the observed effect; but this is a subject which may be reserved for future investigation.

"The observations of July 3, I believe, reveal to us the most potent cause of the caprices of the atmosphere as regards the transmission of sound. We shall, moreover, find them throwing light upon anomalies subsequently observed, which, without their aid, would be perplexing in the highest degree.

* * * * . * *

American siren sent for experiment. "During my recent visit to the United States I was favored by an introduction to General Woodruff by Professor Joseph Henry, of Washington. Professor Henry is chairman of the Light-House Board, and General Woodruff is engineer in charge of two of the light-house districts. I accompanied General Woodruff to the establishment at Staten Island, and afterward to Sandy Hook, with the express intention of observing the performance of the steam-siren which, under the auspices of Professor Henry, has been introduced into the light-house system of the United States. Such experiments as were possible to make under the circumstances were made, and I carried home with me a somewhat vivid remembrance of the mechanical effect of the sound of the steam-siren upon my ears and body generally. This I considered to be greater than the similar effect produced by the horns of Mr. Holmes; hence the desire, on my part, to see the siren tried at the South Foreland. The formal expression of this desire was anticipated by the Elder Brethren, while their wishes were in turn anticipated by the courteous kindness of the Light-House Board at Washington. Informed by Major Elliot that our experiments had begun, the Board forwarded to the corporation, for trial, the noble instrument now mounted at the South Foreland.

Principle of the siren. "The principle of the siren is easily understood. A musical sound is produced when the tympanic membrane is struck periodically with sufficient rapidity. The production of these tympanic shocks by puffs of air was first realized by Doctor Robison, and his device was the first and simplest form of the siren. A stop-cock was so constructed that it opened and shut the passage of a pipe seven hundred and twenty times in a second. Air being allowed to pass intermittently along the pipe by the rotation of the cock, 'a musical sound was most smoothly uttered.' A great step was made in the construction of this instrument by Cagniard de la Tour, who gave it its present name. He employed a box with a perforated lid, and above the lid a similarly perforated disk, capable of rotation. The perforations were oblique, so that when wind was driven through, it so impinged upon the apertures of the disk as to set it in motion. No separate mechanism was therefore required to move the upper disk. When the perforations of the two disks coincided, a puff escaped; when they did not coincide, the current of air was cut off. In this way a succession of impulses was imparted to the air. The siren has been greatly improved by Dove, and specially so by Helmholtz. Even in its small form the instrument is capable of producing sounds of great intensity.

"In the steam-siren, patented by Mr. Brown, of New York, a fixed disk and a rotating disk are also employed, radial slits being cut in both disks instead of circular apertures. One disk is fixed across the throat of a trumpet, 10½ feet long, 5 inches in diameter, where the disk crosses it, and gradually opening out till at the other extremity it reaches a diameter of 2 feet 3 inches. Behind the fixed disk is the rotating one, which is driven by separate mechanism. The trumpet is mounted on a boiler. In our experiments steam of 70 pounds pressure has for the most part been employed. Just as in the siren already described, when the radial slits of the two disks coincide, a puff of steam escapes. Sound-waves of great intensity are thus sent through the air; the pitch of the note produced depending on the rapidity with which the puffs succeed each other; in other words, upon the velocity of rotation.

* * * * * * *

"*October* 8.—* * * The heavy rain at length reached us, but although it was falling all the way between us and the Foreland, the sound, instead of being deadened, rose perceptibly in power. Hail was now added to the rain, and the shower reached a tropical violence. The deck was thickly covered with hail-stones, which here and there floated upon the rain-water, the latter not having time to escape. We stopped. In the midst of this furious squall both the horn and the siren were distinctly heard; and as the shower lightened, thus lessening the local noises, the

[margin: Sound not deadened by heavy rain.]

sounds so rose in power that we heard them at a distance of seven and a half miles distinctly louder than they had been heard through the rainless atmosphere at five miles. This observation is entirely opposed to the statement of Derham, which has been repeated by all writers since his time, regarding the stifling influence of falling rain upon sound. But it harmonizes perfectly with our experience on the 3d July, which proved water in the state of *vapor*, so mixed with air as to form non-homogeneous parcels, to be a most potent influence as regards the stoppage of sound. Prior to the violent showers of to-day the air had been in this condition, but the descent of the shower restored in part the homogeneity of the atmosphere and augmented its transmissive power.

"In the cleansed and cool atmosphere the horn-sound appeared to improve more than that of the siren, slightly surpassing it at times. The horn-note was of lower pitch; hence it might be inferred that the change in the atmosphere favored specially the transmission of the longer waves.

"Up to this time the siren had been performing 2,400 revolutions a minute; the rate was now reduced to 2,000 a minute. The sound immediately surpassed that of the horn. By this experiment the foregoing inference was reduced to demonstration; a highly instructive result, as it showed an interdependence between aerial reflection and the lengths of the sonorous waves.

"At 4 p. m. the rain had ceased, the sun shone clearly out: the air was calm afloat, but west with a force of 2, ashore. At nine miles' distance the horn was heard feebly, the siren clearly; the howitzer at this distance sent us a loud report. All, indeed, seemed better at this distance than at five and one-half miles; from which it follows that at this latter distance the intensity of the sound must have been augmented at least threefold by the descent of the rain.

* * * * * * *

"*October* 10.—* * * We descended the 12-ladder shaft, and from the lower station listened to the gun, the upper horn, the siren, and the lower horn. The sound of the siren was strikingly distinguished from that of the upper horn by its *hardness* and almost explosive force. Its echoes also were much louder and longer continued than those of the horn; and from this alone its greater reach of penetration might be inferred. The noise of the surf, however, at the lower station, interfered seriously with the observations.

"*October* 13.—* * * On steaming toward the axis no echo for some time was generated by the horns, none by the Canadian whistle, but long-drawn and distinct echoes came from the south in the case of the siren. When quite abreast of the station the horn-echoes were also heard, but they failed to approach in intensity those of the siren.

"Near the shore the wind was now north; farther out it was southwest, and we steamed between the two currents. As far as the South Sand Head light-ship all the sounds were heard both through violent rain and through the noise of the paddles. To rain I have never yet been able to trace any deadening power; indeed such rain as we have hitherto encountered produced a distinctly opposite effect, and the reason is now intelligible. The siren on the present occasion was clearest and loudest, though at times the Canadian whistle showed great power. A struggle between the winds continued for some time, the north wind, accompanied by a cool atmosphere, at length prevailing. *[margin: No deadening effect can be ascribed to rain.]*

"Once while halting near the light-ship, when the Foreland was hidden in a dense rain-mist, I, being ignorant of its bearing, immediately found its position from the direction of the sound. *[margin: Superiority of the American siren.]*

"Thomson, the chief lamp-lighter at the South Sand Head, an exceedingly intelligent man, reported that on all occasions the sound of the siren had the mastery; and that opinion on this point was unanimous on board the light-vessel. On Friday and Saturday the sounds, he reported, were but faintly heard, being probably impaired by the local noises. To-day we found during our visit all the sounds very good, that of the siren being particularly intense.

* * * * * * *

"*October* 14.—* * * At 11.30 a. m. a gun was fired at the Foreland; report distinct. Up to this time the Canadian whistle had been adjusted to produce a shrill note; it was not heard. The piercing shrillness of this note, when heard at the South Foreland on October 10, suggested its trial to-day. The opinion that a note of this character, which affects an observer close at hand so powerfully and painfully, has also the greatest range, is a common one, and might be true in connection with homogeneous atmosphere. But in 'flocculent' air the shorter waves suffer most from partial reflection, exactly as the shorter waves of solar light suffer most in their passage through the suspended matter of the atmosphere. The blue of the firmament is, in fact, the echo of these shorter undulations. *[margin: Error in supposing a shrill note to be superior as a signal.]*

"According to arrangement, the Canadian whistle was now changed to its old low pitch. It was immediately heard at the Varne buoy.

* * * * * * *

"During the earlier part of this day the atmosphere, which, throughout, was of extreme optical clearness, favored the transmission of the longer sound-waves, corresponding to the deeper sounds.

Changes in the atmosphere on the same day.
"After a lapse of three hours the case was reversed, the high-pitched siren being heard when both gun and horns were absolutely inaudible. But even this was not permanent. Such changes on the part of the atmosphere have never hitherto been noticed, nor am I aware of a single observation bearing upon this selective stoppage of the sound. Its optical analogies have been already pointed out. The parcels of air and vapor play, to some extent, the same part in scattering the waves of sound as the minute particles suspended in the atmosphere do in scattering the solar light, producing by their preferences in this respect the blue of the sky.

* * * * * * *

Daboll's invention of fog-horns.
"*October* 15.—* * * To the late Mr. Daboll, of the United States, belongs the credit of bringing large trumpets into use as fog-signals. At Dungeness one of his horns had been erected under his own superintendence; and wishing to make myself acquainted with its performance, we steamed *Horn at Dungeness.* thither to-day. On examining the horn, I was struck by its similarity in all essential particulars to the horns employed at the South Foreland. Considerable improvements in the working of the horn have been introduced by Mr. Holmes, but the horn itself is substantially that of Daboll.

* * * * * * *

"*October* 18.—* * * There is no doubt that two days might be chosen on one of which the report of a pocket-pistol would be further heard than the report of an 18-pounder on the other. * * * *

"*October* 23.—* * * In the observations of Mr. Ayres to the west of the Foreland, wind and sound were almost in direct opposition; in those of Mr. Douglass they were by no means coincident. For a time both directions inclosed an angle of about 45°, and subsequently a greater angle. The *Effect of a thunder-shower on the atmosphere.* difference in the results is nevertheless striking. I may here draw attention to the remarkable effect of the rain and thunder-storm observed by Mr. Douglass. He was well in the sound-shadow near Kingsdown coast-guard station. He had sent a fly in advance of him, and the driver had been waiting for him for fifteen minutes without once hearing either trumpets or gun; nor had the coast-guardsman on duty heard any sound throughout the day. In fact, the the atmosphere prior to the thunder-storm was in that flocculent condition to which I have so often had occasion to refer, being composed of non-homogeneous locks of air and vapor. The thunder-storm, which I am assured by Mr. Douglass resembled the descent of a water-spout rather than of an ordinary shower of rain, abolished this condition of things, diminishing the partial echoes, and opening a freer way for the sound through the atmosphere.

"In the case of Mr. Ayres, the mastery of the siren over the gun was very conspicuous; in the case of Mr. Douglass also, though the difference was not so great, the siren was heard a mile farther than the gun. *Superiority of the American siren.*

* * * * * * *

"*October* 31.—* * * This was an exceedingly thick and squally day, with dense clouds and vapor everywhere. In acoustic opacity it was almost a match for the memorable 3d of July.

"Steamed with a view of getting dead to windward of station. The siren was clearly heard through all noises. During one particularly heavy squall, when the wind rose to a force of 8, the siren sent us a forcible sound, the horns at the same time being quite inaudible.

* * * * * * *

"*November* 21.—* * * The result of the day's observations was to prove that the siren suffered far more in being directed from us than the gun; this means that the conical trumpet associated with the siren is far more effectual in gathering up the sound and sending it in the direction of the axis than is the cylinder of the gun. *Result of directing the signals from the observers.*

"The siren, pointed on us, was heard to-day through the paddle noises at a distance of five miles.

"We made various observations in the sound-shadow and near it. The fluctuations in the strength of the sound indicated that we were passing through spaces of interference, the sound being sometimes suddenly augmented and sometimes suddenly deadened.

* * * * * * *

"In the neighborhood of an acoustic shadow—we need not be *in* the shadow—and with a wind of a force of 4 against the sound, there are states of the atmosphere in which even the siren with its axis pointed on the observer could not be trusted for a distance of one and a half miles. *Effect of proximity to an acoustic shadow.*

Horns, whistles, and guns, under those circumstances, are simply nowhere.

* * * * * * *

General review. "A brief review of our proceedings will aid the memory of the reader who has taken the trouble of going over the foregoing pages. Daboll's horn had been highly spoken of by writers on fog-signals. A third-order apparatus of the kind has been reported as sending its sound to a distance of from seven to nine miles against the wind, and to a distance of twelve to fourteen miles with the wind. Holmes had improved upon Daboll, and with an instrument of Holmes of *Commencement.* the first order our experiments were made. They began on the 19th of May, 1873. Whistles were also employed on this occasion, but those tested were speedily put out of court. *Whistles found useless.* At a distance of two miles from the Foreland they became useless. At three miles' distance the horns also became useless. At a distance of four miles, with paddles stopped *Gun effective.* and all on board quiet, they were wholly unheard. The 12 o'clock gun fired with a 1-pound charge at the Drop Fort in Dover was well heard on May 19, when the horns and whistles were inaudible. On this first day we noticed the sudden and surprising subsidence of the sound as we approached the acoustic shadow lying beyond the line joining the end of the Admiralty pier and the South Foreland. On the 20th of May the permeability of the atmosphere by sound had somewhat increased, but the steam-whistle failed to pierce it to a depth of three miles. At four miles the horns, though aided by quietness on board, were barely heard. By careful nursing, if I may use the expression, the horn-sounds were carried to a distance of six miles. The superiority of the 18-pounder gun, already employed by the Trinity House, over horns and whistles, was on this day so decided as almost to warrant its recommendation to the exclusion of all the other signals.

"Nothing occurred on the 2d of June to exalt our hopes of the trumpets and whistles. The horns were scarcely heard at a distance of three miles; sometimes, indeed, they failed to be heard at two miles. By careful nursing, keeping everything quiet on board, they were afterward carried to a distance of six miles. Long previously they had ceased to be of use as fog-signals. Considering the demands as to sound-range made by writers on this subject, the demonstrated incompetence of horns and whistles of great reputed power to meet these demands was not encouraging.

"On the 3d of June the atmosphere had changed surprisingly. It was loaded overhead with clouds of a dark and threatening character; the sounds, nevertheless, were heard at a distance of three and three-fourths miles through the paddle-noises, while with quietness on board they were heard beyond nine miles. *June 3, acoustic transparency.*

"On June 10 the acoustic transparency of the air was also very fair, the distance penetrated being upward of eight and three-fourths miles. A large horn employed on this day was heard at a distance of five miles through the paddle-noises. The subsidence of the sound near the boundary of the acoustic shadow on the Dover side of the Foreland was to-day sudden and extraordinary, affecting equally both horns and guns. We were warned to-day that the supremacy of the gun on one occasion by no means implied its supremacy on all occasions; the self-same guns which on the 20th had so far transcended the horns, being to-day their equals and nothing more. *Supremacy of gun not invariable.*

"The 11th of June was employed in mastering still further the facts relating to the subsidence of the sound east and west of the Foreland, the cause of this subsidence being in part due to the weakening of the sonorous waves by their divergence into the sound-shadow, and in part, no doubt, to interference.

"The atmosphere on the 25th of June was again very defective acoustically. The sounds reached a maximum distance of six and a half miles. But at four miles, on returning from the maximum distance, the sound was very faint. The guns to-day lost still further their pre-eminence; at five and a half miles their reports were inferior to the sound of the horn. No sounds whatever reached Dover Pier on the 11th, and it was only toward the close of the day that they succeeded in reaching it on the 25th. Thus by slow degrees the caprices of the atmosphere made themselves known to us; showing that even within the limits of a single day the air, as a vehicle of sound, underwent most serious variations. *Guns inferior.*

"The 26th of June was a far better day than its predecessor, the acoustic range being over nine and one-quarter miles. The direction of the wind was less favorable to the sound on this day than on the preceding one, plainly proving that something else than the wind must play an important part in shortening the sound-range.

"On the 1st of July we experimented upon a rotating horn, and heard its direct or axial blast, which was found to be the strongest, at a distance of ten and one-half miles. The sounds to-day were also heard at the Varne light-ship, which *July 1, rotating horn.*

is twelve and three-quarters miles from the Foreland. The atmosphere had become decidedly clearer acoustically, but not so optically, for on this day thick haze obscured the white cliffs of the Foreland. In fact, on days of far greater optical purity, the sound had failed to reach one-third of the distance attained to-day. By the light of such a fact, any attempt to make optical transparency a measure of acoustic transparency must be seen to be delusive. On the 1st of July a 12-inch American whistle, of which we had heard a highly favorable account, was tried in the place of the 12-inch English whistle; but, like its predecessor, the performance of the new instrument fell behind that of the horns. An interval of twelve hours sufficed to convert the acoustically clear atmosphere of the 1st of July into an opaque one; for on the 2d of July even the horn-sounds, with paddles stopped and all noiseless on board, could not penetrate farther than four miles.

<small>12-inch American whistle.</small>

"Thus each succeeding day provided us with a virtually new atmosphere, clearly showing that conclusions founded upon one day's observations might utterly break down in the presence of the phenomena of another day. This was most impressively demonstrated on the day now to be referred to. The acoustic imperviousness of the 3d of July was found to be still greater than that of the 2d, while the optical purity of the day was sensibly perfect. The cliffs of the Foreland could be seen to-day at ten times the distance at which they ceased to be visible on the 1st, while the sounds were cut off at one-sixth of the distance. At 2 p. m. neither guns nor trumpets were able to pierce the transparent air to a depth of three, hardly to a depth of two miles. This extraordinary opacity was proved to arise from the irregular admixture with the air of the aqueous vapor raised by a powerful sun.

<small>Extraordinary acoustic opacity with optical transparency.</small>

"This vapor, though perfectly invisible, produced what I have called an acoustic cloud impervious to the sound, and from which the sound-waves were thrown back as the waves of light are from an ordinary cloud. The waves thus refused transmission produced by their reflection echoes of extraordinary strength and duration. This I may remark is the first time that audible echoes have been proved to be reflected from an optically transparent atmosphere. By the lowering of the sun the production of vapor was checked, and the transmissive power of the atmosphere restored to such an extent that, at a distance of two miles from the Foreland, at 7 p. m. the intensity of the sound was at least thirty-six times its intensity at 2 p. m. Nothing requiring any notice

<small>Echoes from transparent air.</small>

here occurred on July 4, when our summer experiments ended.

"On October 8 the observations were resumed, a steam-siren and a Canadian whistle of great power being added to the list of instruments. A boiler had its steam raised to a pressure of 70 pounds to the square inch. On opening a suitable aperture this steam would issue forcibly in a continuous stream, and the sole function of the siren was to convert this stream into a series of separate strong puffs. This was done by causing a disk with twelve radial slits to rotate behind a fixed disk with the same number of slits. When the slits coincided a puff escaped; when they did not coincide the outflow of steam was interrupted. Each puff of steam at this high pressure generated a sonorous wave of great intensity, and the successive waves followed each other with such rapidity that they linked themselves together to a musical sound so intense as to be best described as a continuous explosion. *October 8, experiments resumed; siren.*

"During the earlier part of October 8 the optical transparency of the air was very great; its acoustic transparency, on the other hand, was very defective. Clouds blackened and broke into a rain and hail shower of tropical violence. The sounds, instead of being deadened, were improved by this furious squall; and, after it had lightened, thus lessening the local noises, the sound was heard at a distance of seven and one-half miles, distinctly louder than it had been heard through the preceding rainless atmosphere at a distance of five miles. Thus at five miles' distance the intensity of the sound had been at least doubled by the rain, a result obviously due to the removal by condensation and precipitation of that vapor, the mixture of which with the air had been proved so prejudicial to fog-signaling. We established this day a dependence between the pitch of a note and its penetrative power, the siren generating 480 waves, being slightly inferior to the horns; while generating 400 waves a second it was distinctly superior. The change in the atmosphere had been one favorable to the transmission of the larger waves. The maximum range on October 8 was nine miles. On October 9 the transmissive power had diminished, the maximum range being seven and one-half miles. On both these days the siren proved to be superior to the horns, and on some occasions superior to the gun. *Sound improved by rain.* *Dependence between pitch and penetrative power.*

"On the 10th and 11th, our steamer having disappeared, we made land-observations. We found the duration of the aerial echoes to be for the siren and the gun 9 seconds, for the horns 6 seconds. The duration varies from day to day. *October 10 and 11.*

We sought to estimate the influence of the violent wind which had caused our steamer to forsake us upon the sound, and found that the sound of the gun failed to reach us in two cases at a distance of 550 yards against the wind; the sound of the siren at the same time rising to a piercing intensity. To leeward the gun was heard at five times, and certainly might have been heard at fifteen times, the distance attained to windward. The momentary character of the gun-sound renders it liable to be quenched by a single puff of wind; but low sounds generally, whether momentary or not, suffer more from an opposing wind than high ones. We had on the 13th another example of the powerlessness of heavy rain to deaden sound.

<small>Sounds against the wind.</small>

"On the 14th the maximum range was ten miles, but the atmosphere did not maintain this power of transmission. It was a day of extreme optical clearness, but its acoustic clearness diminished as the day advanced. In fact the sun was in action. We proved to-day that by lowering the pitch of the Canadian whistle its sound, which had previously been inaudible, became suddenly audible. The day at first was favorable to the transmission of the longer sound-waves. After the lapse of three hours the case was reversed, the high-pitched siren being then heard when both gun and horns were inaudible. But even this state of things did not continue, so rapid and surprising are the caprices of the atmosphere. At a distance of five miles, at 3.30 p. m., the change in the transmissive power reduced the intensity of the sound to at least one-half of what it possessed at 11.30 a. m., the wind throughout maintaining the same strength and direction. Through all this complexity the knowledge obtained on July 3 sheds the light of a principle which reduces to order the apparent confusion.

<small>Pitch and penetration.</small>

"October 15 was spent at Dungeness in examining the performance of Daboll's horn. It is a fine instrument, and its application was ably worked out by its inventor; still it would require very favorable atmospheric conditions to enable it to warn a steamer before she had come dangerously close to the shore. The direction in which the aerial echoes return was finely illustrated to-day, that direction being always the one in which the axis of the horn is pointed.

<small>October 15; Daboll's horn.</small>

"The 16th was a day of exceeding optical transparency, but of great acoustic opacity. The maximum range in the axis was only five miles. On this day the howitzer and all the whistles were clearly overmastered by the siren. It was, moreover, heard at three and a half miles with the

<small>October 16; superiority of the siren.</small>

paddles going, while the gun was unheard at two and a half miles. With no visible object that could possibly yield an echo in sight, the pure aerial echoes, coming from the more distant southern air, were distinct and long-continued at a distance of two miles from the shore. Near the base of the Foreland cliff we determined their duration, and found it to be 11 seconds, while that of the best whistle-echoes was 6 seconds. On this day three whistles, sounded simultaneously, were pitted against the siren, and found clearly inferior to it. On the 17th four horns were compared with the siren, and found inferior to it. This was our day of greatest acoustic transparency, the sound reaching a maximum of fifteen miles for the siren, and of more than sixteen for the gun. The echoes on this day were audible for a longer time than on any other occasion. They continued for 15 seconds; their duration indicating the atmospheric *depth* from which they came.

"On October 18, though the experiments were not directed to determine the transmissive-power of the air, we were not without proof that it continued to be high. From 10 to 10.30 a. m., while waiting for the blasts of the siren at a distance of three miles from the Foreland, the continued reports of what we supposed to be the musketry of skirmishing parties on land were distinctly heard by us all. We afterward learned that the sounds arose from the rifle-practice on Kingsdown beach, five and a half miles away. On July 3, which, optically considered, was a far more perfect day, the 18-pounder howitzer and mortar failed to make themselves heard at half this distance. The 18th was mainly occupied in determining the influence of pitch and pressure on the siren-sound. Taking the fluctuations of the atmosphere into account, I am of the opinion that the siren, performing from 2,000 to 2,400 revolutions a minute, or, in other words, generating from 400 to 480 waves per second, best meets the atmospheric conditions. We varied the pressure from 40 to 80 pounds on the square inch, and though the intensity did not appear to rise in proportion to the pressure, the higher pressure yielded the hardest and most penetrating sound. *Rifle-practice heard 5½ miles.*

"The 20th was a rainy day with a strong wind. Up to a distance of five and a half miles the siren continued to be heard through the sea and paddle noises. In rough weather, indeed, when local noises interfere, the siren-sound far transcends all other sounds. On various occasions to-day it proved its mastery over both gun and horns. On the 21st, when the deputy master paid us a visit, the wind was *October 20; siren superior in rough weather.*

strong and the sea high. The horn-sounds, with paddles going, were lost at four miles; the siren continued serviceable up to six and a half miles. The gun to-day was completely overmastered. Its puffs were seen at the Foreland, but its sound was unheard when the siren was distinctly heard. Heavy rain failed to damp the power of the siren. The whistles were also tried to-day, but were found far inferior to the siren.

October 22; siren superior when local noises interfere.
"On the 22d it blew a gale, and the Galatea quitted us. We made observations on land on the influence of the wind and of local noises. The shelter of the coast-guard station at Cornhill enabled us to hear gun-sounds which were quite inaudible to an observer out of shelter; in the shelter also both horn and siren rose distinctly in power, but they were heard outside when the gun was quite unheard. As usual, the sound to leeward was far more powerful than those at equal distances to windward. The echoes from the cloudless air were to-day very fine. On the 23d, in the absence of the steamer, the observations on the influence of the wind were continued. The quenching of the gun-sounds, in particular to windward, was well illustrated. All the sounds, gun included, were carried much farther to leeward than to windward. The effect of a violent thunderstorm and downpour of rain in exalting the sound was noticed by the observers both to windward and to leeward of the Foreland. In the rear of the siren its range to-day was about a mile. At right angles to the axis, and to windward, it was about the same. To leeward it reached a distance of seven and one-third miles.

Increase of sound in rain.

"On the 24th, when observations were made afloat in the steam-tug Palmerston, the siren exhibited a clear mastery over gun and horns. The maximum range was seven and three-quarters miles. The wind had changed from west-southwest to southeast, then to east. As a consequence of this the siren was heard loudly in the streets of Dover. On the 27th the wind was east-northeast; and the siren-sound penetrated everywhere through Dover, rising over the moaning of the wind and all other noises. It was heard at a distance of six miles from the Foreland, on the road to Folkestone, and would probably have been heard all the way to Folkestone had not the experiments ceased. Afloat and in the axis, with a high wind and sea, the siren, and it only, reached to a distance of six miles; at five miles it was heard through the paddle-noises. On the 28th further experiments were made on the influence of pitch; the siren, when generating 480 waves a second, being found

October 28; influence of pitch.

more effective than when generating 300 waves a second. The maximum range in the axis to-day was seven and one-half miles.

"The 29th of October was a day of extraordinary optical transparency, but by no means transparent acoustically. The gun was the greatest sufferer. At first it was barely heard at five miles, but afterward it was tried at five and one-half, four and one-half, and two and one-half miles, and was heard at none of these distances. The siren at the same time was distinctly heard. The sun was shining strongly, and to its augmenting power the enfeeblement of the gun-sound was probably due; wind from east-southeast to east-northeast. At three and one-half miles subsequently, dead to windward, the siren was faintly heard; the gun was unheard at two and three-fourths miles. On land Mr. Douglass heard the siren and horn sounds to windward at two to two and one-half miles; to leeward Mr. Edwards heard them at seven miles, while Mr. Ayres, in the rear of the instruments, heard them inland at a distance of five miles, or five times farther than they had been heard on October 23.

"The 30th of October furnished another illustration of the fallacy of the prevalent notion which considers optical and acoustic transparency to go hand in hand. The day was very hazy, the white cliffs of the Foreland at the greater distances being quite hidden; still the gun and siren sounds reached on the bearing of the Varne light-vessel to a distance of eleven and one-half miles. The siren was heard through the paddle-noises at nine and one-fourth miles, while at eight and one-half miles it became efficient as a signal with the paddles going. The horns were heard at six and one-fourth miles. This was during calm. Subsequently, with a roaring wind from the north-northwest, no sounds were heard at six and one-half miles. At South Sand Head the siren was very feeble, the gun and horns being inaudible. The wind was here across the direction of the sound. On land, the wind being also across, the siren was heard only to a distance of three miles northeast of the Foreland; in the other direction it was heard plainly on Folkestone Pier, eight miles distant; such was the influence of the wind. Both gun and horns failed to reach Folkestone. *[October 30; atmosphere thick but acoustically transparent.]*

"Wind, rain, a rough sea, and great acoustic opacity characterized October 31. Both gun and horns were unheard three miles away. The siren at the same time was clearly heard. It afterward forced its sound with great *[October 31; siren superior under very unfavorable conditions.]*

power through a violent rain-squall. Wishing the same individual judgment to be brought to bear on the sounds on both sides of the Foreland, in the absence of our steamer, which had quitted us for safety, I committed the observations to Mr. Douglass. He heard them at two miles on the Dover side, and on the Sandwich side, with the same intensity, at six miles.

"A gap, employed by me in preparing this report, and by the engineers in making arrangements for pointing the siren in any required direction, here occurs in our observations. They were, however, resumed on November 21, when comparative experiments were made upon the gun and siren. Both sources of sound, when employed as fog-signals, will not unfrequently have to cover an arc of 180°, and it was desirable to know with greater precision how the sound in windy weather is affected by the direction in which the gun or siren is pointed.

Effect of changing the line of direction of siren. "The gun, therefore, was in the first instance pointed on us and fired, then turned and fired along a line perpendicular to that joining us and it. There was a sensible, though small, difference between the sounds which reached us in the two cases. A similar experiment was made with the siren, and here the falling off, when the instrument was pointed perpendicular to the line joining us and it, was very considerable. This is what is to be expected, for the trumpet associated with the siren is expressly intended to gather up the sound and project it in a certain direction, while no such object is in view in the construction of the gun. Hence any deviation from that direction must, in the case of the siren, be attended with a greater weakening of the sound than in the case of the gun. The experiments here referred to were amply corroborated by others made on November 22 and 23.

Aerial echoes. "On both of these days the Galatea's guns were fired both to windward and to leeward. The aerial echoes in the latter case were distinctly louder and longer than in the former.

"In front of the Cornhill coast-guard station, and only one and one-fourth miles from the Foreland, the siren, on the 21st, though pointed toward us, fell suddenly and considerably in power. Before reaching Dover Pier it had ceased to be heard. The wind was here against the sound; but this, though it contributed to the effect, could not account for it, nor could the proximity of the shadow account for it. To these two causes must have been added a flocculent atmosphere. The experiment demonstrates conclusively that there are atmospheric and local conditions which when combined prevent our most powerful instruments from making more than a distant approach to the performance which writers on fog-signals have demanded of them.

"On November 24 the sound of the siren pointed to windward was compared at equal distances in front of and behind the instrument. It was louder to leeward in the rear than at equal distances to windward in front. Hence in a wind the desirability of pointing the instrument to windward. The whistles were compared this day with the siren deprived of its trumpet. The Canadian and the 8-inch whistles proved the most effective, but the naked siren was as well heard as either of them. As regards opacity, the 25th of November almost rivaled the 3d of July. The gun failed to be heard at a distance 2.8 miles; it yielded only a faint crack at two and one-half miles. This, as on July 3, was when the air was calm. A revival of the wind subsequently brought with it a revival of the sound.

marginal note: November 24; comparison of windward and leeward direction of sound.

* * * * * * *

"While the *velocity* of sound has been the subjects of refined and repeated experiments, I am not aware that since the publication of a celebrated paper by Doctor Derham, in the Philosophical Transactions for 1708, any systematic inquiry has been made into the causes which affect the *intensity* of sound in the atmosphere. Derham's results, though obtained at a time when the means of investigation were very defective, have apparently been accepted with unquestionable trust by all subsequent writers; a fact which is, I think, in some part to be ascribed to the *a priori* probability of his conclusions.

marginal note: Intensity of sound. Derham's paper.

"Thus Doctor Robinson, whom I have already quoted, * * * relying apparently upon Derham, says: 'Fog is a powerful damper of sound;' and he gives a physical reason why it must be so. 'It is a mixture of air and globules of water, and at each of the innumerable surfaces where these two touch, a portion of the vibration is reflected and lost.' And he adds further on, 'The remarkable power of fogs to deaden the report of guns has been often noticed.'

marginal note: Dr. Robinson's statement.

"Assuming it, moreover, as probable that the measure of 'a fog's power in stopping sound' bears some simple relation to its opacity for light, Dr. Robinson, adopting a suggestion of Mr. Alexander Cunningham, states that 'the distance at which a given object, say a flag or pole, disappears may be taken as a measure of the fog's power' to obstruct the sound. This is quite in accordance with prevalent notions, and, granting that the sound is dissipated, as assumed, by reflection from the particles of fog, the conclusion follows that the greater the number of the reflecting particles the greater will be the waste of sound. But the number of particles, or, in other words, the density of the fog, is declared by its action upon light; hence the optical opacity will be a measure of the acoustic opacity.

These opinions shown to be erroneous.

"This I say expresses the opinion generally entertained; 'clear still air' being regarded as the best vehicle for sound. We have not, as stated above, experimented in really dense fogs, but the experiments that we have made entirely destroy the notion that clear weather is necessarily better for the transmission of sound than thick weather. Some of our days of densest acoustic opacity have been marvelously clear optically, while some of our days of thick haze have shown themselves highly favorable to the transmission of sound. Were the physical cause of the sound-waves that above assigned, did that waste arise in any material degree from reflection at the limiting surfaces of the particles of haze, this result, as I have already pointed out, would be inexplicable.

Falling rain as an obstructor.

"Again, Derham, as quoted by Sir John Herschel, says that 'falling rain tends powerfully to obstruct sound.' We have had repeated reversals of this conclusion. Some of our observations have been made on days when rain and hail descended with a perfectly tropical fury, and in no single case did the rain deaden the sound. In every case, indeed, it had precisely the opposite effect.

Falling snow.

"But falling snow, according to Derham, offers a more serious obstacle than any other meteorological agent to the transmission of sound. We have not extended our observa-

Observation on the Mer de Glace.

tions at the South Foreland into snowy weather. But I may be permitted to refer to an observation of my own which bears directly upon this point. On Christmas night, 1859, I reached Chamouni, through snow so deep as to obliterate the road-fences. On the 26th and 27th it fell heavily. During a lull in the storm I reached the Montanvert, sometimes breast-deep in snow. On the 29th the entry in my journal is 'Snow, heavy snow; it must have descended through the entire night, the quantity freshly fallen is so great.' Dr. Derham had referred to the deadening effect produced by a coating of fresh-fallen snow upon the ground, alleging that when the surface was glazed by freezing the damping of the sound disappeared.

"Well, on December 29, I took up a position beside the Mer de Glace, with a view to determine its winter motion, and sent my assistants across the glacier with instructions to measure the displacement of a transverse line of stakes planted previously in the snow. I was standing at the time beside my theodolite, with snow 4 feet deep around me. A storm drifted up the valley, darkening the air as it approached. It reached us, the snow falling more heavily than ever I had seen it elsewhere. It soon formed a heap on my theodolite; still through the telescope I was able to pick up at intervals the retreating forms of my men. Here there was a combination of thick snow in the air and of soft fresh snow on the ground, such as Derham had hardly enjoyed. Through such an atmosphere, however, I was able with my unaided voice to make my instructions audible for half a mile, while the experiment was rendered reciprocal by one of my men making his voice audible to me.

* * * * * * *

"The real enemy to the transmission of sound through the atmosphere has, I think, been clearly revealed by the foregoing inquiry. That enemy has been proved to be not rain, nor hail, nor haze, nor, I imagine, fog or snow; not water, in fact, in either a liquid or a solid form, but water in a vaporous form, mingled with air, so as to render it acoustically turbid and flocculent. This acoustic turbidity often occurs on days of surprising optical transparency. Any system of measures, therefore, founded on the assumption that the optic and acoustic transparency of the atmosphere go hand in hand must prove delusive. *[Non-homogeneous vapor the cause of acoustic opacity.]*

"There is but one solution of this difficulty: it is to make the source of sound so powerful as to be able to endure considerable loss by partial reflection and still retain a sufficient residue for transmission. Of all the instruments hitherto examined by us the siren comes nearest to the fulfillment of this condition. Its establishment upon our coast will, in my opinion, prove an incalculable boon to the mariner. *[Superiority of the siren.]*

* * * * * * *

"We had, as already stated, been favored with thunder, hail, rain, and haze, but not with dense fog. All the more anxious was I to turn the recent excellent opportunity to account. On Tuesday, December 9, I therefore telegraphed to the Trinity House, suggesting some gun-observations. A prompt reply informed me that such observations would be made in the afternoon at Blackwall or in its neighborhood. *[No experiments had been made in dense fog.]*

* * * * * * *

"Slowly, but surely, we thus master this question. And the further we advance the more we are assured that our reputed knowledge regarding it has been erroneous from beginning to end. Fogs, like rain, have no such power *[Fogs do not deaden sound.]*

to deaden sound as, since the time of Derham, has been universally ascribed to them.

* * * * * * *

Experiments in Hyde Park, London.

"An assistant placed at the end of the Serpentine sounded the whistle and pipe for fifteen minutes without interruption. An observer at the bridge noticed the fluctuations of the sound. Sometimes the whistle was heard alone, sometimes the organ-pipe. Sometimes both whistle and pipe began strongly and ended by sinking almost to inaudibility. Extraordinary fluctuations were also observed in the peal of bells, to which reference has been already made. In a few seconds they would sink from a loudly ringing peal into utter silence, from which they would rapidly return to loud-tongued audibility. The intermittent drifting of fog over the sun's disk, by which his light is at times obscured, at times revealed, is, as already stated, the optical analogue of these acoustical effects. In fact, as regards such changes, the acoustic deportment of the atmosphere is a true transcript of its optical deportment.

* * * * * * *

December 31.

"On December 31 I went to the end of the Serpentine, at noon, to listen to the Westminster bell. Not one of the twelve strokes was audible, nor were the chimes heard. On several of the first days of this year I placed myself beside the railing of St. James's Park, near Buckingham Palace, and waited at noon for the stroke of the bell; it was quite unheard. These days were moist and warm, the air was calm, and the clock-tower in sight. On January 19 I placed myself in the same position; fog and drizzling rain obscured the tower; still I heard not only the strokes of the big bell, but also the preceding chimes of the quarter-bells. The air was calm at the time.

January 22.

"During the exceedingly dense and 'dripping' fog of January 22 I placed myself near the same railings, and heard every stroke of the bell. On the same day I sent an assistant to the end of the Serpentine, and when the fog was densest he heard the Westminster bell striking loudly eleven. Toward evening this fog began to melt away, and at 6 o'clock I went to the end of the Serpentine to observe the effect of the clearing of the atmosphere upon the sound. Not one of the strokes reached me. At 9 o'clock, and at 10 o'clock, my able assistant, Mr. Cottrell, was in the same position, and on both occasions failed to hear a single stroke of the bell. It was a case precisely similar to that of December 13, when the dissolution of the fog was accompanied by a decided acoustic thickening of the atmosphere. All this shows what instructive results are to be obtained in connection with the transmission of sound through the atmosphere from a mode of observation accessible to all.

"This opportune fog enabled us to remove the last of a congeries of errors which, ever since the year 1708, have attached themselves to this question. As regards phonic coast-signals, we now know exactly where we stand; and, through the application of this knowledge to maritime purposes, a meteorological phenomenon, which was bewailed in London as an unmitigated evil, may in the end redound to the advantage of the public. *Errors removed.*

"Since the publication of the first notices of this investigation various communications have reached me, to one or two of which I should like to refer. The Rev. George H. Hetling, of Fulham, has written to me with a circumstantiality which leaves no room for doubt, that he has heard the Portland guns at a distance of forty-four miles through a dense fog. *Portland guns heard forty-four miles through dense fog.*

"The Duke of Argyll has also favored me with the following account of his own experience. Coming as it does from a disciplined scientific observer, it is particularly valuable. 'This fact [the permeability of fog by sound] I have long known, from having lived a great part of my life within four miles of the town of Greenock, across the Frith. Ship-building goes on there to a great extent, and the hammering of the calkers and builders is a sound which I have been in the habit of hearing with every variety of distinctness, or of not hearing at all, according to the state of the atmosphere; and I have always observed on days when the air was very clear, and every mast and spar was distinctly seen, hardly any sound was heard, whereas on thick and foggy days, sometimes so thick that nothing could be seen, every clink of every hammer was audible and appeared sometimes as close at hand. This has been long a very familiar experience with me.' *Statement of the Duke of Argyll.*

"It is hardly necessary for me to say a word to guard myself against the misconception that I consider sound to be assisted by the fog itself. Fog I regard as the visible result of an act of condensation, which removes the real barrier to transmission, that barrier being aqueous vapor so mixed with air as to render it acoustically flocculent or turbid. The fog-particles appear to have no more influence upon the waves of sound than the suspended matter stirred up over the banks of Newfoundland has upon the waves of the Atlantic. *The real barrier to transmission.*

* * * * * * *

Superiority of the siren.

"An absolutely uniform superiority on all days cannot be conceded to any one of the instruments subjected to examination; still, our observations have been so numerous and long-continued as to enable us to come to the sure conclusion that, on the whole, the steam-siren is beyond question the most powerful fog-signal which has hitherto been tried in England. It is specially powerful when local noises, such as those of wind, rigging, breaking waves, shore-surf, and the rattle of pebbles have to be overcome. Its density, quality, pitch, and penetration render it dominant over such noises after all other signal-sounds have succumbed.

"I do not hesitate to recommend the introduction of the siren as a coast-signal.

Change in mounting siren desirable.

"It will be desirable in each case to confer upon the instrument a power of rotation, so as to enable the person in charge of it to point its trumpet against the wind, or in any other required direction. This arrangement has been made at the South Foreland, and it presents no mechanical difficulty. It is also desirable to mount the siren so as to permit of the depression of its trumpet 15° or 20° below the horizon.

Position of fog-signal.

"In selecting the position at which a fog-signal is to be mounted, the possible influence of a sound-shadow, and the possible extinction of the sound by the interference of the direct waves with waves reflected from the shore, must form the subject of the gravest consideration. Preliminary trials may in most cases be necessary before fixing on the precise point at which the instrument is to be placed.

* * * * * * *

Siren with compressed air.

"The form of the siren which has been long known to scientific men is worked with air, and it would be worth while to try how the fog-siren would behave supposing compressed air to be substituted for steam. Compressed air might also be tried with the whistles. * * *

Robinson's conditions not fulfilled by any fog-signal.

"No fog-signal hitherto tried is able to fulfill the condition laid down by Dr. Robinson, * * * namely, '*that all fog-signals should be distinctly audible for at least four miles under every circumstance.*' Circumstances may arise to prevent the most powerful sounds from being heard at half this distance. What may with certainty be affirmed is, that in almost all cases the siren, even on steamers with paddles going, may be relied on at a distance of two miles; in the great majority of cases it may be relied upon at a distance of three miles, and in the majority of cases at a distance greater than three miles.

"Happily, the experiments thus far made are perfectly concurrent in indicating that at the particular time when fog-signals are needed, that is, during foggy weather, the air in which the fog is suspended is in a highly homogeneous condition; hence it is in the highest degree probable that in the case of fog we may rely upon these signals being effective at much greater distances than those just mentioned. *[Signals most efficient during fog.]*

"I say 'probable,' while the experiments seem to render this result certain. Before pronouncing it so, however, I should like to have some experience of warmer fogs than those in which the experiments have hitherto been made. That the fog-particles themselves are not sensibly injurious to the sound has been demonstrated; but it is just possible that in warm weather the air associated with the fog may not be homogeneous. I would recommend the experiment necessary to decide this point to be made on some of the fogs of the early summer.

"I am cautious not to inspire the mariner with a confidence which may prove delusive. When he hears a fog-signal he ought, as a general rule, at all events until extended experience justifies the contrary, to assume the source of sound to be not more than two or three miles distant, and to take precautions accordingly. *[Distance at which a signal should be heard.]*

"Once warned, he may, by the hearing of the lead or some other means, be enabled to check his position. But if he errs at all in his estimate of distance, it ought to be on the side of safety.

"Unless very cogent practical reasons can be adduced in its favor, I should strongly deprecate a lengthened interval between the siren-blasts. My own small experience has shown me how harassing to the mariner are some of our revolving lights with a long period of rotation. No light, in my opinion, ought to be obscured for more than 30 seconds, and the interval between the two blasts of our fog-signal ought not to be longer. *[Intervals between blasts.]*

"With the instruments now at our disposal, wisely established along our coasts, I venture to believe that the saving of property in ten years will be an exceedingly large multiple of the outlay necessary for the establishment of such signals. The saving of life appeals to the higher motives of humanity."

* . * . * . *

General Duane, of the Corps of Engineers of the Army and light-house engineer of the New England coast, forwarded to the Light-House Board on January 12, 1872, a *[Report from General Duane.]*

report which corroborates the results of Professor Tyndall's experiments, some of which were foreshadowed in his treatise on Sound, published in 1867. This report contains much practical information in regard to fog-signals, and it is to be regretted that it has not been published. The following are extracts from General Duane's report:

* * * * * *

Fog-signals on the Maine coast. "There are six steam fog-whistles on the coast of Maine. These have been frequently heard at a distance of twenty miles, and as frequently cannot be heard at the distance of two miles, and this with no perceptible difference in the state of the atmosphere.

Signal sometimes inaudible in one direction. "The signal is often heard at a great distance in one direction, while in another it will be scarcely audible at the distance of a mile. This is not the effect of wind, as the signal is frequently heard much farther against the wind than with it. For example, the whistle on Cape Elizabeth can always be distinctly heard in Portland, a distance of nine miles, during a heavy snow-storm, the wind blowing a gale directly from Portland toward the whistle.

Belt impenetrable by sound, surrounding the signal. "The most perplexing difficulty, however, arises from the fact that the signal often appears to be surrounded by a belt, varying in radius from one to one and a half miles, from which the sound appears to be entirely absent.

Sound lost for a certain distance, and then recovered. "Thus, in moving directly from a station, the sound may be audible for the distance of a mile, is then lost for about the same distance, after which it is again distinctly heard for a long time. This action is common to all ear-signals, and has been at times observed at all the stations, at one of which the signal is situated on a bare rock twenty miles from the main-land, with no surrounding objects to affect the sound. All attempts to re-enforce the sound by means of reflectors have hitherto been unsuccessful. Upon a large scale sound does not appear, on striking a surface, to be reflected after the manner of light and heat, but to roll along it like a cloud of smoke.

* * * * * *

Conditions of the atmosphere. "From an attentive observation during three years of the fog-signals on this coast, and from the reports received from captains and pilots of coasting-vessels, I am convinced that in some conditions of the atmosphere the most powerful signals will be at times unreliable.

* * * * * *

Reflection affecting sound. "Now, it frequently occurs that a signal, which under ordinary circumstances would be audible at the distance of fifteen miles, cannot be heard from a vessel at the distance of a single mile. This is probably due to the reflection mentioned by Humboldt.

"The temperature of the air over the land where the fog-signal is located being very different from that over the sea, the sound, in passing from the former to the latter, undergoes reflection at their surfaces of contact. The correctness of this view is rendered more probable by the fact that when the sound is thus impeded in the direction of the sea, it has been observed to be much stronger inland.

"When a vessel approaches a signal in a fog a difficulty is sometimes experienced in determining the position of the signal by the direction from which the sound appears to proceed, the apparent and true direction being entirely different. This is undoubtedly due to the refraction of sound passing through media of different density. *[Difficulty in determining position of the signal]*

"Experiments and observation lead to the conclusion that these anomalies in penetration and direction of sound from fog-signals are to be attributed mainly to the want of uniformity in the surrounding atmosphere, and that snow, rain, fog, and the force and direction of the wind have much less influence than has generally been supposed." *[Reason for the anomalies found to exist]*

* * * * * * *

While this report is passing through the press, Sir Frederick has also kindly sent me a copy of his memorandum to the Elder Brethren of the Trinity House concerning the report of Professor Tyndall, and it will be found below.

MEMORANDUM BY SIR FREDERICK ARROW, THE DEPUTY MASTER OF THE TRINITY HOUSE, UPON DR. TYNDALL'S REPORT ON THE EXPERIMENTS AT SOUTH FORELAND.

"At the close of a series of important investigations, undertaken with the desire of adding to the safety of navigation round our seaboard, to which a committee of the Elder Brethren, acting in concert with the corporation's scientific adviser, have devoted many months of careful attention, it will be convenient, now that the report of Dr. Tyndall has been presented to the board, to consider how far the conclusions arrived at may be practically and usefully applied.

"Before entering, however, upon the subject-matter of the report, it is due to the members of the committee, who have freely sacrificed their time and comfort during a protracted period, that the board should record its acknowledgment of their careful prosecution of a long and arduous duty.

And if this acknowledgment is due to the members of the committee, much more is to be accorded to their distinguished scientific guide, Dr. Tyndall, who at great inconvenience to himself, with serious encroachment on his valuable time, and frequently some personal discomfort, has applied himself to this investigation with characteristic patience and perseverance. Step by step, after repetitions almost wearying to those unversed in such trained and dispassionate habits of procedure, the conditions affecting the traveling-power of sound have become clearer and clearer, old errors have been corrected, and a great advance has been made toward accurately estimating the value of sound-signals; and the important knowledge has been acquired that the seaman's greatest enemy affords in itself aid to mitigate its worst evils.

"To Mr. Douglass also and his assistants much credit is due for the very thorough and efficient manner in which they performed their duties in connection with this inquiry, and Mr. Douglass's assistance as a practical observer on board the yachts has throughout the experiments been of great service.

"Observations at sea were commenced on the 19th of May, 1873, after some months previously employed by the engineering department in mounting the requisite instruments, (the selection of which was based upon the report of the committee who had visited the fog-signal establishments of the North American seaboard,) and in making such arrangements as were suggested by the experience of the Elder Brethren and Mr. Douglass or by the views of Dr. Tyndall on the subject. From that time to the 21st of February, 1874, on shore and at sea the observations have been going on at short intervals, and frequently for weeks together.

"Foremost among the practical results is the important fact before alluded to, viz, that fog does not impede the transmission of sound, (as has long been supposed;) indeed, it is shown that a foggy atmosphere is a highly favorable condition for the traveling of the sound-wave; further, during heavy, blinding rain or snow storms the passage of sound through the air is not obstructed; indeed, the observations of the committee in the former case record an increase in the power of the sound either during or immediately after a rain-storm; while the evidence of Dr. Tyndall of his Alpine experience with regard to falling snow may be accepted as proving the latter. It may safely be concluded, therefore, that whenever the state of the weather is such as to render sound-signals necessary, the atmospheric conditions are most favorable for their efficient application, and it may also be concluded that, under the conditions of weather above referred to, the range of the signals will be much greater than the limit laid down in the report as the result of the general observations.

"Turning to the action of the wind upon sound, the report confirms all previous experience, and shows that the most powerful sound fails to penetrate the opposing force of a strong wind to any considerable distance; but it is satisfactory to be assured that even against a moderate gale and unfavorable conditions for sound-transmission signals may be relied on for sending sound to a distance of two or three miles, and, under ordinary conditions of fog, considerably farther. Having regard, however, to the variability of the sound-range of the same instrument on different days, attributable to the varying conditions of the atmosphere, it is not possible to assert positively that any signal has an absolute range which may be relied upon at all times. The practice of the Elder Brethren of not publishing in their notices of fog-signals a maximum range of audibility, or of accepting isolated instances as every-day occurrences, is therefore amply justified. Happily long ranges are not very necessary, inasmuch as the mariner does not need to hear a sound-signal at ten, fifteen, or twenty miles. Dr. Tyndall quotes, in relation to this part of the subject, an opinion expressed by our late scientific adviser, Professor Faraday, that 'a false promise to the mariner would be worse than no promise at all.' The Brethren need scarcely be reminded that in so saying our dear and venerated friend was simply giving utterance to the standard axiom of the Trinity House, as old, perhaps, as the corporation itself, viz, that safety is only to be found in certainty, and that anything which does not secure the latter condition is a foe, rather than a friend, to the mariner.

"Bearing in mind, therefore, the liability to atmospheric interference under ever-varying conditions, as shown in the report, attention must next be directed to the instruments used in the investigation and the conclusions arrived at with regard to each of them, so far as the inquiry has now advanced.

"The instruments tried were the American siren, Holmes's trumpets, American, Canadian, and English steam-whistles, and three guns. The effects obtained from all these instruments have varied in a remarkable manner at different times, but for general efficiency there is no doubt that the American siren takes the first place. It has shown the greatest penetrative power, especially where local noises have to be overcome, but at present it has the drawback of being worked by steam at the high pressure of 70 pounds, which is not only a serious element of danger, but entails considerable expense for fuel and labor. Mr. Douglass, however, tells us that the caloric-engine will work the siren, and he confidently anticipates being able to do away with steam altogether, and so to render this instrument a safe, economical, and efficient signal for general adoption.

"The air-trumpet has also shown itself to be an efficient instrument, superior to the whistles and sometimes equaling the siren. Its chief advantage is that it is blown by means of the caloric-engine at something over 20 pounds' pressure, and can be worked without skilled labor, and, avoiding the danger attendant upon the employment of steam or gunpowder, combines safety with economy; and its clear, musical note may be an element of distinctiveness capable of being developed so as to make it ultimately of some service in this respect. In actual practice there are one or two drawbacks to the use of reed-instruments, such as the difficulty of tuning, liability of reeds cracking, &c. Such contingencies are not likely to arise in regard to the siren, and if the economy and simplicity of working by means of the caloric-engine can be also applied to the more powerful siren, it seems clear that the result will be highly advantageous as a most useful combination of power, safety, and economy well suited for fog-signal purposes. Nevertheless the satisfactory performance of the trumpets during the late trials fully justifies their present employment as fog-signals.

"Not so much can be said in favor of whistles. Throughout the trials their marked inferiority to the other instruments has been recorded. The American whistle, yielding a harsh roar, when close at hand was deafening, but its sound failed to penetrate to any useful distance.* The Canadian whistle appears to have been better, but it also failed in general effective power, although occasionally it was heard a great distance, even obtaining superiority over the other instruments, but this was of very rare occurrence. As a rule, the whistles were behind the siren, trumpet, and gun, and seem to have been dependent, more than the other instruments, on exceptional atmospheric conditions for

---

* This statement does not agree with our experience in the practical use of a large number of steam fog-whistles on the seacoast of the United States.—E.

yielding their best results. The general conclusion seems to be that for practical purposes the steam-whistles, as at present tried, are not proved to be advantageous as fog-signals.

"With respect to the usefulness of guns, it appears from the report that they possess certain disadvantages, viz, the short duration of the sound, the liability of that sound to be quenched by local noises, and its comparative inability to cope with an opposing wind. Dr. Tyndall nevertheless ranks the gun as a first-class signal, an opinion which long experience of its use confirms. The gun, as a signal, is well known to mariners, while the flash is also said to be of service in thick weather. With regard to the guns used in the experiments, it appears that the short $5\frac{1}{2}$-inch howitzer, with a 3-pound charge, is superior to the long 18-pounder or the 13-inch mortar with the same charge, the howitzer having generally yielded a louder and more effective report. The subject of a special gun for fog-signal purposes is now under consideration, and it seems probable that both in effective power and facility of working, the gun may be rendered considerably more serviceable than hitherto.

"From the foregoing observations it will be seen that at present there are three kinds of instrument practically available for future service as fog-signals, viz, the siren, the trumpet, and the gun, and as further experience is gained with regard to these instruments, it may reasonably be expected that great improvement will be made in them and that the future results to be obtained from them will surpass those now recorded.

"It will be well now to consider briefly one or two points in connection with the selection of a site for, and the instrument to be used as a fog-signal, when the locality has been determined upon. In the investigations the question as to the height above the sea at which it is desirable that a signal should be placed has received some attention, and the results show that it is advantageous that such signals should be placed at a considerable height above the sea-level in order to avoid the interference caused by the noise of waves breaking on the shore, the rattle of pebbles, &c. The comparative trial made between a pair of horns on the summit of the South Foreland cliff and another pair 200 feet below, close to the sea-surface, proved with scarcely an exception that the higher horns were superior to the lower ones. From this it would appear advisable, where possible, to place the signal high above the sea, but there are positions on our coasts where fog-signals are necessary, and yet where no considerable elevation could be obtained, notably Dungeness, Orfordness, &c. For such positions, therefore, a large and powerful siren would be very suitable as being able to overcome the noises of the sea-shore. For river-banks, light-vessels, and other places undisturbed by interfering noises, a smaller instrument of the same description, or a trumpet, would prove serviceable, and the gun would be a fit signal for such places as are of some elevation clear above the sea, without adjacent outlying rocks, and which vessels may approach 'close to.' It is not intended in the above suggestions to lay down any system which shall be invariably followed in the allocation of fog-signals, but, having regard to the performances of the instruments referred to as shown in the report, the foregoing observations may be regarded as indicating to some extent how the respective merits of the instruments may be most usefully applied.

"While upon the subject of fog-signal sites allusion may be made to the caution conveyed in the report that in selecting the position for a fog-signal the possible influence of sound-shadow must be taken into account. This is a point to which reference was made in the report of your committee to America, it being therein recorded that in the experiments carried out at Portland Bay, United States,* the effect of a sound-shadow was distinctly experienced, and your committee stated in their report that 'in selecting the site for a fog-signal care must be taken that no outlying point or cliff shall interfere with the arc of sound.' It is satisfactory to find the conclusion of that committee on this point entirely borne out by the investigations of another.

"Another important conclusion to be drawn from this report in regard to the question of sites is that no signal should be required to mark dangers extending seaward more than a mile or a mile and a quarter. The minimum effective range of a signal being $2\frac{1}{2}$ miles, vessels approaching such dangers and coming into the sound-range would have room to maneuver and be able to keep at a safe distance. This is, of course, taking the minimum range of the signals, as stated in the report, but it is more than probable, as has been stated previously, that in foggy weather, the sound-range being extended farther than the minimum limit referred to, a larger range may be allowed.

"Another important consideration has to be borne in mind, viz, the direction in which sound should be projected. As

---

* The experiments referred to have been carried on from time to time for some years, under the direction of Professor Henry.—E.

the sound of a signal is ascertained to be most effective in the line of its axis, it follows that the instrument should be capable of such adjustment that its strongest sound may be projected directly against the opposing wind.

"It is to be observed that on another point these experiments confirm the opinions expressed by your committee to America, for, with regard to the question of distinctions, it is clearly shown that it is not possible to rely upon distinctiveness of note alone, for the mariner would not appreciate such a distinction; indeed the siren, horns, and whistles have invariably been spoken of by sailors in the vicinity as "the fog-horns." Between the report of the gun and the sound of the siren or the trumpet there is a perfectly intelligible difference; but for further purposes of distinction for the latter instruments, variation of the length of the silent interval between each blast offers the most satisfactory means. With regard to this point it will be seen that Dr. Tyndall has, with some reservation, expressed an opinion which hardly seems to harmonize with the experience of the Elder Brethren. Dr. Tyndall would restrict the silent interval to a length of 30 seconds, and in support of his opinion draws an analogy between the action of the eye and the ear, which does not commend itself to actual nautical experience. The board will probably not be disposed to waive a clear advantage in power and great scope for distinctiveness, in order that the longest interval of silence should not exceed 30 seconds, especially with the knowledge that guns fired at intervals of a quarter of an hour have proved of great service to the mariner hitherto.

"A general review of the entire report shows that a considerable amount of knowledge has been gained, both as to the influence of the atmosphere in the transmission of sound and to what extent the appliances we possess may be relied on for producing such sounds as will be of practical service to the mariner. We have learned something of our ignorance in regard to sound-transmission. We now know that the varying conditions of the atmosphere render no judgment infallible, and that conclusions founded on the experience of to-day, are not trustworthy for estimating the results of the morrow. We know, moreover, that after bringing forward all the aid which science can at present give to guide the mariner in thick weather, there is still a large element of uncertainty and mystifying influence with which he has to combat, and which renders it incumbent on him to use the greatest caution and prudence in thick weather, to regard and make use of the sound-signals as means for assuring the vessel's position, and not as aids for running at high speed; and, above all, never to trust so implicitly to sound-signals as to neglect the use of the seaman's best friend and truest guide, the lead.

"The subject of fog-signals has by means of this investigation received a great impetus. It may fairly be said that we have taken a considerable step in advance, and it only remains to follow it up. As we go forward our experience will widen, and although it is more than probable that a few years of practical experience and testing of fog-signals will materially modify our present views, and improve considerably the instruments we have, yet we now know *how* to go forward and in what direction to head our efforts. It is to be hoped that before very long our coasts will be guarded by a complete chain of sound-signals, all effective and useful to the mariner. No unnecessary delay need now occur before proceeding to supply the light-ships and the important stations already selected by the board, and when they are all established the lights rendered useless at a quarter of a mile by fog will be superseded by sound-signals capable of warning the mariner at a distance of three miles.

"It is almost unnecessary to add that in thus giving practical effect to the spirit of the recommendations of this valuable report, the Elder Brethren will have the satisfaction of knowing they are acting in the highest interests of humanity and conferring an inestimable boon on the nautical community at large."

SOUTH FORELAND.

Location. — The great electric lights at South Foreland, two in number, are three miles east of Dover Pier, on the high chalk-cliffs overlooking the Strait of Dover, from which can be seen Gris-nez and other French lights. They are about 1,000 feet apart, the high light 372, the low one 275 feet above the sea, and form a range or lead as a guide to clear the Goodwin Sands, one of the greatest dangers in British waters. A general plan of the establishment is shown in Plate I.

Engine-house. — A fire-proof engine-house, a plan of which is shown in Plate II, is placed midway between the towers, and contains the magneto-electric machines, the engine-room, boiler-room, coal-room, and two repair-shops. Near by are the dwellings of the engineer who superintends the establishment, and those of some of the keepers, there being six at this station.

Height of focal planes.

Dwellings.

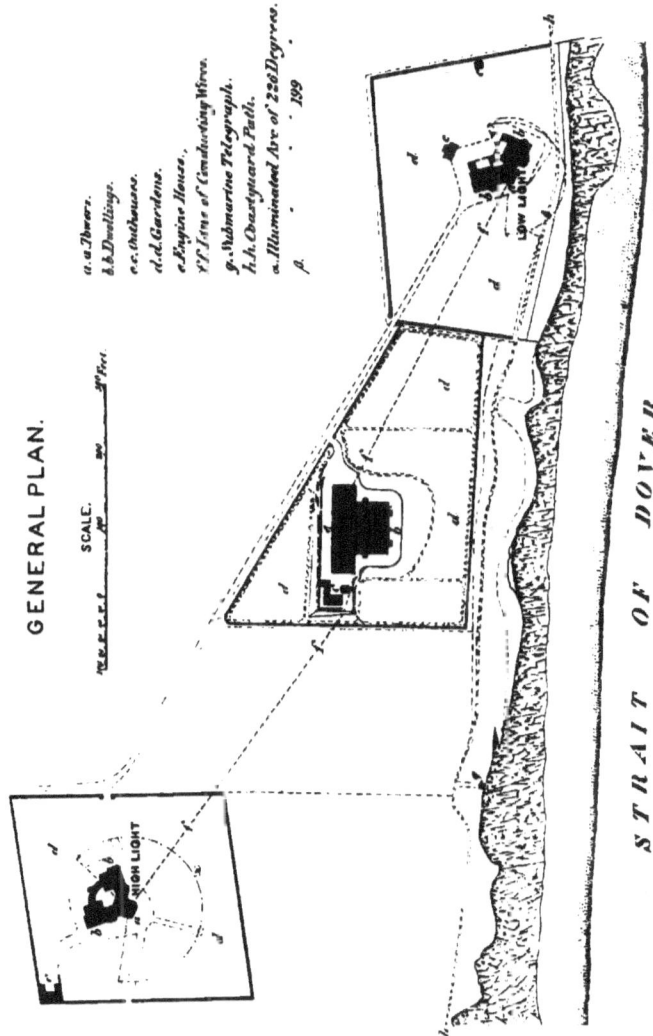

The electric current is generated by means of large magneto-electric machines, two of which are provided for each light, though habitually in clear weather but one machine is used for each. These machines are driven by means of belting connected with a steam-engine, a duplicate of which is kept for use in case of accident or repairs. The boilers, which are of the ordinary locomotive class, are also in duplicate. About 56 pounds of coke per hour are consumed during the night; during the day the fires are banked. One of these electric machines is of French manufacture, having been made by the *Compagnie l'Alliance* of Paris. The others are English-made machines of Professor Holmes's patent, and are considered by the Elder Brethren to be superior, though the French appeared to be the simpler in construction and is the one shown to visitors in explaining the operation of generating the electric current. This operation is fully illustrated in the description, which will be found further on, of the magneto-electric lights at La Hève near the mouth of the Seine. Magneto-electric machines. Manner of operating the same. Coke used. Machines of French and English manufacture. Description given at La Hève.

It may be well to state here, however, that each machine is composed of ninety-six helices mounted upon six gunmetal wheels, each having sixteen helices. Helices.

Between these wheels are placed the magnets, eight in each division, forty of which are composed of six layers or leaves riveted together, and sixteen (the end ones) similarly constructed but having only three leaves or layers. These magnets, which are mounted in frames, are stationary, while the helices revolve at the rate of four hundred revolutions per minute. Magnets.

The power absorbed by the machine alone, disregarding friction, is four indicated horse-power, and the actual power required to work one of the machines, including the friction of engine and shafting, is six indicated horse-power. Power required for operating the machines.

The power of a magneto-electric machine is according to the gross attractive power of its magnets, each magnet having a certain lifting or attractive power, (expressed in pounds.) In the machines at South Foreland each of the six-plate magnets will lift 108 pounds, and each three-plate magnet will lift 54 pounds, making the attractive power of the magnets in one machine to be $40 \times 108 + 16 \times 54 = 5,184$ pounds. This may be considered as expressing the power of the machine. The proportion of the lifting power to the actual weight of a magnet is a good indication of its value, and, generally speaking, a magnet which will lift two and one-half times its own weight is a good one. Each six-plate magnet at South Foreland has a weight of $43\frac{1}{2}$ pounds, and Power of the machines. Value of magnets.

will lift 108 pounds. The total weight of all the magnets in one machine is 2,088 pounds, the total attractive power being, as stated, about 5,184 pounds.

Cables connecting machines and lamps. The machines are connected with the electric lamps placed in the lenses of the tower by underground cables.

Thanks to Professor Tyndall for explanations, &c. The manner of operating these machines, the arrangements of the lenses, lamps, &c., were carefully explained to me by Professor Tyndall, who had kindly accompanied me through the station, and who spared no efforts to make my inspection a thorough and minute one.

Carbon points. Each lamp contains two pieces of carbon, each of which is about 10 inches long by three-eighths of an inch square. These are placed end to end, one above the other, and are kept at the proper distance apart by an automatic apparatus.

The current leaps across the small space separating the 'carbons,' and a series of sparks is formed, but so rapidly that the eye cannot separate them, and a most brilliant light is produced. By the automatic apparatus the carbon pencils are moved toward each other as fast as they are consumed, and the only danger of irregularity of the lights, *Danger from the presence of foreign matter in the carbons.* if the machines and cables are in good order, arises from the presence of foreign matter in the carbons. I was told by the keepers that the carbons in use give them trouble in this particular, the lights being sometimes extinguished. This is only for an instant, however, as all that is necessary to relight them is to bring the carbons in contact, after which they are replaced in their proper positions.

Arrangement for managing the light. The arrangements for bringing the electric light to the focus of the lens, and for feeding the carbons as fast as they are consumed, are simple and ingenious, and the duties of the keepers, beyond watching for the occurrence of imperfections in the carbons, are very light. These carbons are made from coke-dust; their rate of consumption is 34 inches per night for each light, at a cost of one penny per inch, exclusive of waste and breakage.

Lenses. The lenses in use at South Foreland are of about the same size as ordinary third-order lenses, (39 inches interior diameter,) and were especially designed for the electric light and this locality by Chance Brothers & Co., of Birmingham.

Required arc of illumination. Utilizing the rear light. The arc of illumination required being but little more than 180°, the rear light is ingeniously utilized to re-enforce the other by means of totally reflecting prisms, and it was observed that a much greater development was given to the catadioptric prisms below the central belt than can be done in the use of the large burners of the oil-light.

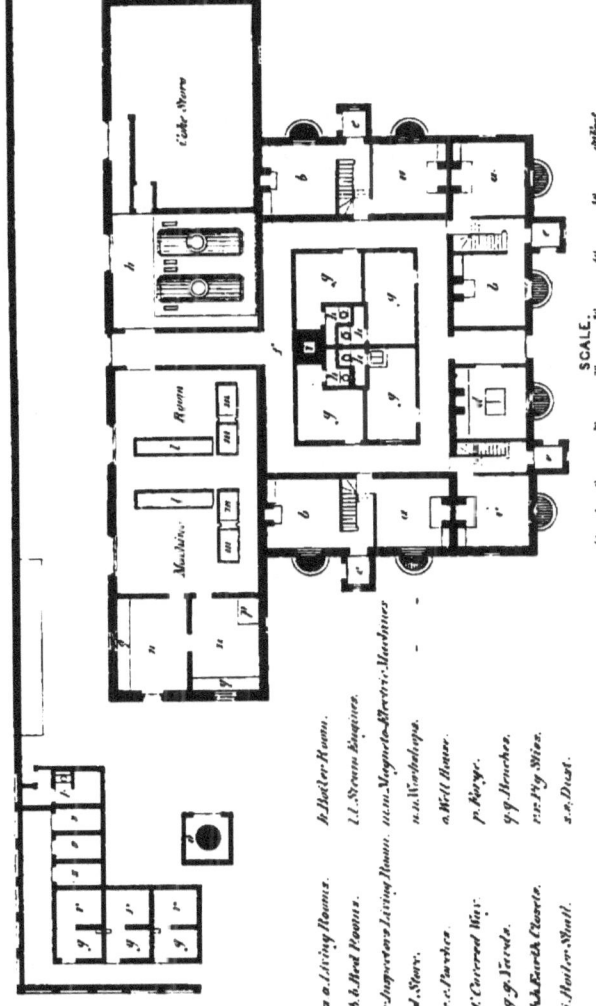

Both lights are fixed, and illumine somewhat more than half the horizon. In both the portion of light which would otherwise be wasted landward is led round by subsidiary apparatus to intensify the illuminated arc; but the low light presented the peculiarity that it was desired to light the sea brightly from the horizon to 300 yards from the base of the tower, and as the height of the focal plane is 290 feet above the sea, this requires that a portion of the light should dip below the horizontal plane no less than $17° 23' 32''$, while the major portion should go to the sea-horizon. This requirement was met in the following manner: The light from both the top and bottom systems of reflecting prisms was directed wholly to the horizon, while the central refracting portion of the cupola was specially arranged to give the required dip. The latter consists of sixteen refracting segments, eight above and eight below the focal plane, there being no central refracting belt as is usually the case. These refracting segments are so arranged that the intensity of the light viewed from the point distant 300 yards, or from the horizon, is sensibly the same; also, that no portion of the sea shall depend for its illumination on one prism only, as otherwise that point might be placed in total darkness, owing to a lantern-bar or other obstacle intercepting the light.

<small>Description of optical apparatus.</small>

<small>Main apparatus.</small>

The arrangement adopted for using the light which would otherwise be wasted toward the land is symmetrical about the middle line of the apparatus, each half consisting of a portion of a holophote, and of a frame of six vertical prisms. The former condenses the light of the spark into a beam of horizontal parallel rays, which pass to the side of the apparatus and are there distributed over half the illuminated arc by the vertical prisms. The rays emerging from these prisms do not pass through a single focus as in lights previously constructed on the azimuthal condensing plan, but each prism has a special focus so situated that the light is equally distributed over the arc illuminated. Probably the plan of a common focus was adopted in some earlier cases as simpler for calculation, and also in order to facilitate the examination of the apparatus for adjustment, but the examination can be conducted with equal accuracy with special foci, and no trouble should be spared to render such an apparatus as perfect as possible. The outer prism of the set is of a special and difficult construction, required because the arc it has to illumine is situated at so great an angle from the direction of the rays it receives. Prisms of a similar construction had been previously suggested by Mr.

<small>Subsidiary apparatus.</small>

Thomas Stevenson, engineer of the Scottish lights, and have been used by Mr. James Chance in several other instances. In cases like the South Foreland, many plans for attaining the end in view suggested themselves. It might be proposed that some arrangement similar to the subsidiary one at Souter Point (which will be described further on) should be used. Such an arrangement would be unsuitable in the case of South Foreland, for various reasons, among which may be mentioned that the light would pass through three optical agents instead of two. At Souter Point, the light being revolving, the only feasible plan was to take the spare light downward through the pedestal. Again, the central cupola might have been continued entirely round the spark and the spare light from the land-side of the cupola reflected seaward by large vertical prisms. This plan was rejected on account of its being unnecessarily expensive, and for other reasons. The advantages of these and other plans were considered by Mr. Chance before he adopted the arrangement shown on the drawing.

Moderator-lamp for use in case of accident. To supply any contingency necessitating its use a "moderator" oil-lamp is placed under the electric lamp and can be quickly substituted for it. This lens is shown in elevation and plan in Plate III, in which a and a' show the electric lamp in position and withdrawn for removing the carbon pencils; $b\ b$, the carbon pencils; $c\ c$, the electric wires; d, the bed-plate; e, the burner of the oil-lamp; f, the telescopic supply and overflow pipes for the oil; g, the oil-lamp; h, elevation of one of the half-holophotes; $h'\ h'$, the two half-holophotes in plan; $i\ i$, $i'\ i'$, vertical condensing prisms.

In substituting the oil for the electric light, the bed-plate d is removed and the oil-burner e is run up to the focus by *Power of electric light.* means of a rack and pinion. The power of the uncondensed beam from each of the electric lights at South Foreland, *i. e.* of the naked light without the condensation of the rays produced by the lenses, is equal to the combined light of 2,000 candles, while the corresponding power of the four-wick sea-coast light-house oil-light is 328 candles. Estimating the power of the condensed beam from the South Foreland lenses as ninety times the power of the naked light, (which is the result of Mr. Chance's calculation of the condensing-power of the South Foreland lenses,) we have for the power of the beams from each of the electric lights at South Foreland, 180,000 candles!

Observations at light from Dover pier. I was shown the method of lighting the electric lamp, and in the evening observed the two lights from Dover Pier. While the upper light was decidedly the superior, the lower

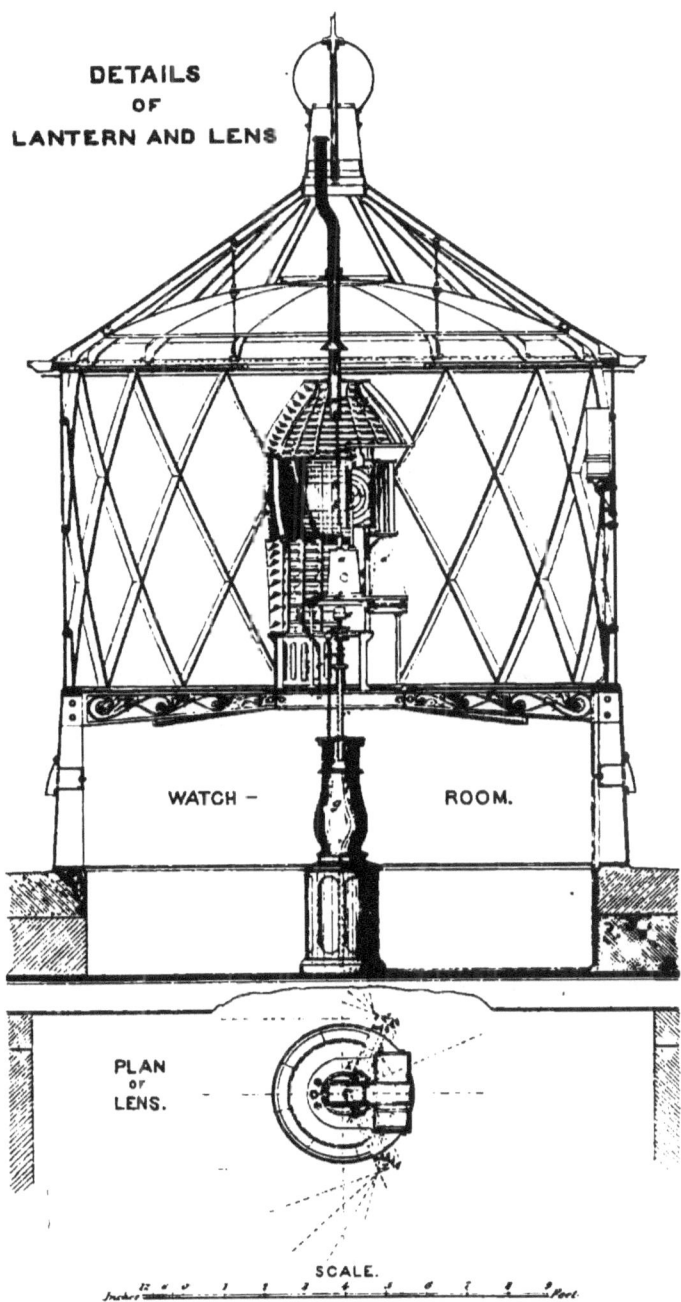

waxing and waning very perceptibly, which was caused, as Mr. Douglass informed me, by some imperfection in the machinery; both of them surpassed anything I had ever seen, and I could not convince myself that they were three miles off. Even at this distance, the shadows of objects on the pier were very distinct.

The towers supporting these lights are not high, (their site being an elevated one,) and they are attached to keepers' dwellings. The buildings are all constructed in the most substantial manner, and each dwelling is sufficient for the accommodation of two keepers and their families, with room for the supernumeraries who are kept at the station for instruction. The steam-engineer (the principal official of the establishment) and the principal light-keepers are competent to manage the engine and magneto-electric machines, and they attend to this duty in turn. No laborers or firemen are employed. The dwellings at this station, as at all the stations I visited on the English coast, were kept extremely neat. Rooms for accommodation of any visiting officer of the Trinity House are fitted up in the dwelling attached to the engine-house, and everything is provided for his comfort, even to a full set of table-furniture. *Buildings. Machines managed by steam-engineer and principal keepers. Rooms for accommodation of Trinity House officers.*

Attached to the central station (the engine-house) are the different store-rooms for the supplies, which include lubricating and colza oils, the latter for use in the lenses in case of failure of the electric light or of other accident; but I believe it has rarely if ever been needed. Some of the oil-butts are of tin, the others of galvanized sheet-iron. *Store-rooms. Oil-butts.*

The colza (rapeseed) oil is of a pale sherry-color, very clear and limpid, with a strong and peculiar vegetable odor. *Colza-oil.*

In all the houses the steps and stairs, as well as the paving of all the halls and corridors, were of stone, rubbed once a week with Bath-stone, which produces a color like that given by a wash of hydraulic cement.

In the towers, and also at the engine-house, speaking-tubes are arranged to communicate from one to the other, and for the purpose of calling relief. I found such tubes at all the stations I visited. Each keeper's watch is four hours. *Speaking-tubes.*

Both watch-rooms and lanterns at South Foreland are considerably larger and more commodious than our own. *Watch-rooms and lanterns, size of.*

At each dwelling is an earth-closet, placed in an outbuilding, in which, instead of earth, the ashes produced at the station are used, and I was informed that the use of earth-closets at light-stations is universal, and gives entire satisfaction. *Earth-closets.*

Fire, means for extinguishing. Means for extinguishing accidental fires are provided by the engine which drives the magneto-electric machines. The pumps are connected by pipes to each of the towers and dwellings, the water being drawn through the chalk, from a well at a depth of 200 feet, during high tide in the Strait of Dover, when the water backs up into the well. Reservoirs are provided for use at low water.

Meteorological observations. Two keepers are designated for each tower, who, in addition to their other duties, make daily observations with the barometer and with wet and dry bulb thermometers, keeping memoranda for the use of some department of the government.

The keepers. The two principals, who are assistants to the engineer, I found to be very intelligent men who seemed thoroughly to understand the magneto-electric machines, and who gave me a very accurate account of their operation. One of them was by trade a watch-maker, and the other a stone-mason. The latter told me, with evident pride, that he had laid all the stone at the Bishop Rock, near the Scilly Islands, one of the most exposed stations in the English service, and had been for some years the principal keeper of that light, a position he was obliged to resign, the close confinement affecting his health. Each of these men had been more than fifteen years in the service.

Cost of maintenance of light. The annual cost of maintenance of a single electric English light is about £800, (or $4,000,) about double that of a first-order single oil-light station, while the light produced by the former is between six and seven times that of the most powerful lens with the four-wick Douglass oil-burner.

Cost of substituting electric-light apparatus, &c., for oil. The approximate cost of substituting at a double-light station, the magneto-electric lights as used at South Foreland for the oil-lights commonly used is as follows:

Building works.......................	£7,360	$36,800
Lantern and dioptric apparatus	3,088	15,440
Electric-apparatus	5,356	26,780
Miscellaneous	650	3,250
Total........................	£16,454	$82,270

I am indebted for my detailed description of the excellent optical apparatus at South Foreland to the manufacturers, Messrs. Chance, Brothers & Co., of Birmingham.

THE ROMAN PHAROS IN DOVER CASTLE.

Visit to Dover Castle. Through the kindness of Colonel Collinson, of the Royal Engineers, I had an opportunity of visiting the castle at Dover, and of attending a review of the three regiments of Kent County militia, and of the garrison of three regiments of regular troops. Colonel Collinson showed me many objects of great interest—the grand old castle, from which were distinctly visible the coasts of France, the towers of the cathedral of Boulogne, and the light-house at Calais, on the other side of the Channel; the rooms occupied by Charles I and by Queen Elizabeth; the church of Saint Mary, Within the Castle, founded A. D. 161,* and the modern exterior forts; but nothing was more interesting to me, considering the nature of my mission to Europe, than the old pharos within the castle walls, the present condition of which is represented in Fig. 1.

Fig. 1.

Roman pharos in Dover Castle.

The antiquity of this light-house, which has not probably been used as such since the Conquest, no doubt exceeds that of any light-house in Great Britain, and it is supposed to have been built in the reign of the Emperor Claudius, about A. D. 44.*

Upon it burned for many centuries those great fires of wood and coal formerly maintained on several towers still standing on the coasts of Great Britain. These earliest guides to mariners at length gave way to reflectors; they, in their turn, being replaced in the year 1819 by that great triumph of scientific skill, the Fresnel lens. Date of invention of the Fresnel system.

The pharos, like its sister light-house, the Tour d'Ordre at Boulogne, is built of brick, in color and shape like those in the Roman structures found elsewhere in Great Britain; they are of a light-red color, about 14 inches long, and not more than an inch and a half thick. This latter dimension Construction of the pharos. Description of the Roman bricks.

---

* Hasted's History of Kent.

is but little more than the thickness of the joints, which are filled with a mortar composed of lime and finely-powdered Roman brick. The preservation of this famous relic of the Romans in England is doubtless due to the fact that some centuries ago the tower was turned into a belfry for the church of Saint Mary, and was surrounded by walls of stone. These are now nearly destroyed by time, and the old Roman work is again exposed.

Visits to Trinity House.

While the Trinity House steam-yacht Vestal, in which I was to take my first cruise among the English lights, was fitting out for her annual voyage to the northeast coast of England, I made frequent visits to the Trinity House, where I was always cordially welcomed, and thus I acquired much information regarding the English light-house system.

Kindness received from Sir Frederick Arrow.

I received much kindness in many ways from Sir Frederick Arrow, and on the 21st of May accompanied him to a dinner at the Mansion House, to which, through his good offices, I had the honor to be invited by the Lord Mayor and Lady Mayoress, by whom it was given in honor of the return of the Master of Trinity House, the Duke of Edinburgh. About three hundred guests were present, and it was a highly enjoyable and interesting occasion.

Dinner at the Lord Mayor's.

Acknowledgment made by the Duke of Edinburgh for attentions to Sir Frederick Arrow and Captain Webb while in the United States.

In his response to a toast to the Trinity House, His Highness spoke of the gratitude of the corporation for the services rendered Sir Frederick Arrow and Captain Webb of the Elder Brethren during their stay in this country, (to which I have before referred,) and commissioned me to convey his thanks to my associates of the Light-House Board of the United States. While the services referred to were insignificant as compared with those rendered to me while in England, I was much gratified by the highly complimentary terms in which he mentioned our country, and particularly our light-house establishment.

Interview with Captain Doty.

The Doty lamp.

While in London at this time Captain Doty, patentee of a burner for light-house illumination, addressed me a note requesting an interview, which request I complied with, and he showed me his lamp, which has been patented in several countries, including the United States. It combines the outer "cone" or "jacket," the central "button" and adjustable gallery, but has not the conical "tips" peculiar to the Douglass lamp, which also comprises the above improvements. Captain Doty claims that the Douglass or Trinity House lamp is an infringement of his patent, and the ques-

Claim of Captain Doty that the Trinity House lamp is an infringement of his patent.

tion has been before the courts, with what results I am not fully informed, but, as I found in my subsequent inspection, the English are rapidly changing their light-house lamps for those of the Douglass pattern, and their illuminant from colza to mineral oil. *English using new lamps and mineral oil.*

Both the Doty and the Douglass lamp are especially designed for burning the latter, though they are equally adapted to the use of vegetable and animal oils. *Lamps adapted to all oils.*

Captain Doty claims, among other things, to have first suggested the great economy in the use of mineral oil, and that the invention of his lamp made its use practicable in light-house illumination. He stated that the French government had issued a general order to change all the lights on the coast of France from colza to mineral oil; that the Scotch were rapidly introducing such oil for use in their light-houses, and that his lamp had been adopted by France, Scotland, and Sweden. Judging from my observations at the trial, (which, however, was not comparative,) I believe Captain Doty's lamp to be an excellent one. Of his claim to priority of invention I was not sufficiently informed to judge; but, from what I afterward learned in Paris from M. Lepaute, who showed me the lamp invented by his father in 1845, I am inclined to doubt if such claim can be sustained, though I do not question the fact that Captain Doty is entitled to much credit for having directed attention to the advantages to be gained by the adoption of mineral oil for light-houses. *Further claims of Captain Doty. Opinion regarding the Doty lamp. Doubt as to Captain Doty's claim of priority of invention.*

One evening, after dining with Professor Tyndall at his club, I went with him to observe, from the terrace on which stands the Duke of York's column, the competitive gas and electric lights on Westminster clock-tower. The electric light I have already described in my account of the South Foreland light-station; the gas-light will be fully described in my treatment of the subject of Irish light-houses. *Observation of the trial of gas and electric lights on Westminster clock-tower.*

It may be well, however, to state here that in the experimental gas-light on Westminster clock-tower three Wigham burners (each composed of 108 jets, but so arranged as to burn 28, 48, 68, 88, or 108 jets as desired) were placed one above another, at a distance of three feet from center to center. Before the lower one was a refracting belt of a first-order dioptric apparatus for a fixed light; before the upper two burners were placed two refracting panels of a first-order apparatus for revolving light, each panel being for an arc of 45°. These panels were arranged for rotating before the flame and producing in combination with the refracting *Wigham burner. Disposition of burners. Arrangement for producing fixed light varied by flashes.*

of belt, a fixed light varied by flashes. Both of these lights were magnificently bright, but their nearness to our place of observation was unfavorable to a comparative test, and caused the gas-light to appear vertically elongated, an effect which my subsequent observations of the gas-light in actual use in light-houses on the coast of Ireland and east coast of England convinced me was not a necessary feature of the system. The reddish tinge which prevailed in the gas much more than in the electric light would probably enable it to more successfully penetrate fog.

Mr. Douglass found the power of this gas-light, burning 108 jets, the beam being uncondensed by lenticular apparatus, to be equal to that of 1,199 candles when consuming 300 feet of cannel-coal gas per hour of the illuminating value of 25 candles.

TRINITY HOUSE DEPOT AT BLACKWALL.

On the 11th of June I went by rail to visit the principal depot of the Trinity House at Blackwall, on the Lower Thames. This depot is much the same as our own at Staten Island, New York Harbor, but at the former are repaired the numerous light-ships employed on the coast above and below the mouth of the Thames, while our light-ships are repaired at private yards.

The grounds are rather limited in extent, and some of the buildings are old and inconvenient, but facilities for all kinds of work and for storage are as good as can be obtained until the additional area which is desired can be purchased.

The lamp-shop is very complete in its appointments, and a large number of men are employed there. Many of them are constantly engaged upon the manufacture and repair of the catoptric apparatus for light-ships, of which the English have a great number.

In many of the light-houses reflectors are still in use, being considered better than lenses for some localities, especially for range or "leading" lights.

The apparatus for light-ships is hung, as in our own service, upon a universal joint, and an ingenious improvement has been recently adopted, by which the reflector is adjusted by passing the shaft of the gimbal through slots which allow it to be moved backward or forward, and the face of the reflector maintained in a vertical position.

In the new Douglass lamp, which is being rapidly supplied to all English light-houses, the light may be increased

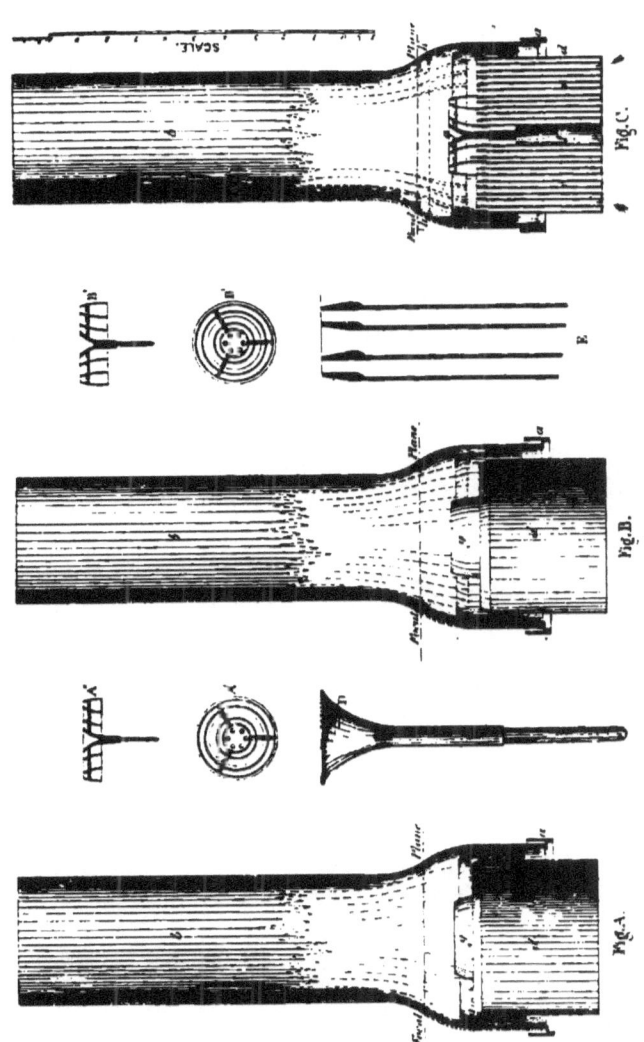

as desired in thick and foggy weather, being in this regard similar in effect to the electric and gas lights, although it is not possible to make the proportionate increase so great with oil-light as with those just mentioned.

These lamps are of different orders. Among those I saw at Blackwall was the six-wick lamp for mineral or colza oil, (in which mineral oil was to be used,) designed to be placed in one of the towers at Haisborough, on the east coast of England, for experimental comparison with the gas-light, which will receive mention when I come to describe that station. *Six-wick lamp.*

This lamp is called by Mr. Douglass the "lamp of single and double power," from its capacity for increasing or diminishing the light to suit the state of the atmosphere. For example, in the case of the six-wick burner, shown in Plate IV, the ordinary fair-weather light is produced by the flame from the outer three wicks only, but, as the weather becomes thick or foggy, the inner three may be successively lighted, increasing the power of the flame from 342 to 722 candles—more than double its fair-weather power. *Description.*

These lamps burn either animal, vegetable, or mineral oil. The burner of one for six wicks is shown in Plate IV, in which figures A and B are elevations showing the adjustment for burning mineral and colza oils respectively, and figure C is a section of the latter; a is the chimney-holder; $b\ b$ the chimney; c the exterior deflector; d the outer wick-case; e the inner cases; f the central air-space; g the interior deflector; A' and A'' plan and section of interior deflector for mineral-oil; B' and B'' the same for colza-oil; D the central button, and E an enlarged view of the burner-tips.

In burning mineral oil the wicks are raised about one-sixteenth of an inch above the tips of the burners, and the exterior deflector is kept in the position shown at A in Plate IV. *Use of mineral oil.*

In burning colza the wicks are raised about five-sixteenths of an inch above the tips of the burners. The oil overflows as in our lamps, and the exterior deflector is placed as shown at B, Plate IV. *Use of colza.*

About one and one-fourth inches below the tops of the wick-cases are small holes, kept closed by caoutchouc valves when colza is used, but when mineral oil is burned the holes are opened and the oil is maintained at that height.

Both the exterior and interior deflectors are readily removed for the purpose of trimming the wicks. *Removal of deflectors.*

The "buttons" and "tips" are the same for all sizes of *Buttons and tips.*

lamps, their number corresponding to the order of the lamp, *i. e.*, the "tip" and "button" of a one-wick lamp is applicable to the inner wick-case of a two, three, four, five, or six wick lamp, and a tip of any specified number will fit the corresponding wick-case of any order of lamp.

Zone of maximum intensity. Mr. Douglass states that a notable feature in the flames of the improved lamp is the increased power of the beam in the direction of the sea-horizon over that from an old one of the same initial power, such increase being due to the narrow zone of maximum intensity found in the flame of the new burner, and which is fully utilized for the longest range by the refracting portions of dioptric apparatus. In the old flames the zone of maximum intensity does exist, but the difference in power between it and the portions of the flame of minimum intensity is not great.

Adoption of level of maximum intensity. This level of maximum intensity, shown at $h\,h$ in Plate IV, is now being adopted by the Trinity House for the sea-horizon focus of the refractors of dioptric apparatus and its height above the tips of the burners in the several lamps is as follows:

	Millimeters
Height above the tips of the burners. Burner of one wick	13
Burner of two wicks	14
Burner of three wicks	15
Burner of four wicks	16
Burner of five wicks	17.5
Burner of six wicks	19

Gain to light at the zone of maximum intensity. With these adjustments of the foci some light is necessarily cut off from the lower catadioptric prisms by the "exterior deflector," but the increase of light from the refractor is very great, it having been found that the power of the light sent to the sea-horizon or maximum range is from 25 to 30 per cent. more than can be obtained from one of the old flames; that is to say, by taking two flames, one old and one new, of the same total initial candle-power as measured by the photometer, the beam of maximum intensity in the direction of the sea-horizon from a dioptric apparatus with the new flame in focus will be 25 to 30 per cent. greater than with the old flame.

Gas-burner. I was shown a new kind of gas-burner, an invention of Mr. Douglass. The top is perforated with a circle of very small holes, and this is combined with a perforated button similar to that used for the oil-lamp for light-houses. Outside is placed the adjustable jacket or cone, and the adjustable chimney is also used.

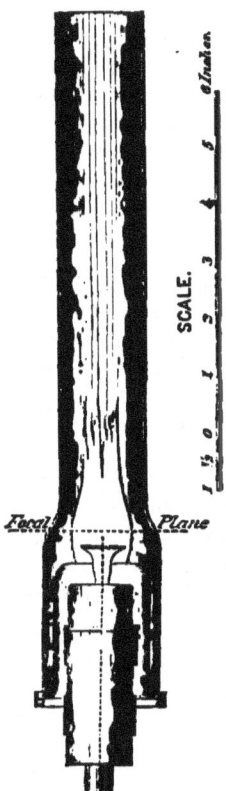

Fig. 2.

Douglass Gas-burner.

The light from this gas-burner, which is shown in Fig. 2, was not, in the experiments at Blackwall, equal to that from either colza or mineral oil, on account of the inferior quality of the gas at that place; but I was told that it was a marked improvement over the common gas-burner.

Mr. Douglass told me that the power of this Argand gas-burner when burning London gas of sixteen candles is about the same as that of the Argand oil-burner, viz, twenty-three candles, and the consumption of gas is about 30 per cent. less than the best burner before used in England.

Actual power of gas-burner.

The buttons for the gas-burner when made of brass last but a little while. They have been made of platinum, but this is expensive, and when I last heard from Mr. Douglass he was experimenting with buttons made of lava, which promised good results. He states that there is no difficulty in producing a gas-burner of the size of the six-wick oil-lamp, or even larger, on the same principle. One of five rings of flame has been tried, and the illuminating power is precisely the same as with the small burners *pro rata* of gas consumed.

Buttons when made of brass soon destroyed.

Lava proposed.

Gas-burner the size of a six-wick lamp easily produced.

The following extract from a letter of recent date, written me by Mr. Douglass, may prove of importance to us in the future, and I here place it on record:

Extract from letter concerning a recently-patented gas-burner.

"I have just found that a patent has been taken out by Mr. Silber (the patentee of a mineral-oil burner) for a gas-burner nearly identical with my own, but, fortunately, the date of the patent is nearly a month after I reported to the Trinity House the results of experiments made with burners I had made for some new light-houses. It is probably important that you know this, as it is not unlikely that a patent has been taken out for the burner in America. I will send you a copy of the paper as soon as it is published."

As I have before remarked, the light-house establishments of Europe have greatly increased the power of their lights

Increase of power of light and decrease of expenses.

with a concurrent decrease of expense, changes which have been produced by the introduction of mineral-oil and the improvements in burners.

Report of Mr. Douglass. I can best illustrate this important subject by quoting one of the earliest reports in regard to this matter made by Mr. Douglass to the Trinity House, dated 30th of March, 1871, which I find in a "Return to an Order of the House of Commons dated 26th of June, 1870, for a copy of 'Correspondence between the general light-house authorities and the Board of Trade, relative to proposals to substitute mineral oils for colza oil in light-houses.'"

"TRINITY HOUSE, *March* 30, 1871.

"Referring to the various and lengthened experiments which have been made at this House for the purpose of determining the suitability of paraffine and petroleum oils for the illumination of light-houses, and the most efficient lamp for consuming these mineral oils, I now beg to submit the following report:

Samples of oils obtained. "For the purpose of ascertaining the relative merits of paraffine and petroleum oils, a sample of the best burning paraffine was obtained from Messrs. Young & Co., and samples of the best burning petroleum-oil were obtained from two respectable manufacturers of that article. The samples were as follows, viz:

Comparative table.

	Specific gravity.	Flashing-point.	Net price per gallon.
Messrs. Young & Co.'s paraffine-oil	.811	136°	1s. 6d.
Trinidad Petroleum Company's petroleum-oil.	.820	130°	Exact price not stated; said to be about 2s.
Carless, Capel & Co.'s petroleum-oil	.796	116°	2s.

Result of trial in ordinary lamp. "It was found that the three samples, when consumed in an ordinary single-wick paraffine lamp, gave nearly the same photogenic results; and, being fairly equal as to safety, the paraffine of Messrs. Young appeared, on the score of economy, to be the most desirable material for the purpose; it was therefore decided to carry out all further experiments with this oil.

Colza used. "The colza-oil used in the experiment was taken from the corporation's stock at Blackwall; it had a specific gravity of .915, and was of excellent quality for illuminating purposes.

First experiments. "The first experiments were made by consuming the oil in the ordinary Trinity House Argand single-wick, and first-order four-wick lamps for colza-oil. No alteration of the lamps was found to be necessary to effect this, but the best result was found to be obtained when the surface of the oil in the wick-case was lowered somewhat below the level of the tip of the burners. A good, bright, steady flame was maintained in each lamp, and the following are the results:

"In recording these results, the photometric value of each light is expressed in English units or standard sperm candles consuming 120 grains per hour. The value of colza-oil is the contract-price for this article for the current year, (3s. 4d. [83 cents] per gallon,) and the value of paraffine is the contract-price of Messrs. Young for the current year, (1s. 6d. [37½ cents] per gallon:)

Value of colza.

Value of paraffine.

Argand burner—Paraffine and colza oils.

	Lamp consuming paraffine.	Lamp consuming colza-oil.
Illuminating power of the light	8.4 units	13.9 units.
Consumption of oil per burner per hour	.0130 gallon	.0115 gallon.
Consumption of oil per unit of light per hour	.00123 gallon	.00083 gallon.
Cost of light per burner per hour	1.86d	.46 0d.
Cost of light per unit per hour	.022d	.033d.

Result with the Argand burner.

First-order four-wick burner—Paraffine and colza oil.

	Lamp consuming paraffine.	Lamp consuming colza-oil.
Illuminating power of light	209.7 units	260.0 units.
Consumption of oil per burner per hour	.2076 0 gallon	.2582 4 gallon.
Consumption of oil per unit of light per hour	.00099 gallon	.00086 gallon.
Cost of light per burner per hour	3.73d	10.33d.
Cost of light per unit per hour	.018d	.038d.

Result with four wick burner.

"In these experiments the lamps were kept burning for six hours without any trimming of the wicks; the illuminating power of the lights was determined every hour by a Bunsen's photometer, and the powers given are a mean of those powers.

"The first lamp improved was the single-wick or Argand; the alterations effected to this lamp were as follows, viz: The tips of the wick-case were closed so as to fit more closely to the wick, and beveled for the purpose of admitting the ascending currents of air freely to the lower part of the flame; a perforated button was introduced at the center of the burner; the wick-case was lengthened for the purpose of economizing the consumption of wick, and an alteration was made in the form of the glass chimney. The results of a six hours' trial with this improved burner,

Improvement in Argand lamp.

which has now been in use at the Milford leading-lights for the last three months with good practical results, are as follows:

Improved Argand burner.

Result with improved Argand burner.

	Lamp consuming paraffine.	Lamp consuming colza-oil.
Illuminating power of the light	20.6 units.	13.9 units.
Consumption of oil per burner per hour	.0109 gallon.	.0115 gallon.
Consumption of oil per unit of light per hour	.00053 gallon.	.00083 gallon.
Cost of light per burner per hour	.190d.	.461d.
Cost of light per unit per hour	.009d.	.033d.

"The above experiments with this lamp, as now improved, show the comparative cost of light produced by colza-oil and paraffine to be as 33 to 9, 11 to 3, or 55 to 15. (See tables.)

Improvement in four-wick burner.

"The experiments with the Argand burner, which had been so successful, were followed by similar experiments with the first-order or four-wick burner. The alterations effected to this burner were as follows, viz: The tips of the wick-cases were closed so as to fit closely to the cotton wicks, and were beveled in the same manner as the Argand lamp, for the purpose of admitting the ascending currents of air freely to the lower part of each ring of flame; and the wick-cases were considerably lengthened for the purpose of economizing the consumption of cotton-wick. Glass chimneys of various forms were tried, but no better result has been obtained than with the chimney usually used with the four-wick burner for consuming colza-oil. After a series of trials it was found that the best photogenic result was obtained with the surface of the oil in the wick-case three inches below the tips of the burner; it was, therefore necessary to keep the oil uniformly at this level during the burning of the lamp, which was done by placing at a short distance from and level with the burner one of the flow-regulating cisterns formerly used for light-house lamps, which receives from an upper reservoir a supply of oil, and maintains the supply to the burner at the required level by a self-acting ball-cock. I have devised another and, in my opinion, a more perfect arrangement for regulating the flow of oil in these lamps, which has been tried with perfect success.

Oil-regulator.

"I beg to submit herewith drawings of the improved Argand and first-order four-wick lamps, on the latter of which is the self-acting regulator. This regulator consists of a stand-pipe placed at the side of the burner; the top of this pipe, which is open, is at the level at which the oil is intended to flow in the wick-cases; a portion of the supply of oil from a reservoir placed above the level of the burner, or from the cylinder of a pressure-lamp, flows up the standpipe and overflows at the top, descending by another pipe surrounding the stand-pipe to a cistern below the burner. A glass thimble is screwed on to the top of the regulator, through which the state of the flow can be observed. The necessary adjustments of supply are made from time to time by the ordinary regulating-valve placed in the supply-pipe below the burner. The tops of the inner and outer tubes of the stand-pipe are rendered telescopic by a piece of pipe fitted to and sliding on them externally; by means of these sliding pieces the flow of the lamp can be altered at any time so as immediately to adapt it for burning any descriptions of hydrocarbon. As the invention appears to me to be of importance for regulating the flow of all lamps used for burning mineral oils, I have had it provisionally protected.

"As it is found to be necessary in burning paraffine that its level in the wick-case be considerably below the top of the burner, it may reasonably be expected that the tips of the burners will be destroyed by the heat of the flame much sooner than with burners consuming colza-oil, where the latter is constantly overflowing the burners and keeping them cool. I have provided for this increased destruction of burners by fitting each with removable tips, as shown on the accompanying drawings. With this arrangement, and by keeping a supply of spare tips at each station, the tips of the burners may be renewed at any moment by the light-keeper in charge, thereby avoiding the necessity for returning the burner to the workshops for repairs. *Removable tips.*

"The following are the mean results of several six-hour experiments.

Improved first-order four-wick burner.

	Lamp consuming paraffine.	Lamp consuming colza-oil.	Result with improved four-wick lamp.
Illuminating power of light	280 units	200 units	
Consumption of oil per burner per hour	.21240 gallon	.25824 gallon	
Consumption of oil per unit of light per hour	.00081 gallon	.00096 gallon	
Cost of light per hour per burner	3.87d	9.02d	
Cost of light per unit per hour	.014d	.033d	

"The above experiments with this lamp, as now improved, show the comparative cost of light produced by colza-oil and paraffine when consumed in a first-order burner to be as 33 to 15. From these results it will be observed that the superiority in illuminating power of the paraffine over the *Comparative cost of colza and paraffine.*

colza oil is much greater when consumed in the Argand burner than when consumed in the large four-wick burner; consequently the contrast between the cost of light produced by the two oils is greater when the oils are consumed in the Argand burner than when consumed in the four-wick burner.

"Further experiments were made for ascertaining the relative illuminating power of the paraffine and colza lamps during the time required at a light-house on the longest winter night. Each lamp was kept burning for sixteen hours without any trimming of the wicks, and the following are the photometric results of several experiments:

	IMPROVED ARGAND BURNER.			IMPROVED FIRST-ORDER FOUR-WICK BURNER.	
Hours.	Paraffine lamp.	Colza lamp.	Hours.	Paraffine lamp.	Colza lamp.
	Units.	*Units.*		*Units.*	*Units.*
1	20.6	13.9	1	280	269
2	20.6	13.9	2	274	263
3	20.6	13.9	3	280	261
4	20.6	13.9	4	278	216
5	20.4	13.6	5	278	245
6	20.4	13.3	6	278	244
7	20.0	12.7	7	274	245
8	20.0	12.7	8	277	246
9	19.6	12.7	9	271	240
10	19.6	12.7	10	266	214
11	19.2	12.7	11	265	213
12	19.0	12.3	12	277	227
13	18.7	12.1	13	267	231
14	18.6	12.0	14	266	224
15	18.5	11.9	15	256	2.0
16	18.5	11.8	16	230	200
Mean	19.7	12.8	Mean	271	230

Deductions. "From these results it is apparent that the paraffine lamps will burn throughout the longest winter night in this country without any trimming of the wicks, and give during *Sustained intensity.* this time a light of nearly uniform photometric value. At the end of sixteen hours the illuminating power of the Argand lamp, burning paraffine, was only 10 per cent. less, and in the four-wick lamp only 10.7 per cent. less, than at the commencement of the trials, while the illuminating power of the lamps burning colza-oil gradually decreased soon after the commencement of the trials, and at the end of sixteen hours the illuminating power was reduced 15.1 per cent. in the Argand burner, and 25.6 per cent. in the four-wick burner.

State of wicks. "At the termination of these trials the wicks of the paraffine lamps were not much fatigued; the tips were charred only one-eighth of an inch in depth, and to all appearances the lamp was fit for burning many hours longer. The wicks of the colza-lamp were much distressed; they were charred five-sixteenths of an inch in depth, and evidently nearly worn out; trimming would have been absolutely necessary if burned for three or four hours longer.

"In conclusion, it may be generally stated as the result of the lengthened experiments which have been made at this House with paraffine as an illuminant for light-houses—

"1st. *The cost of light is 72.7 per cent. less when produced by the Argand or single-wick lamp, and 60.5 per cent. less when produced in the first-order or four-wick lamp, than colza-oil.* Comparative cost.

"2d. *The lamps burning paraffine will give a light of more uniform illuminating power throughout the night, without trimming, than the lamps burning the colza-oil.* Paraffine light more uniform.

"3d. *The lamps burning paraffine are more readily ignited; they burn with greater certainty, and require less attention than lamps burning colza-oil.* Paraffine lamps easier to tend.

"4th. *The lamps burning paraffine may be arranged for increasing the power of the light when the state of the weather requires it, as is now done with the electric light and coal-gas.* Increase of power possible.

"5th. *Paraffine can be stored and used at light-houses with safety, provided that ordinary care is used.* Paraffine may be stored with safety.

"I am, &c.,

"JAS. N. DOUGLASS."

The tables given in the foregoing report showed—

1st. *That light for light with a first-order lamp, the cost of the paraffine was about one-half that of the colza light.* Cost with first-order lamp.

2d. *Light for light with the fourth-order lamp, the cost of paraffine was about one-fourth that of colza.* Cost with fourth-order lamp.

These results were confirmed by a comparison of the figures in the following table of results obtained by Dr. Macadam, scientific adviser of the board of commissioners of northern (Scottish) lights, Mr. Douglass, engineer of the Trinity House, and Professor Tyndall, scientific adviser of the Trinity House, by quite separate and distinct experiments. Confirmation of above.

Comparative statement of experiments for testing the values of colza and paraffine as illuminants for light-houses.

	Dr. Macadam.		Mr. Douglass.		Dr. Tyndall.	
	Colza.	Paraffine.	Colza.	Paraffine.	Colza.	Paraffine.
FIRST-ORDER LAMP:						
Illuminating power of light expressed in standard candles, consuming 120 grains per hour	261.3	261.3	209.	280.	244.85	294.03
Percentage of increase of light in favor of paraffine	......	......	......	4.09	......	2.134
Consumption of oil by the lamp during one hour in imperial gallons	.2077	.1967	.2582	.2321	.1928	.2226
Consumption of oil per candle per hour in imperial gallons	.0079	.00075	.00096	.00083	.00067	.00076
Cost of light per hour in pence	8.39	3.55	9.02	3.87	7.71	4.01
Cost of light per candle per hour in pence	.032	.014	.033	.014	.027	.013
Percentage of saving in cost in favor of paraffine	......	56.25	......	57.57	......	51.85
FOURTH-ORDER LAMP:						
Illuminating power of light expressed in standard candles, consuming 120 grains per hour	11.33	19.83	13.9	20.6	11.63	18.40
Percentage of increase of light in favor of paraffine	......	75.02	......	48.2	......	58.9
Consumption of oil by the lamp during one hour in imperial gallons	.0133	.0125	.0115	.0109	.0146	.01335
Consumption of oil per candle per hour in imperial gallons	.00117	.00063	.00083	.00053	.00126	.00072
Cost of light per hour in pence	.592	.218	.451	.126	.6132	.2461
Cost of light per candle per hour in pence	.047	.011	.633	.609	.053	.013
Percentage of saving in cost in favor of paraffine	......	76.6	......	72.72	......	75.47

The foregoing tables were among the earliest results of experiments with paraffine oil and petroleum.

Result of improvements made by Mr. Douglass. The improvements made by Mr. Douglass from time to time in lamp-burners have resulted in what is apparently a perfect light-house lamp, which can hardly be surpassed in economy and efficiency except by the supply of pure oxygen to the flame.

Details of lamp. Each of the Douglass lamps now made for the English light-houses has—

Burner-tips. 1st. A series of burner-tips so constructed that they may be removed when burned out, instead of substituting new burners, as is done under the old system.

These "tips," which slightly compress the wicks, give to them a perfectly cylindrical form, and their exterior surfaces being slightly conical, the air from below is sent directly into the flame without danger of forming eddies where the gases burn imperfectly.

Interior deflector. 2d. A curvilinear perforated "button," (interior deflector,) which sends the interior current of air into the flame.

Exterior deflector. 3d. An outer "cone," (exterior deflector,) by which a second current of air is thrown into the flame at its most advantageous zone.

Outer air-current. 4th. A space between the cone and the chimney, by which a third or outer air-current is produced.

This air-current is injected into the flame above that admitted by the "exterior deflector," and is for the purpose of increasing combustion. A portion of this current being drawn up along the surface of the chimney, prevents the heat from melting or otherwise injuring the glass, in which a milkiness resulting from disintegration sufficient to impair the light, is found to be produced by long-continued heat.

Another result produced by this current is that the base of the chimney is kept at a degree of temperature sufficiently low to admit the removal of the chimney with the naked hand, rendering the use of tongs unnecessary. *Chimney kept cool.*

5th. An adjustable gallery by which the chimney is raised and lowered at will, it being found that the height of the shoulder of the chimney has a marked effect upon the flame. *Adjustable gallery.*

6th. Very soft and compressible wicks adopted for this lamp only after many repeated experiments. *Soft wick.*

Each of these inventions, some of which were original with Mr. Douglass, and some had previously been in use, is considered of importance, but a combination of the whole is essential to produce the remarkable results given in the following table deduced from the most recent experiments of Mr. Douglass and kindly sent me by him since my return to this country: *Combination of these inventions.*

Comparative statements showing the mean illuminating power and consumption of oil with the old and improved light-house burners of the Corporation of Trinity House. *Comparative results of experiments with Mr. Douglass's burner.*

	Old concentric burners consuming colza.				Improved concentric burners consuming colza.					
Number of wicks:										
Full power....	4	3	2	1	6	5	4	3	2	1
Half power...					3	3	2	2	1	
Mean diameter of outer wick, (in inches)	3.32	2.40	1.65	.82	5.00	4.16	3.32	2.40	1.65	.82
Illuminating power of flame in standard sperm candles, (or units,) consuming 120 grains per hour:										
Full power....	200	167	58	21.9	722	514	328	208	82	23
Half power...					343	225	173	146	51	
Consumption per burner per hour in fluid-ounces, 1.25 to American wine-gallon:										
Full power....	39.77	23.09	7.70	2.77	70.21	51.81	32.12	20.17	7.87	2.20
Half power...					36.12	24.68	17.43	14.16	4.80	
Consumption per unit per hour in fluid-ounces...	.148	.138	.134	.128	.106	.101	.098	.097	.096	.096
Mean height of flame, (in inches).	3.75	2.5	2.0	1.5	5.5	5.0	4.5	3.2	2.8	2.2
Increased illuminating power of improved burner, per cent					21.93	24.55	41.38	63.47		
Saving in oil with new burner for each unit of light, per cent...					13.77	29.86	28.57	24.81		

Consumption of colza. An inspection of the above table shows the consumption of colza-oil in the new burner to be very nearly one-tenth of an ounce for each unit of light, that unit being the light from a standard sperm-candle consuming 120 grains per hour, and Mr. Douglass informs me that the improved burners have raised the power of light produced from colza to such a degree that this oil is now practically equal in illuminating power and consumption to the best mineral-oil found in Great Britain, the latter having a specific gravity ranging from .810 to .820, flashing above 130° Fahrenheit and distilling between 212° and 572° Fahrenheit, so that the difference between mineral and colza oil for light-house illumination is principally one of economy. In England the cost of colza-oil is about 2s. 9d. (68 cents) per imperial gallon, that of mineral-oil of the quality above stated being 1s. 7d. (39 cents) per imperial gallon, so that in that country, for all orders of light-house lamps, *the cost of maintaining mineral-oil sea-coast lights is about one-half that of maintaining colza-oil lights.*

Substitution of mineral for colza oil. For this reason, as well as because mineral-oil lamps are much more cleanly, more easily lighted, and require no trimming during the night, thus making their efficiency less dependent on the watchfulness of keepers, the English are rapidly substituting mineral-oil lamps in their light-houses for those for colza formerly used.

Price of lard-oil. In the United States the average price of our illuminant, lard-oil, of which we annually use about 100,000 gallons, is 89 cents per gallon; the cost of mineral-oil of the quality required by the British and French contract-specifications is about 35 cents per gallon.

The following table gives the comparative values in standard candles of the new English Douglass and the American light-house lamps:

Number of wicks	6	5	4	3	2	1
English lamps, value in candles	722	514	328	208	82	23
American lamps, value in candles			210	133	44	15.3
Percentage in favor of English lamps			56	56	86	50

An examination of the foregoing tables and the facts just stated shows:

1st. *That by the adoption of the triple current of air burners of the English and French into light-houses, we should gain more than 50 per cent. in the power of our lights.*

2d. *By the adoption of American mineral-oil instead of the lard-oil now used in our light-houses, we should save $54,000 per annum in cost of oil.*

This is on the supposition that the American refiners will produce, if stimulated by the use of the former by the Government, an article equal to that used in Europe for light house illumination.

3d. If it should be found that such American oils cannot be obtained, precisely the same excellent quality of mineral-oil (Scotch) as is used in the light-houses of France and Great Britain can be imported at a cost, including freight,* not exceeding 36 cents per gallon, and the saving in this case would not be less than $53,000 per annum in cost of oil.

Since my return from Europe the Trinity House has kindly sent to the Light-House Board several improved four-wick burners, for both colza and mineral oil; interior deflectors for cutting off one, two, or three wicks; one Argand lamp, complete, for burning colza or mineral oil; and one Argand gas-burner, as well as a supply of wicks, chimneys, &c., for use with the lamps and burners sent.

In regard to the purchase of mineral-oil, I give below some extracts from the specifications which the Trinity House furnishes to bidders : <small>Specifications for mineral-oil.</small>

"1. The mineral-oil required to be supplied under this contract is to be of the best possible quality, the greatest care is to be taken in its preparation, and it must be as free as possible from oil of vitriol. <small>Quality.</small>

"2. If the oil be *petroleum*, it must have a specific gravity not less than .785, nor greater than .790, at 60° Fahrenheit; its flashing-point not lower than 125° to 130° Fahrenheit, and it shall distil between 212° and 482° Fahrenheit. <small>Specific gravity if petroleum.</small>

"3. If the oil be *paraffine*, it must have a specific gravity not less than .810 nor greater than .820 at 60° Fahrenheit; flashing above 130° Fahrenheit, and distilling between 212° and 572° Fahrenheit. <small>Of paraffine.</small>

"4. A sample of five gallons of each of the oils proposed to be supplied is to accompany the tender; such sample will be tested by burning to ascertain its action on the wick, and its specific gravity, flashing-point, and chemical composition (to be determined by fractional distillation) being ascertained, the several results shall, if approved, be considered binding in all subsequent deliveries of oil under this contract. <small>Samples</small> <small>Tests.</small>

"5. The contractor will be furnished with tinned iron cans to hold the required amount of oil for each light-house; <small>Cans furnished.</small>

---

* The contract-price of oil in England is, as stated above, 1s. 6d. per imperial gallon, or about 39 cents. The freight to New York I have ascertained will not be over 4 cents, making its cost here of the imperial gallon 43 cents, or, as this is one-fifth larger than the American wine-gallon, about 36 cents per wine-gallon.

every can will contain about five gallons, and will weigh net about 21 pounds. The cans will be delivered to the contractor at the Trinity Buoy-Wharf, Blackwall, in perfectly tight and sound condition, and the contractor must assure himself thereof at the time of his receiving them, as no plea of unsoundness will be admitted after acceptance, in case of leakage or fracture.

"6   *   *   *   *   *   *

"7   *   *   *   *   *   *

Test when oil is delivered.
"8. Samples will be drawn at pleasure from any portion of each parcel delivered, and shall be tested in accordance with the tenor of paragraph 4. In the event of the test being satisfactory, such parcel shall be formally received, but if not, the contractor shall, at his own expense, remove the whole quantity within ten days after receiving notice of rejection, and must immediately replace the same with another supply, which latter shall be tested in like manner: Records will be kept of the results in each case of testing samples."

Tests at depot.
All oils are thoroughly tested twice: first by the officers of the depot at Blackwall, and afterward in samples, by the engineer and Elder Brethren at the Trinity House.

Process of testing colza.
The process of testing the colza-oils received from the contractors is thus described:

Samples from bidders.
Each of the parties tendering bids sends a sample of twenty gallons marked with a letteronly, so that, while testing, the manipulator has no knowledge of whose oil is under test. A sample of the previous year's supply is also introduced with the others; of the identity of this also the manipulator is ignorant.

Special qualities required.
The oils are tested with regard to the consumption; the effect of combustion upon the wick; the illuminating power and specific gravity; also with regard to congelation.

Lamps used.
For consumption and combustion, the samples, including that of the previous year's supply, are tried in Argand lamps kept burning for sixteen hours, that being considered as about equal to the longest winter night. During this time particular notice is taken of any diminution of the flame, (which must be kept $1\frac{1}{2}$ inches high,) or any other irregularity of combustion, and at the expiration of the time the residue in each lamp is carefully measured to ascertain the exact quantity consumed.

Measure of illuminating power.
The illuminating power of each oil is measured by a photometer, (Bunsen's, used also in our service,) and the unit is, as with us, the standard candle consuming 120 grains per hour. The observation is taken at the full power of the light when first lighted, and again at the end of the trial.

The specific gravity of each oil is measured by an oleometer in the usual way. *Measure of gravity.*

To ascertain with regard to congelation, samples of each oil are placed in half-pint clear glass bottles and subjected to a temperature of 25° Fahrenheit, at which temperature they should remain fluid for sixteen hours. *Congelation.*

A portion of the original twenty gallons supplied by the firm who obtains the contract is retained, and a sample from every cask delivered by them during the period of the contract is comparatively tested with the original sample. *Test of oil when delivered.*

Trinity House acts as agent for the purchase of oil for the light houses of some foreign countries and of several of the English colonies. The storage capacity for oil at Blackwall is about 210 tons, (about 54,000 gallons.) The great cast-iron tank for colza-oil is divided into compartments, and is much like our oil-tank at Staten Island. *Trinity House agent for foreign light-house establishments. Storage capacity at the depot. Oil-tanks.*

The oil is drawn from two levels, and it is the universal practice in the English light-house service to draw from more than one level wherever large oil-vessels are used, some of the 100-gallon oil-butts having as many as three cocks at different heights. *Cocks.*

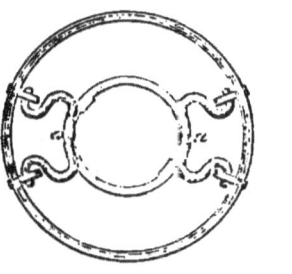
Fig. 3.

The level of the oil in each tank is shown by means of a float attached to a chain, which passes over a pulley at the top, to a weight moving upon a graduated scale on the outside. The oil is emptied into the tanks through pipes leading from the floor above, and is delivered from the tanks for issue to the light-houses into sheet-iron "drums" (cans) holding five gallons each, and which are 11 inches in diameter and 15 inches in height. *Float to show oil-level.*

Five-gallon oil-can.

Five-gallon cans.

A recess, made by sinking the top below the sides, contains the handles *a a*, (see Fig. 3,) and a screw-plug for emptying.

These cans are purchased by contract, but are tinned *Purchase of cans.*

on the inside by the workmen at the depot. They are in size and construction admirably designed for transport. Three of them can be easily carried by two men, and they stow in the supply-vessels and boats with little loss of space. They are also strong and durable. I think it a better plan for delivering oil to light-houses than ours of shipping it in barrels on the eastern coast and in square tin cans to the Pacific coast light-houses.

Plan of delivery of oil in these cans a good one.

Especial attention is paid to the testing of lamp-chimneys, which are distributed to the light-houses from Blackwall depot.

Lamp-chimneys.

A gauge, b, of metal is fastened to a frame, a, (see Fig. 4,) in which the chimney is placed, and its accuracy of shape and thickness is determined by turning it around its axis. Those which do not fit the gauge closely are rejected.

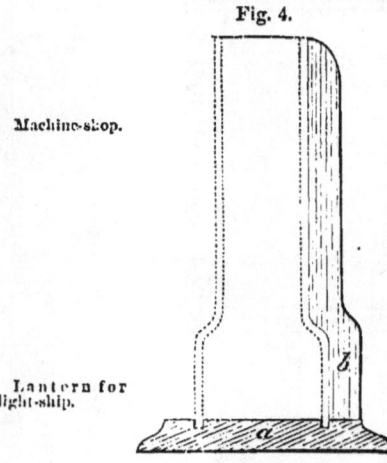

Fig. 4.

Chimney-gauge.

Machine-shop.

The machine-shop is well fitted up with machinery, including lathes, drilling and screw-cutting machines, and all kinds of work, excepting heavy forging and casting of iron, are done here; even the manufacture of the furniture used in the light-house towers and keepers' dwellings.

Lantern for light-ship.

Among great varieties of work I saw constructing in the smithery a lantern for a light-ship, made entirely of iron, the sash-bars being of malleable cast iron.

Lamps trimmed without lowering lantern.

This lantern was not intended to be lowered to the deck for trimming the lamps, as is usual, but was designed to remain in its place on the mast, which is to be of hollow steel about two feet in diameter. Small doors open from the hollow of the mast on to the deck, and also into the lantern when placed in position, an inside ladder affording a means of communication with the lantern at all times.

There is about 20 inches space between the sides of the lantern and the reflectors, to enable the keepers to trim the lamps, and an outside foot-rail is provided to enable them to clean the exterior of the lantern-glass.

Question of ability of light-ship to carry mast and lantern.

Whether so heavy a mast and lantern can be carried on light-ships in all weathers is yet a matter of experiment, but there can be no doubt that facility of trimming without lowering the lantern is much to be desired.

There was also constructing a set of machinery for a fog-bell, the driving-weights of which were square and furnished with rollers on the sides to avoid friction against the boxes. *Fog-bell machinery.*

Above the buoy-shed is a convenient photometric gallery for testing lamps, oils, and lenses. It is about 80 by 12 feet, the interior is painted black, and the sky-lights are so arranged that all daylight can be excluded. *Photometric gallery.*

At Blackwall is always kept a light-ship for relief in case of accident to any of the numerous light-ships on the coast near the Thames, and this is also the place of repair not only for these vessels but for the steam-tenders. *Relief light-ship.*

Trinity House ceased some years ago to build iron light-ships, on account of their fouling so rapidly and the consequent necessity of bringing them into port once in two years, while the wooden light-ships require to be brought in but once in seven years. Further reasons for giving up the use of iron light-ships are that they are cold and damp, and when run into, sustain much more danger than the wooden vessels. *Iron light-ships considered inferior.*

The approximate cost of a light-ship, either of iron or wood, including everything, is, in England, £5,500, or about $27,500. *Cost of light-ship.*

The size of English light-ships is about the same as our own, although the latter have much greater beam and depth of hold, and are more rounded at the sides, which probably enables them to ride easier and with less shock to the catoptric apparatus and the lantern. *Size.*

Those placed in exposed situations are usually held by single mushroom-anchors, weighing two tons, and each light-ship has 210 fathoms of 1½-inch cable. *Moorings.*

The Seven Stones light-ship, off Land's End, is anchored in 42 fathoms of water and has 315 fathoms of cable attached to her mushroom. *Anchors of the "Seven Stones."*

In addition to these moorings every light-ship has on board spare anchors and cables in readiness to be let go at a moment's notice in case the vessel should drag her anchor, but such an occurrence has not happened for several years. *Additional anchors.*

The chain-cables used by all the vessels of the Trinity House are made with great care. The requirements and the mode of testing are as follows: *Chain-cables; requirements of contracts.*

All the cables, mooring-chains, rigging and crane chains, and articles appertaining thereto, excepting the stay-pins and steel pins, are manufactured from fine fibrous iron of approved quality, and must bear a tensile strain of not less than *Quality of chains.*

23 tons per square inch of original area, with a contraction at fracture of not less than 45 per cent. of the original area.

Quality of stay-pins. The cast iron in stay-pins is of the best tough gray metal, and must bear a compressive strain of not less than 52 tons per square inch of original area, with a reduction in length of not less than 10 per cent.

Steel pins. The steel pins for retaining the joining shackle-bolt are of the best and toughest manufacture, and must bear a tensile strain of not less than 35 tons per square inch of original area, with a contraction at fracture of not less than 45 per cent. of original area.

Workmanship. All cables, mooring-chains, rigging and crane-chains, and articles appertaining thereto, are required to be of the best possible workmanship. They are proved at a proving-machine, licensed by the Board of Trade, at the expense of the contractor, and the proving is carried out in the presence of the engineer or other designated officer.

Proof strains for testing chains. The chains are subjected to the following proof-strains, viz: Open-link cable and mooring-chains, and close-link rigging and crane chains are proved to 8.47 tons per square inch of each side of the link, or 466 pounds per circular one-eighth inch of the diameter of the iron, and stud-chains are proved to 11.46 tons per square inch of each side of the link, or 630 pounds per circular inch of the diameter of the iron. Any links which may appear to be defective are cut out and replaced, and the chain re-proved at the expense of the manufacturer. If more than three links are found to be defective in any length after proving, such length is considered liable to rejection. All forelock-shackles, connecting-shackles, and spare swivels, are proved to the same strain as the chain with which they are intended to be used.

Further tests. In addition to the above proof, and for the purpose of ascertaining the exact quality of the iron and welding, the engineer selects, as he may think fit, from any lengths of each of the sizes of chain ordered, three test-pieces, four feet long of each size. These pieces are cut out of the chains, and, together with a shackle of each size, also selected by the engineer, are stamped with a Trinity House stamp, and are sent by the contractor to such public testing-works as may be designated by the Trinity House, to be tested and reported on by the engineer of the Trinity House, at the contractor's expense. The quality of the iron is ascertained by testing the iron in one link cut off each of the pieces of chain, also the iron in each of the shackles. The remainder of each of the four-foot lengths of chain is then tested to ascertain the quality of the welding, when the ultimate breaking-stress must not be less than 16 tons per square inch of each side of the link, or 880 pounds per circular one-eighth inch of the diameter of the iron. In the event of these tests proving satisfactory, the lengths of chain from which the test-pieces were taken are to be made good by the contractor, and the lengths reproved at his expense. In the event of any portion of the material or welding, when tested, proving inferior in quality to that specified, the chains and shackles are rejected, and the above tests are repeated at the expense of the contractor on other pieces of chains and shackles selected by the engineer in the same manner from other chains and shackles manufactured by the contractor, and submitted for approval.

At the depot are a great many buoys of all kinds, most of them of timber, but I believe none are now made of that material. Some of the different kinds are the "nun," "can reversed," "can," "egg-bottom," "convex bottom," "flat bottom," "hollow bottom," "spherical," and "conical." *Buoys. Names.*

The English can-buoy corresponds in shape to our nun-buoy, except that the larger end is in the air. The cylindrical buoys used by the English are much like our can-buoy, and are said to satisfy all conditions required. An English nun-buoy is conical at both ends and is used to mark wrecks. An English flat-bottom buoy is, as are some of the others, water-ballasted, *i. e.*, they have a cross diaphragm at a proper distance from the bottom, and the water is allowed to flow in and out of the lower compartment thus made through eight holes an inch in diameter, placed at equal distance around its sides. The water cannot be discharged unless the buoy is careened for some time, and it is therefore as completely ballasted as if the water had no means of exit. *Can-buoy. Cylindrical. Nun. Flat-bottom water-ballasting.*

When those buoys are required for deep water where the weight of the mooring-chain is sufficient for ballast, water-ballast is not used, and the holes are plugged with hard wood. With the increased buoyancy thus obtained the same line of flotation as in shoal water is approximately attained. *When buoys are required for deep water.*

This is a matter of importance, preserving, as it does, a uniformity of appearance in each class of buoys, irrespective of the depth of water in which they are moored.

The buoys said to be the best for strong tideways are the can, cylindrical, and flat-bottom; for exposed channels and coasts the egg-bottom is used. *Best buoys for strong tideways; for channels.*

Another which is used with satisfactory results is a patent hollow-bottomed one, called "Herbert's" buoy. *Herbert's buoy, theory of.*

The theory of the action of this buoy is that the air confined in the bottom forms an elastic spring upon which the buoy rebounds in gentle and easy motions, causing but moderate friction to the mooring-chain, little or no pull upon the sinker, and a corresponding relief from agitation or friction to the globe and staff above.

Size of buoys. I was especially struck with the great size of some of the buoys which I saw at Blackwall and at other places, many of them being 20 feet in length.

Mooring of English buoys. The English ordinarily moor their buoys by a single chain and sinker or mushroom, but in some instances double moorings are used. The chains of light-ships, after three years' service as such, are converted into buoy-chains. The proportion of chain used in mooring buoys is generally three times the depth of water.

Shifting buoys. Small buoys are shifted twice a year, but the buoys above 8 feet remain at their stations for two or three years and are painted at their anchorage periodically.

System observed in buoying English waters. The following is the system observed in buoying channels:

Sides of channels. The side of the channel is to be considered starboard or port with reference to the entrance to any port from seaward.

Entrance to. The entrances of channels or turning-points are marked by conical buoys with or without staff and globe or triangle, cage, &c. Singled-colored can-buoys, either black or red, mark the starboard side, and buoys of the same shape and color, either checkered or vertically striped with white, mark the port side; further distinctions are given when required, by the use of conical buoys with or without staff and globe, or cage, globes being on the starboard and cages on the port hand.

Middle ground. Where a middle ground exists in a channel each end of it is marked by a buoy of the color in use in that channel, but with annular bands of white and with or without staff and diamond or triangle, as may be desirable; in case of its being of such extent as to require intermediate buoys, they are colored as if on the sides of a channel. When required, the outer buoy is marked by a staff and diamond, and the inner one by a staff and triangle.

Wrecks. Wrecks are marked by green nun-buoys.

Marks. Each buoy is plainly marked with a running number *and the name of the locality where it belongs.*

The white stripes or checkers of buoys are about 20 per cent. less in size than the black and red, it being found that the characteristic distinctions of the buoys are better observed at a distance by this inequality.

Bell-buoys. Bell-buoys, of which there are many in use, are constructed of iron, and have four hammers or clappers, each hung by a Y, which prevents jamming and obviates the use of guides. They cost about £200 ($1,000) each.

Buoy-shed. At Blackwall the buoys are kept under a commodious buoy-shed, with convenient arrangements of rails, &c., for moving them. The sinkers are square and of iron. The prices of iron buoys most recently obtained are as follows: *Prices of buoys.*

Eight-foot drum-buoy, price £32 10s., ($162.50.)
Eight-foot spherical buoy, price £145, ($725.)
Eight-foot water-ballasting buoy, price £82 10s., ($412.50.)
Thirteen-foot water-ballasting buoy, price £198 6s. 8d., ($991.66.)
Bell-buoy, price £224 12s., ($1,123.)

Towers in which tests are made. At the depot are two towers, each having a fixed lens of the second order, and in them are tested, under the direction of Professor Tyndall or Mr. Douglass, the different lamps, lenses, and oils, the effect of fog, &c.

The principal point of observation is Greenwich, distant about two miles.

Light-house depots on the coast of England. There are several light-house depots on the coast of England, at Yarmouth, Coquet Island, and other places, but Blackwall is the principal depot for manufacture, supply, and repair.

Superintendents of Light-House Districts. The immediate agents through which the authority of Trinity House is exercised are called *superintendents*, and each has some special duties assigned him, either the sole care of the service in some specified part of the coast or the charge of some special branch, such as the supply and store houses at Blackwall. The tenders are under their orders. *Tenders under orders of superintendents.* They wear a uniform on all occasions when on duty.

Keepers. Light-keepers are appointed by the corporation. The rules require that applicants shall be between the ages of 19 and 28. They must produce certificates of character and physical ability, and (from a schoolmaster) of ability to read, write, and perform simple operations of arithmetic. As vacancies occur successful applicants are taken on probation, *i. e.*, are appointed *supernumerary* light-keepers. *Supernumeraries.*

Instruction to keepers on probation, (supernumeraries.) They are then sent to the depot at Blackwall and placed under the orders of the superintendent there. They are carefully trained in the use and care of lamps and all light-house apparatus, including meteorological instruments; the keeping of the light-house journal and accounts, and the general management of affairs at a light-house.

Certificates given.   A certificate of the lowest grade is given for competency in their duties.

A second course of instruction includes the use of tools in carpentry and plumbing, that he may be able to effect ordinary repairs; also the management and general knowledge of the steam-engine.

A third course includes instruction in the management of the magneto-electric machine and lamp.

A fourth course includes the use and management of fog-horn apparatus.

Separate certificates are given for each course.

Number of supernumeraries at Blackwall and South Foreland.   There are always eight of these candidates for light-keepers at Blackwall, and two at South Foreland, the latter for instruction in the management of electric lights, and to the great care exercised in their selection, and the thoroughness with which they are instructed before they enter upon their duties as keepers, is to be attributed the excellent condition of the lights, towers, dwellings, and grounds that I observed at every station which I afterward visited.

Supernumeraries uniformed and paid.   Supernumeraries are supplied with uniforms, and are paid at the rate of £45 ($225) per annum; but on obtaining the four certificates and giving satisfactory proofs of steadiness and sobriety, they become entitled to an assistant keeper's pay.

The rates of pay are as follows:

Grade of keeper.	Gross rate per annum.		Deduct insurance.	
	£	s.	£	s.
Principals who have served as such above 10 years, if insured	72	0	3	0
Principals who have served as such above 10 years, if uninsured	70	10	0	0
Principals above 5 and under 10 years, if insured	68	0	3	0
Principals above 5 and under 10 years, if uninsured	66	10	0	0
Principals under 5 years, if insured	66	0	3	0
Principals under 5 years, if uninsured	64	10	0	0
Assistant keepers who have served as such above 10 years, if insured	58	0	3	0
Assistant keepers who have served as such above 10 years, if uninsured	56	10	0	0
Assistant keepers above 5 and under 10 years, if insured	56	0	3	0
Assistant keepers above 5 and under 10 years, if uninsured	54	10	0	0
Assistant keepers under 5 years, if insured	54	0	3	0
Assistant keepers under 5 years, if uninsured	52	10	0	0

Keepers pensioned.   When no longer able to do service, keepers are pensioned, the pension computed on an estimated allowance of £18, in addition to the above scale.

Term of service at rock and screw-pile stations.   Keepers and assistants at rock and screw-pile stations remain on shore, in rotation, one month each.

The regulations in regard to the care of lamps and premises and keeping watch are much the same as our own, but where mineral-oil is used, the following instructions are added:

"The oil is to be placed in the metallic cisterns provided for the purpose; and these are to be kept perfectly closed by means of the cover and tops with which they are provided. In drawing oil from the cisterns it is to be drawn into a proper can, provided with an oil-tight screwed cover and an air-tight screwed cap to the spout. After charging the lamps the can is to be returned to the store with the covers, spout, and top screwed tight. All oil required for the service of the establishment is to be taken from the store during day-light, and keepers are not, under any circumstances, to enter the oil-room with a lighted lamp or candle." *Care of mineral oil.*

IRON LIGHT-HOUSES OFF THE MOUTH OF THE THAMES.

After leaving Blackwall we proceeded down the Thames in the Trinity House steam-yacht Vestal on a cruise of inspection of the lights on the east coast of England, during which we visited nearly all of them between the Thames and the Scottish border. On this journey it was my good fortune to accompany Admiral Collinson, C. B., and Captain Weller, of the Elder Brethren, and I shall long remember the great kindness and attention of which I was the recipient from both of these gentlemen. *Cruise in the Vestal.*

We passed the Mucking light-house, situated in the Thames, below Gravesend, and the Maplin Sand light-house, off the mouth of that river. Both of these are screw-pile structures; the latter was, I believe, the second of that kind in the world, having been lighted in 1841, and was one of the earliest applications of that useful invention of Mitchell, of which we have many examples, there being more than fifty light-houses built on that plan in the United States. *Mucking light-house. Maplin Sand light-house. Maplin Sand light-house the second screw-pile light-house built. Number of screw-pile light-houses in the United States.*

It may be mentioned here that the first screw-pile light-house was built at the mouth of the river Wyre, on the northwest coast of England, two or three years before the light-house on the Maplin Sand. The screws were three feet in diameter, the piles five inches; and above the ground, instead of iron, as at Maplin, wooden columns were used. This light-house was destroyed in 1870. *First screw-pile light-house built. When destroyed.*

The Maplin Sand light-house, a view of which is shown in Plate V, is a hexagonal structure, with one central and eight exterior piles. The piles were driven vertically, but above the water-line they bend toward the center and incline in a pyramidal form to the lantern-floor. The screws are four feet in diameter, the piles five inches, and they support cast-iron columns 12 inches in diameter. The col- *Description of Maplin Sand light-house. Piles. Screws. Columns.*

umns are very strongly braced, and the structure had an appearance of great strength.

Gunfleet light-house. We stopped at the Gunfleet light-house, situated on a sand of that name, north of the mouth of the Thames, and thirty-one miles from the Nore light-ship. It is exposed to the full force of the North Sea.

Piles. There are one central and six exterior piles supporting columns of about 12 inches in diameter, strongly braced.

Sockets. The sockets for the columns are not cast in one with the sockets for the braces, but the latter are bolted against the face of the piles by tap-bolts.

Form of the structure. Unlike Maplin Sand light-house, the piles were not driven vertically, and are inclined from the bottom to the top in the form of a pyramid. The piles, braces, and sockets are of a very massive character, and give an appearance of great durability and of the strength which the site demands.

Keepers' dwelling. The dwelling for the keepers (below the lantern-floor) is but one story in height, and is smaller and less convenient than in similar structures in the United States. The sides and roof are made of corrugated iron with wrought-iron angle-plates. Below the floor of the dwelling additional space is furnished by placing a store-room in an inverted pyramid, to which access is had by a ladder from the gallery. The dwelling is divided into a living room, (also used as a kitchen,) a bed-room, and an oil-room. It was stated that the sea rarely rises to the bottom of the house, and the object of the peculiar form given was to allow the wind and spray to be warded off without imparting shocks to the structure. I should judge the device to be one of questionable utility, and that but little more expense would have been incurred by raising the building a few feet higher and placing another full story for the accommodation of the keepers.

Additional space furnished.

Reason for the pyramidal form of the structure.

Keepers. There are two keepers, one less than we would have in the United States, and it will be observed throughout this report that the British lights are maintained by a less number of keepers for each than for the same order of light in our service.

Catoptric apparatus and lantern. The lantern, which is large and commodious, contains a revolving catoptric apparatus composed of fifteen reflectors and Argand burners in sets of five, placed on a frame of three sides, and this being a red light, panes of red glass, in frames hung on hinges, were placed in front of each reflector. This structure seems admirably adapted to the locality, and I should think the question of replacing by similar structures some of the great number of light-ships which mark

Red light, how managed.

Advisability of replacing light-ships by screw-pile light-houses.

Plate V

MAPLIN SAND LIGHT-HOUSE.

the channels through the shoals obstructing navigation on the east coast of England, would have attracted attention, and there are probably some special reasons why it has not been done.

While the first cost of a screw-pile light-house in an exposed locality is greater than that of a light-ship, the cost of maintenance as well as of repairs is much less; and besides, the danger which sometimes occurs of light-ships being dragged from their stations and leading vessels into the very dangers from which they are intended to warn them, is avoided. *Comparative costs of screw-pile light-houses and light-ships.*

These considerations have induced us to replace our light-ships by screw-pile light-houses except in the case of shifting shoals like those off the island of Nantucket.

ORFORDNESS.

We did not visit these light-houses, but as viewed from seaward they are substantial structures. Seen in one, in either direction, they guide clear of certain dangers; and besides this, they mark out, by means of red sectors of light, other dangers. This was the first instance I saw of what the Elder Brethren call "red cuts," which I shall fully describe when I come to treat of the lights at Souter Point and Coquet Island. We ran from the white into the red light, and the line of division was quite distinct. *Object of the light-houses at Orfordness.* *"Red cuts."*

YARMOUTH.

On the 12th of June we arrived at Yarmouth, where there is an extensive supply and buoy depot on the river Yare. It consists of a very fine buoy store-house, a store-house for chain-cables, a cooper-shop for wooden buoys, smith and paint shops, a store-house for oil, slips for the repair of light-ships, and quarters for the superintendent, foremen, and clerks. There is also a fire-proof store-house for the signal-rockets used on the light-ships. *Depot at Yarmouth.* *Buildings.*

The buoy store-house is well built of masonry, paved with wooden blocks, and traversed by a railway. A trussed traveling-crane, of excellent construction, supported on girders resting on piers projecting from the side-walls, gives great facility for moving the buoys; large sliding-doors open toward the river, and on the wharf is placed a ten-ton crane for hoisting the buoys into the steam-yacht, Beacon, which, with its steam-launch or pinnace, is constantly engaged in the service of the district under the superintendence of Mr. Emerson. *Buoy store-house.* *Yacht Beacon.*

Buoys repaired. At this depot are painted and repaired the buoys for the neighboring coast and channels, and a large number are kept in store for relief and to supply losses.

Wooden buoys brought in each year. The wooden buoys (not *spar*-buoys; of these I believe the English have none) are brought in once a year for painting, *Iron buoys painted at their moorings.* but the iron buoys are painted without being unshackled from their moorings, and are but rarely changed. The *Marking buoys.* buoys are marked with their numbers and *names;* these last being the names of the spits, channels, &c., which they are intended to point out. A practice very different from our own is that of painting these numbers and names on heavy canvas strips of two thicknesses, which are fastened to the buoys by means of bolts and nuts, or by lashings. These strips are frequently changed, (without lifting the buoys,) as it is considered of great importance that the names and numbers of the buoys shall always be plain and distinct.

Method of determining if buoys and light-ships in view from the depot are in place. There are a great many buoys and several light-ships in view from the high lookout-tower above the dwelling of the superintendent, and Mr. Emerson has contrived a simple and ingenious mode of detecting if any of them have been driven from their positions.

A large telescope, (shown in Fig. 5,) provided with spider-lines, is movable on a vertical axis, *c*, fixed upon a platform on which are marked cross-lines *c a, c a, c a*, and the names of the buoys and light-ships, which indicate the precise direction in which they should be found by means of a pointer, *c b*, attached to the pedestal of the telescope; thus the slightest drifting from their proper positions is at once discovered.

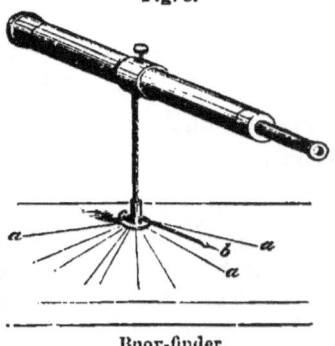

Fig. 5.

Buoy-finder.

In regard to buoys, I should mention that the buoy-list of the English differs in several respects from our own. There is noted in regard to each buoy its *name, size, description, (kind,) color, material, weight of sinker, fathoms of chain, depth of water, when first laid, and date of last removal.*

The following is an example:

RAMSGATE CHANNELS.

Name.	Size.	Description.	Color.	Material.	Mooring.		Depth of water, lowest.	First laid.	Removals.	Remarks.
					Anchor.	Chain.				
	Feet.				Cwt.	Fms.	Feet.			
West side.										
Dike...............	6	Can......	Black................	Wood...	10	10	10	1844	1859	
Quorn..............	6	Can......	Black and white checkered	Wood...	10	10	7	1844	......	Color altered from white, 1859.
North Fairway.....	9	Cl. H. B.	Red and white striped	Iron	14	10	6	1783	......	Name and color altered, 1869.
Checkered Fairway.	7	Can......	Red and white checkered	Wood...	12	12	15	1783	......	Color altered, 1869.
South Fairway.....	7	Can......	Red and white striped	Wood...	14	10	14	1865	......	Color altered, 1869.

The following abbreviations are used for designating the kind (description) of buoys: H. B., hollow bottom; C. B., convex bottom; C., can; Cr., can reversed; F. B., flat bottom; E. B., egg-bottom; Sph., spherical.

A spare light-ship is kept at Yarmouth depot, and I saw one repairing in the dock. The bottom was exposed, and was provided with bilge-pieces or "bilge-keels" to prevent rolling, as shown in Fig. 6.

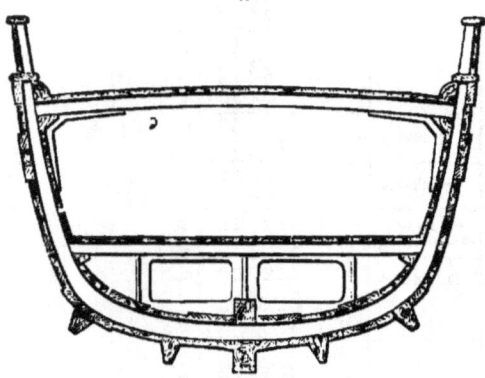

Fig. 6.

Bilge-keels for Light-ships.

I was informed that this was the common practice in the English service.

HAISBOROUGH.

Position. There are two first-order sea-coast fixed lights at Haisborough, in the county of Norfolk, the northern or high light being 140 feet, the southern or low light 94 feet, above the sea. They are about a mile apart and form a range or lead for vessels passing through Haisborough Gat.

Competitive trial of gas and Douglass's four-wick oil-lamp. At the time of my visit to this station there had been in progress for some months an important competitive trial between a Douglass four-wick lamp in the low tower and a gas-lamp, patented by Mr. Wigham, of Dublin, in the high tower. The apparatus for the latter was manufactured by Edmundson & Co., engineers, of London and Dublin.

Gas in use in Irish light-houses. The commissioners of Irish lights have introduced the use of gas into several of their light-houses, as we were informed by Professor Tyndall when he was in the United States last year, and in the remarks which he made at a session of the Philosophical Society of Washington, he mentioned the great "flexibility" of this kind of light when used for light-house illumination.

Flexibility of gas-light.

Lamp. I will treat more particularly of this invention of Mr. Wigham when I describe the Irish lights that I visited with him, but I will state here that I found the lamp to consist of a horizontal circular disk (hollow) about one foot in diameter, supported upon a stand and into which the tubes supplying the burners, in sets or frames, were connected by joints made tight by means of quicksilver.

The lamp is designed to burn 28 jets in clear weather. They are arranged in concentric rings, the diameter of the inner row being about the same as that of the outer burner of the ordinary four-wick lamp, *i. e.*, four inches. *Arrangement of gas-jets.*

In case the atmosphere becomes hazy, an additional exterior row of 20 jets is placed in two frames of 180° each, each frame being supported by a short supply-tube set into a cup containing a quicksilver joint. *Increase of number of jets.*

During this operation the lights from the 28 jets forming the nucleus are turned low, and when the cocks are reopened the flame from these lights the exterior row. As required by increasing density of fogs or thick weather, additional rows of jets are successively placed in each case, increasing the number by 20, so that from 28, the number in the nucleus, the various powers are 48, 68, 88, and 108 jets, the latter being used only in very thick weather or dense fog.

There is no chimney *surrounding* the flame, but above it, at a distance of about 12 inches, is suspended a chimney of mica, into which the flame is carried by the draught through the cowl of the lantern. The mica chimneys vary in diameter, and are changed to accord with the number of jets used. *Chimney of mica.*

The entire operation of changing from one set of jets to the next higher or lower, or from the lowest to the highest, or the reverse, and also changing the mica chimneys, occupies but a few seconds, not more time, I should think, than the trimming of a four-wick lamp. *Time occupied in changing power of light.*

The diameter of the flames corresponding to the different powers of the lamp are respectively $3\frac{3}{4}$, $5\frac{1}{2}$, $7\frac{1}{2}$, $9\frac{1}{4}$, and $10\frac{1}{2}$ inches for the 28, 48, 68, 88, and 108 jets. It will therefore be observed that a great part of the larger flames is necessarily exfocal, increasing the divergence of the light, and the increase of intensity when seen at any point within the arc of visibility is no doubt due to the great *thickness* of the flame. *Diameters of the flames.*

The heat inside the lantern, when the larger flames are turned on, is very great, but I was told that it was not sufficient to injure the lenticular apparatus nor to seriously annoy the keepers. *Heat produced by the largest flames.*

As in the electric light at South Foreland, an oil-lamp is always at hand in the watch-room, and in case of accident to the gas-lamp, it can be removed and the former lighted in less than two minutes. *I did not learn that occasion for *Oil-lamp on hand in case of accident.*

its use had yet occurred, and I should think it even less likely to occur here than in the case of the electric light.

Coal used. In the gas-house near the tower common Newcastle coals are used for heating the five retorts, and cannel-coal yields the supply of gas.

Gas, how conducted to the receivers. The gas issuing from the retorts is, after being caused to pass through water, conducted through several layers of slaked lime contained in flat boxes, thence through a system of pipes, depositing the tar-product *en route*, and finally is carried into the receivers (of 4,900 cubic feet capacity) from which the light-house lamps are supplied.

Meters. Separate meters are used for registering the quantities of gas consumed in the dwellings and in the light-house, and each amount is reported monthly to the Trinity House. The consumption of gas in the 48-jet burner, in a night of 7½ hours, was 830 feet, or 2.3 feet per burner per hour.

Consumption in April and May, 1873. In the months of April and May preceding my visit the consumption of gas in the tower had been 21,980 and 26,450 feet respectively.

Number of keepers. There are two keepers at each light-house at Haisborough, (that being the rule for all English sea-coast lights, except rock-stations,) and in addition to them is employed a laborer from the neighboring village, to make the gas, but his attendance at the station is only required every other day. He is paid a weekly salary of fifteen shillings, (about $3.75.)

Laborer employed.

Fuel saved by use of tar. A large saving in fuel is effected by consuming the tar which is produced in the manufacture of the gas.

A general plan of the buildings at the light-house is shown in Plate VI.

Plan of tower. The tower is built of brick and stuccoed; it and the dwellings, out-buildings, and walls surrounding the premises are kept scrupulously clean and neat.

Stairs. The interior of the tower is cylindrical; the stairs, like those in the towers at South Foreland, are circular, and apparently self-supporting, one end only being built into the wall, as in our Treasury at Washington, and in several other buildings I have seen in America. This method of stair-building I found to be universal in Europe, in private as well as public buildings. I think our most recent towers with conical interior and iron stairs winding around the interior of the cone, superior to any I saw in Europe. European towers are, however, superior to any constructed by us until within a few years, on account of the greater amount of light and the airiness of towers with a free and open interior.

Comparison of American and European towers.

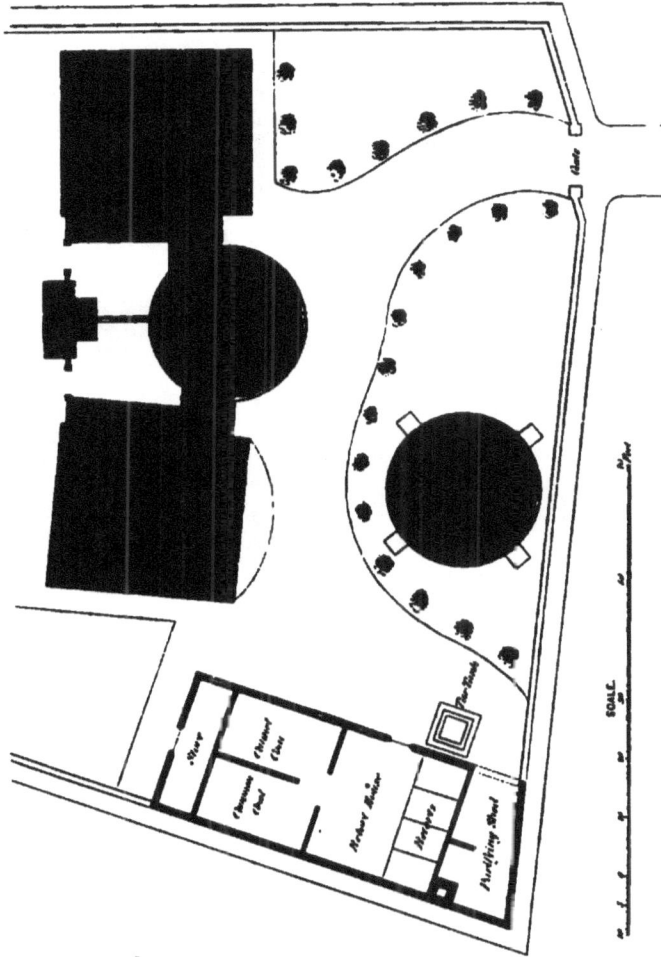

HAISBOROUGH HIGH LIGHTHOUSE.
GAS LIGHT.

PLATE VI.

The practice abroad is, in most cases, to make the interior wall cylindrical and the exterior one conical, leaving an unnecessarily large air-space between the two near the base, while in our latest towers, such as those at Bodie's Island, Saint Augustine, and others, a small air-space sufficiently large for the purpose intended is left between the walls, (both of which are conical,) and the space gained by this mode of construction is thrown into the interior.

The amount of masonry in our present system is the same as in the former, and is calculated to resist by its weight the overturning effect of the severest gales. Masonry in old and new systems the same.

At Haisborough the oil-cellars are placed below the floor of the towers, the cisterns or cans, each holding about 100 gallons, being arranged around the walls. There are no arrangements for pumping the oil to the top of the towers, and it is carried by the keepers by hand. The filling-room below the lantern is provided with brass measures of different sizes, from a gallon downward, and every morning the keeper notes the consumption of oil the previous night, and makes monthly returns of the amounts to the Trinity House. Oil-c Measures to show nightly consumption of oil.

These measuring-vessels, the brass-work of the lamp, and the hand-rails of the stairs, are always neatly burnished. All metal-work burnished carefully.

The English lanterns in all the recent light-houses have diagonal sash-bars, as it is considered that the upright bars obstruct a large portion of the light in certain directions. I will more fully treat of the latest lanterns which I saw, and particularly of the advantages of the diagonal sash-bars, when I come to describe the light-house at Holyhead. Diagonal sash-bars.

The glass for the lanterns at Haisborogh is half an inch thick, the panes are lozenge-shaped, and the surfaces are curved to conform to the diameter of the lantern. Glass for lanterns.

No special means are used to prevent large sea-fowl from breaking the lantern-glass, and I was told that the necessity of such means does not exist in England as it does with us, particularly on our southern coast.

In the lantern-floor there is provided a basin, covered when not in use, into which is led rain-water from the roof for use in washing the interior of the lantern. Basin in lantern-floor.

The air, which supports the combustion of the lamp, is not let directly through the sides of the lantern, as in our service, but is admitted below and passes through the grating which forms the lantern-floor. The object in this is to give the air a uniform temperature, and great importance is attached to this in the English service. (See Plate VII.) Egress of air.

The windows of the tower are arranged without admitting the rain, according to an excellent plan which is shown Method of ventilation.

in Plate VIII, in which it will be observed that the upper sash is hinged at *a*, and swings, as shown by the dotted arc. To the lower part of the sash is fastened a rod, *b*, which passes through a sleeve, *c*, which is movable about an axis, and through which a set screw passes by which the window can be fastened at any desired angle. One only of the lower sashes opens, as is shown in the plate.

Cost of changing from oil to gas light. The cost of changing from oil to gas at Haisborough was about £1,700. ($8,500,) the gas-holder and other parts of the apparatus being designed to serve both lights.

Painting the dwellings, &c., periodically. In the English service the towers and dwellings are generally painted white (to make them serve better as day-marks) once in four years, by painters permanently employed by the Trinity House, and who for this purpose visit the stations in rotation.

The lantern, watch-room, &c., are painted by the keepers once a year. The hand-rails, when of iron, are painted with bronze paint, and when they are of brass, which is often the case, they are kept neatly burnished.

Wind-vanes and lightning-rods. At the summit of the lanterns are always placed wind-vanes and lightning-rods.

Flag-staffs. Flag-staffs are provided at each station, placed either on the tower or in the grounds surrounding it. The Trinity-House flag is displayed whenever the tenders are seen approaching; also on Sundays and holidays. I observed that a neat pavement of pebbles, about 15 inches wide, was laid at the foot of each wall, to protect the soil from the wash from the wall in rainy weather.

Rooms furnished keepers. In regard to the dwellings, each keeper is furnished with a living room, three bed-rooms, a scullery, wash-room, a place for coals, and a garden.

Books at the station. There are at Haisborough, as at all other light-stations in the English service, certain books furnished by the Trinity House in which are kept the records of the stations. Among them I observed an Order-Book, in which any officer of the corporation enters the orders or directions given by him to the keeper while on his visit to the station. It is his duty to observe whether previous orders of himself or others have been properly executed.

Another book is called the Visitors' Book, and in it are recorded the names and professions of the persons visiting the station.

Libraries. Small libraries are provided at each station for the use of the keepers and their families. They always include a Bible and Prayer-book, and are otherwise composed of books suitable for persons of their class.

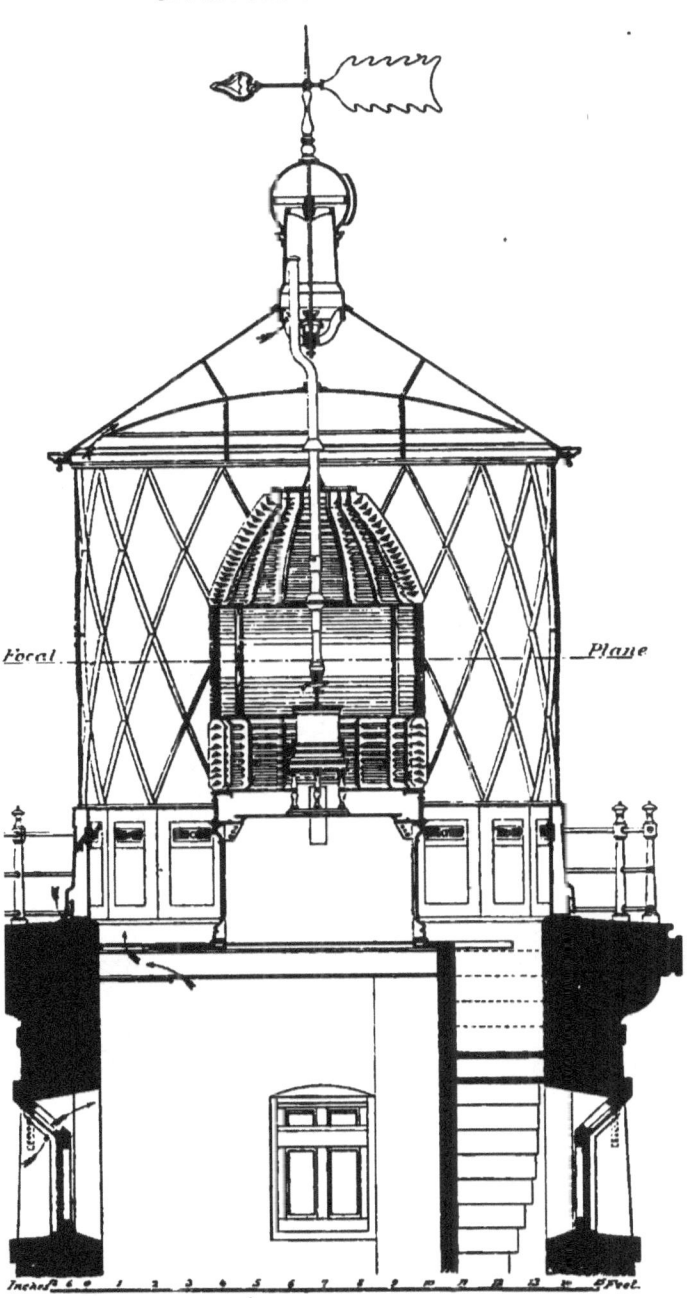

These libraries are interchanged between the stations on the annual visits of the supply-vessels.

Medicine-chests also are furnished to each station. *Medicine-chests.*

I did not observe any room set apart especially for a work- *No work-rooms at English stations.* shop as at our large stations, though keepers are furnished with necessary tools, and their education at Blackwall in mechanical operations would, with the general intelligence possessed by the keepers, make this provision more useful even than in our own service.

A certain amount of standing furniture is provided in *Furniture.* each dwelling. It includes iron bedsteads, chairs, tables, a desk, &c.

When a keeper is removed from one station to another (either to a better one as a reward of merit, or to an inferior one as a punishment) his family is transported at public expense.

The keepers at Haisborough, as at all the other stations *Uniform worn by the keepers.* which I visited, wore the neat uniform of the Corporation of Trinity House.

From Haisborough we steamed out to the Newarp light- *Observation of lights at night.* ship, (to be noticed farther on,) and returned after dark to observe from the sea the comparative intensities of the gas and oil lights.

The gas-light is in the northern light-house, the oil-light in the southern, on a point of land nearer the sea, at an elevation 46 feet below the former, the respective heights of focal planes being, as before stated, 140 and 94 feet above the sea. The lower tower is lighted by one of Douglass's four-wick lamps.

The Vestal was stopped at a distance of six and a half miles from the lights, and at a point equidistant from both. The night was clear, and the opportunity for fair-weather observations was excellent. The Trinity House officers on board had directed the keeper of the upper (the gas) lighthouse to burn the ordinary number of jets, viz, 48, till 9 o'clock. At that time the number of jets was to be reduced to 28, and the changes were to be as follows:

At 9 p. m. reduce to 28 jets; at 9.10 p. m. increase to 48 *Memoranda of changes in lights.* jets; at 9.20 p. m. increase to 68 jets; at 9.30 p. m. increase to 88 jets; at 9.40 p. m. increase to 103 jets; at 9.50 p. m. reduce to 28 jets; at 10 p. m. reduce to 68 jets; at 10.10 p. m. reduce to 48 jets; at 10.20 p. m. reduce to 28 jets; at 10.30 p. m. increase to 48 jets.

The comparative brightness of the lights was estimated *Manner of observing the lights.* by observing them with the naked eye, and also through different thicknesses of red glass. The method of using the latter was to place successive layers of small plates of glass into frames made for the purpose, until one or both of the lights when seen through them could barely be discerned, and I found that the eye could thus much more readily detect differences between the intensities of the lights than when viewing them without the use of the glass media.

Question of using a neutral-tinted glass through which to compare the lights.

I am not, however, satisfied as to the advisability of using red glass, since it is probable that those flames which have more of that color in their composition would be placed at a disadvantage, and I would prefer a neutral-tinted glass.

Appearance of the lights.

When we first observed the lights from our position the gas-light (48 jets) was not equal to the oil-light; between 9 p. m. and 9.10 p. m. (28 jets) it was still more inferior; between 9.10 and 9.20 (48 jets) the same difference was observed as before; between 9.20 and 9.30 (68 jets) we pronounced the two lights equal.

Fog obscures both lights.

At this time a dense fog rolled in from seaward, obscuring both lights, and we steamed toward them till we got within (as we afterward found) two miles of them, both continuing eclipsed. About midnight the fog rolled away, and the lower (oil) light came gradually into view, but when it had apparently attained its full power we could still see no sign of the upper (the gas) light.

Fifteen minutes afterward the upper light dimly appeared and slowly increased in brightness till about half past 12, when both lights were fairly free from the fog, and in the opinion of all the party the upper (gas) was very much superior to the lower (oil) light.

Superiority of the gas-lights.

As the time covered by the instructions given to the keepers had long since expired, it was not until our return to London that we learned the number of gas-jets burning, which was then shown to be 108, the number corresponding to the instructions of the keepers for times of dense fog.

Number of jets burning.

It was fortunate for our experiment that the fog shut in during our observations. That the oil-light was first to be seen was no doubt due to the fact that the fog rolled in over the land from seaward, (though this was not apparent to us, there being no perceptible breeze,) and that light, being on a point projecting into the sea, was first free from it.

Judgments arrived at by observation of lights.

As far as determined by our experiments at Haisborough, I have no doubt the following judgments were correct:

1st. In fair weather the gas-light of 68 jets was equal to the first-order light from the oil-lamp of four wicks as improved by Mr. Douglass.

2d. Neither the light of 28 nor of 48 jets was equal, but that of 108 jets was decidedly superior to the oil-light.

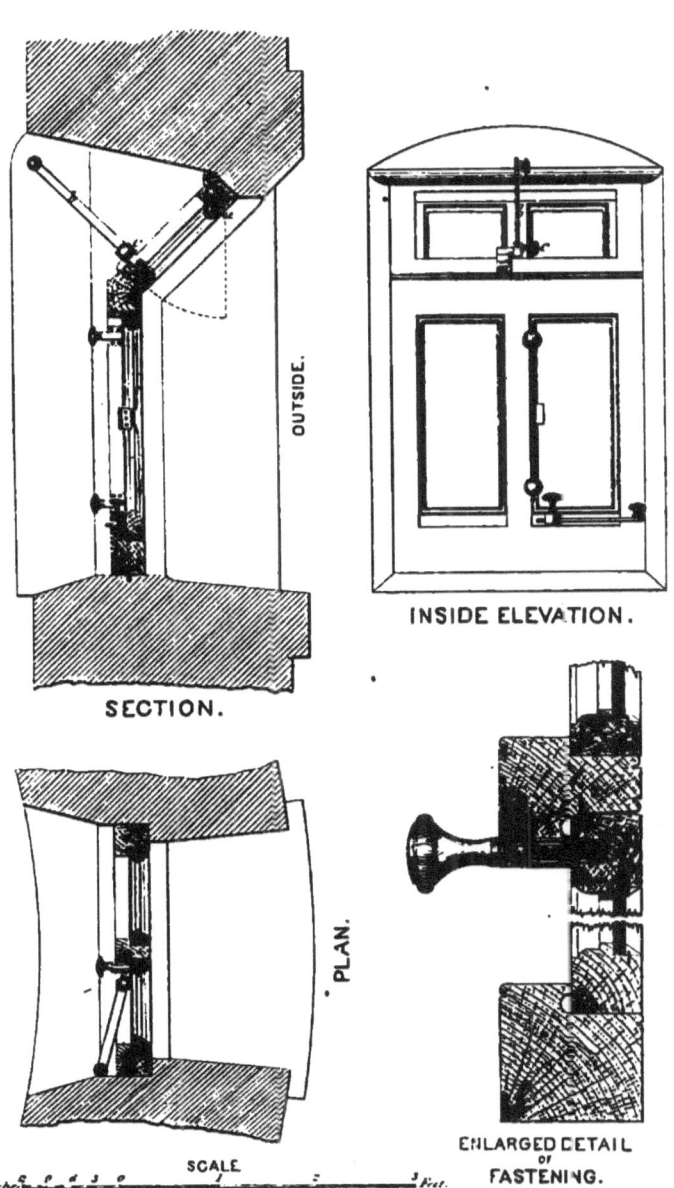

3d. In a dense fog, to an observer at a distance of two miles, neither the gas-light of 108 jets nor the oil-light from a four-wick burner gives any indication, however slight, of its existence.

It has been observed by Professor Tyndall, in his experiments in Dublin, that a steam-cloud of even a few feet of thickness is sufficient to totally obscure the rays of the sun itself, and as either the four-wick oil-lamp or the 28-jet gas-light is sufficiently powerful to illuminate the horizon corresponding to the ordinary elevations of sea-coast light-houses, *i. e.*, at distances from eighteen to twenty nautical (twenty-one to twenty-three statute) miles in clear weather, it is evident that the increased intensity of sea-coast lights is desired for those intermediate states of the atmosphere between dense fog on the one hand and clear weather on the other. *A slight cloud of steam sufficient to obscure the sun.*

In these conditions of the air, including all the varieties of haziness and "thick" weather, up to dense fog, (and also in snow-storms,) light-houses are most useful; for these the light-house engineers of Europe are striving to find the most powerful lights, and to meet this want the electric light (which I have described at South Foreland, and which I shall more clearly exhibit in an account of my visit to the light-houses at the mouth of the Seine) and the gas-light (such as I have described at Haisborough and afterward saw on the coast of Ireland) have been introduced within a few years. *Conditions of the atmosphere for which the most powerful lights are desired.*

While both these lights can be maintained at moderate power in fair weather, they have this advantage: they can be increased almost without limit when it becomes hazy and thick; this can be done without any increase in the size or cost of the lenticular apparatus, since the electric light requires a lens much smaller than that required for an oil-lamp, and as I shall illustrate farther on, 324 jets can be burned in Wigham's triform gas-light without increasing the size of the lenticular apparatus or the diameter of the flame beyond the maximum (108 jets) which I saw at Haisborough. *Power of electric and gas lights can be increased when needed.*

It is this power of being increased, (according to the conditions of the weather, from 28 jets by steps of 20 at a time, till 324 jets the beam from which, even when uncondensed, is equal to more than the united beam from 6,000 candles can throw their rays in a solid beam through the lenticular apparatus,) which gives to the gas-light of Mr. Wigham its great "flexibility," to adopt the term so happily used by Professor Tyndall when speaking of this light, *Flexibility of the gas-light.*

and I believe in this regard it is superior to the electric light.

Economy in use of gas. In the gas-light in clear weather only a sufficient quantity of gas is used to carry the light distinctly to the horizon. The large quantity required for "thick" weather remains stored in the reservoir till wanted, and the expensive light is burned *only when needed*, whereas in the electric light, though the engine-power is doubled in "thick" weather, yet the ordinary fair-weather expense of the engines is much greater than the fair-weather cost of the gas light-houses; and, further, the gradations of power to which the gas-light is subject are much more varied than in the electric light, and the former can be suited by intelligent keepers to any state of the atmosphere.

Absence of flexibility in oil-light. Of course the oil-light which we use in the United States has no "flexibility" and burns the same in fair weather as in foul, in the twilight of the evening as in the darkness of the night. This is a fact of very great importance in this country, and particularly in high latitudes in Great Britain. In the long twilights of the last summer, while between the mouths of the Tyne and Tweed, I found no difficulty in reading on the deck of the Vestal at half past 10 o'clock, and indeed it could hardly have been said to be dark during the entire night. In these long twilights and in clear nights great economy can be attained in the use of illuminating power, which can be stored up, as it were, to be used only when the weather demands that it shall be put forth in all its strength.

Question of the relative penetrating powers of the different lights not yet determined. It is to be observed in this connection that the relative penetrating powers of the oil, the electric and gas lights, have not yet been sufficiently tested *at a distance and in all sorts of weather.* This is a matter of great importance, and should be made the subject of an exhaustive series of experiments.

Almost any illuminant is good enough for fair weather, but the light which will be finally adopted by all nations will be that which will send its rays to the greatest distance in storm and thick weather.

Illustration of the gas referees of London. The gas referees of London, to whom the English Board of Trade have referred the matter of light-house illumination by gas, very cleverly illustrate this desideratum as follows:

"Suppose the case of two regiments armed in the main with short-range rifles, but each comprising a body of marksmen twenty in number in one regiment, and forty in the other, armed with rifles of the longest range.

"At 1,200 yards the power of these regiments would be represented solely by the numbers of their long-range riflemen—the power of the one at that distance being double that of the other, although at close quarters their destroying-power would be equal.

"Every flame of gas or oil may be said to be a sheaf of rays of various lengths or penetrating power, so that two lights which are equal near to their source may become unequal when viewed from a distance; and an analogous effect to that of distance will be produced by mists and fog, obstacles with which it is most desirable that light-houses should be able successfully to contend."

THE NEWARP LIGHT-SHIP.

This light-ship marks one of the sands which form a perfect labyrinth off the coast of Norfolk and Lincolnshire. It is built of wood, is registered as 212 tons builders' measurement, and has three masts carrying fixed lights; the fore and mizzen being 24, and the main-mast light 34 feet above the sea. It is anchored in 17 fathoms of water by an anchor weighing 45 cwt., having 210 fathoms of 1½-inch chain, and carries besides, two bower-anchors of 20 and 14 cwt. respectively, with 150 fathoms of chain each. *Position. Material and size. Lights. Moorings.*

The ship carries a Daboll fog-trumpet, which is sounded by means of an Ericsson hot-air engine with an 18-inch cylinder, placed below the deck and near the bow of the vessel. Both the smoke-funnel and the trumpet are placed forward of the foremast. The latter, which is removable, is kept below deck when not in use; when sounding it revolves once a minute. *Daboll trumpet.*

A Chinese gong is provided for use in case of accident to the trumpet or engine, and it was sounding when we left the vessel, but we ran out of its range at a very short distance; I thought it inferior to the bells used in our light-ships. *Chinese gong provided.*

The trumpet was also sounded after we left the vessel, and although I judged it to be pitched at too high a note, according to the conclusions arrived at in our American experiments, we heard it with remarkable distinctness. At a distance of two miles it sounded very loud and clear; at six miles the sound had sensibly decreased, but it was quite audible when the Vestal was under way, and it was not until we had gone eight miles that it ceased to be heard. *Trumpet sounded. Heard at eight miles.*

There was no wind to interfere with the sound, but my experience on this occasion satisfied me that, for localities where fogs are as prevalent as at the stations occupied by *Fog-signals necessary on light-ships.*

S. Ex. 54——8

our light-ships off our northern coast and in Long Island Sound, powerful fog-signals, operated by steam or hot-air, would be extremely useful to the immense commerce depending on these vessels for safety.

Arrangements of lights on the masts. In order to assist in determining at night the direction in which light-ships are riding at their anchors, the lights on the mizzens are placed at lower elevations than those on the main masts.

English light-ships' crews. In the English service each light-ship has the following crew: one master, one mate, three lamp-lighters, and six able seamen, one of whom may be a carpenter. No applicants under thirty-two years of age are admitted.

Table of rates of pay. The following table shows the uniform rates of pay in the service:

Rate per month.	Gross.			Deduct insurance.			Increase after 3 years service for good conduct.		
	£	s.	d.	£	s.	d.	£	s.	d.
Masters, uninsured	6	13	4	0	0	0	0	0	0
Masters, insured	6	15	10	0	5	0	0	0	0
Masters who have served as such five years and upward	5	2	6	0	5	0	0	0	0
Masters who have served as such under five years	4	10	6	0	5	0	0	0	0
Carpenters who have served as such five years and upward	3	11	0	0	5	0	0	2	6
Carpenters who have served as such under five years	3	8	6	0	5	0	0	2	6
Lamp-lighters who have served as such five years and upward	3	7	6	0	5	0	0	2	6
Lamp-lighters who have served as such under five years	3	3	0	0	5	0	0	2	6
Seamen who have served as such five years and upward	3	0	0	0	5	0	0	2	6
Seamen who have served as such three years and under five years	2	17	6	0	5	0	0	2	6
Seamen who have served as such under three years	2	17	6	0	5	0	0	0	0

Table of rations. The master furnishes the provisions per the following table:

Meat, 10 pounds per week each man.
Bread, 7 pounds per week each man.
Flour, 2 pounds per week each man.
Peas, 1 pint per week each man.
Potatoes, 7 pounds per week each man.
Suet, ½ pound per week each man.
Tea, 2 ounces per week each man.
Sugar, ¾ pound per week each man.
Beer, 3 gallons per week each man.

When on shore 1s. 7d. per day is allowed each man in lieu of provisions.

Uniform. The master and mate are furnished a regulation uniform-suit, and the crew a cap, one shirt, and one pair of trousers annually. When unfit for longer service they receive a pension, computed on length of service, varying from 4½d. to 1s. per day. Their pay and allowances for this service is much better than those of either the royal navy or the merchant-service. *Pensions.*

Either the master or mate must remain on board. One-third of the crew is on shore at a time, the relief occurring once a month. They must remain near the shore-station, with the officer (master or mate) on shore at the time, and execute such service as may be required, and, if the vessel goes adrift, join her as soon as possible. If vessels are observed in distress guns are fired, and if at night, rockets are thrown until assistance approaches. Careful and regular observations with meteorological instruments are taken on light vessels as well as at light-houses. *Master or mate and two-thirds of crew to remain on board. Meteorological observations to be taken.*

The officer in charge on light-ships, at light-houses, and in the Trinity House yachts are required to assemble the men under his orders every Sunday, (when they have no opportunity of attending church,) and to read the church-service for the day and a sermon or homily from a volume provided by the Trinity House for the purpose; when one of the Elder Brethern is present on Sunday the church-service is read by him. *Officer in charge to hold church-service. Or any member of the Elder Brethren, if present.*

THE COCKLE LIGHT-SHIP.

This light-ship, which we visited on our return from the north, is placed at the northern entrance to Yarmouth Road. It is 155 tons measurement, is built of wood, and is anchored in seven fathoms of water. It carries a revolving white light which is produced from nine reflectors arranged on three faces; three reflectors on each. The interval between flashes is one minute. *Position. Material, measurement, and moorings. Light flashing.*

This is one of several light-ships of this kind which I saw in my journeys around the English and Irish coasts, and they are, no doubt, much more useful in attracting the attention of the mariner than light-ships with fixed lights. *Preferable to fixed lights.*

We passed the Cockle at night, and I had a good opportunity of seeing the light. It lies about six miles from the Newarp, and the crew state that in light winds and clear weather the Daboll trumpet of the latter can be heard with great distinctness, but that fog "kills" the sound to a great degree. *Daboll trumpet on board the Newarp heard from the Cockle, six miles.*

SPURN POINT.

Spurn Point is a low sand-spit, projecting into the mouth of the river Humber. There are two towers at this station, and they form a double range or "lead," the outside one *Range-lights.*

being effected by the eclipse of the inner light by the outer tower.

Number of range-lights on the English coast. The number of stations on the east coast of England with two towers, many of them with first-order lights, forming ranges or leads, is noticeable, and is accounted for by the intricacies of the channels between the sands and shoals off the coast, and their distance from the land.

Construction of high tower. The high or main light-tower was built in 1776 by John Smeaton, the builder of the Eddystone light-house, and is as unlike the graceful light-house towers of the present day as can well be imagined.

The rooms are very large. The lower story only is arched over, and is used for an oil-room, while the upper rooms serve for the families of the keepers, and one is used as a *Chapel for keepers.* chapel for the keepers, coast-guardsmen, and fishermen who live at the Point.

Lens. The lens is of the first order, and a part of the arc of illumination is covered by white light, while certain dangers *Red cut.* are marked by a red sector. The red glass in this light covers the required arc of the lens, and is fastened to its frame, but in order to sharpen the "cuts" between the red and white light, narrow strips of red glass are placed in the lantern opposite the edges of the red glass outside the lens, as is shown in Fig. 7, in which $a\,a\,a$ is the shade of red

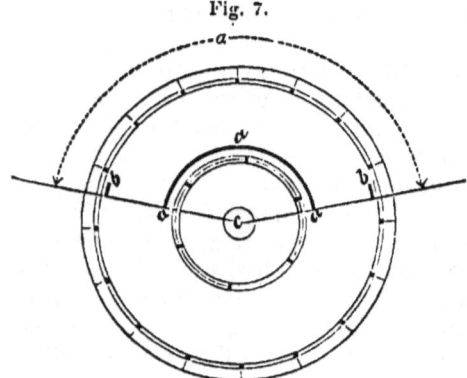

Fig. 7.

Red cut at Spurn Point.

glass of the entire height of the lens; $b\,b$ vertical strips of red glass for the purpose of intensifying the "red cut;" c the lamp, a the sector of red light.

Copy from British light-house list. The following, copied from the British light-house list, will suggest the purpose of this mode of covering by red light any desired area, and of marking by the cuts between the red and white lights any line upon the sea within the circle of illumination:

"SPURN POINT.—*A sector of red light is thrown from the high light-house bearing from N. W. by N., (cutting two cables N. E. of the Sand Hayle buoy,) round northerly to S. E. by E. ¾ E., on which bearing it will cut one cable north of Grimsby Pier. In other directions the light is white.*"

The oil in use at this station when I visited it was colza, but I believe it is the intention to change this and all the other lamps in the service of the corporation of Trinity House for the use of mineral-oil. Colza-oil used. Mineral to be substituted.

The lantern was a fine one of gun-metal, the sash-bars being diagonal. The diameter is 14 feet. Lantern.

The service-room was fitted up with closets and shelves, and the articles for daily use are neatly stowed away in boxes marked "Cleaning-cloths," "Skins," "Powder and brushes," "Cotton wicks," "Polishing-powder for brass and copper," &c. Service-room.

The lens is supported by the lantern-floor, which is carried upon eight double iron brackets, and the interior of the watch-room is finished in corrugated iron. Lens.

The station, comprising the lower dwelling, &c., is surrounded by a high wall similar to that at the Longstone, (mentioned farther on,) but for a different purpose, viz, to keep out the drifting sand. Wall around the station.

The lower light-house is a comparatively new structure, and was built in shoal water inside the main tower, as a substitute for the tower which Smeaton built *outside* the high light which was some years ago undermined and destroyed by the sea. Lower light-house.

The apparatus in this tower is of the fourth order. Apparatus.

FLAMBOROUGH HEAD.

This part of the Yorkshire coast is high and bold, resembling the coast of California. Appearance of coast.

I was interested, not only in the inspection of the fine new tower on this remarkable headland, but in seeing near by a well-preserved example (built in 1674) of one of those great coal-burning light-towers whose use preceded the invention of either of the systems of illumination now in use; from this tower was witnessed the naval battle between the Serapis and the Bon Homme Richard, which was fought off this headland on the 23d of September, 1799. Old tower for burning coal still standing.

118 EUROPEAN LIGHT-HOUSE SYSTEMS.

Height of focal plane. The focal plane of the new tower is 214 feet above the sea, and 87 feet above its base. The tower is surmounted by a first-order revolving lens, showing, alternately, one red and two white flashes.

Characteristics of lens.

Quality of lens. This lens, made by Chance, Brothers & Co., of Birmingham, is a fine piece of workmanship, and the Trinity House officers state that all of the optical apparatus furnished by this firm give great satisfaction.

Areas of red and white panels. The area of each of the red panels, is to the area of each of the white panels of the lens, as 21 to 9, thereby producing an equalization of the distances at which the flashes can be seen.

Lamp. The lamp was one of the latest, combining all of Douglass's improvements, and burned mineral-oil, though it is suited also for burning colza.

The lantern is of the same character as that at Haisborough, which I have described, and there were no points of special interest at this station that were not mentioned in connection with that light, except that on the edge of the bluff there is a fog-gun station, in charge of a special set of keepers, (two,) who have dwellings and gardens separate from those at the light-station.

Fog-gun station.

Fog-gun, how used. The gun, an 18-pounder, is in a small masonry building having an embrasure on the sea-side; it is fired at intervals of fifteen minutes in foggy weather, the charges being three pounds. About one thousand rounds are fired annually, and they are kept in ready-filled cartridges in barrels in the magazine. The gunners have no other duties.

WHITBY.

These two first-order lights are on the coast of Yorkshire, and, like the Haisborough lights, form a range or head which clears a dangerous rock.

Towers.
Red cut. The towers are about 250 yards apart. A red cut shown from the northern tower covers certain other dangers to be avoided by vessels.

Fastening for red panes. The mode of fastening the red panes, so that they can be easily removed for cleaning the lens, is very simple, as will be seen in Fig. 8, consisting of a turning-plate, which, when shut, rests on a slight projection.

Lamp-guard. Another simple contrivance in use was a movable metallic guard, which is slipped over the burner before the wick is trimmed, so as to catch the cuttings. (See Fig. 9.)

The stairs leading from the watch-room to the lantern were noticeable, the step, newel-post, and ornamental bracket being cast in one piece.

The smoke-pipe leading from the watch-room stove was **Smoke-pipe.**
of *brass* neatly burnished.

The dwellings for the keepers (each light having two) **Keepers' dwellings.**
were placed on opposite sides of each tower, and the rule is
general that each keeper has a dwelling quite separate and
detached from any other.

Fig. 8.

Fastening for red panes.

The dwellings at Whitby—and this is also the rule—are
only one story in height.

There was nothing noticeable in regard to the lenses,
except the large amount of rear light (*i. e.* through an arc
of 180°) not utilized, and I was informed that formerly the **Reflector formerly used for throwing the rear light to seaward.**
land side of the lens was occupied by a metallic reflector,
which, reflecting the heat as well as the light from the flame,
caused the wick to burn so much more freely on the rear
side than on the other as seriously to impair the light.

Fig. 9.

Lamp-guard.

Totally reflecting glass prisms, such as are now used in
light-houses, would not produce this effect, but they have
not as yet been supplied.

I learned at Whitby that one of the light-keepers has dis- **Method of preparing glass**
covered in his experience that dipping the lamp-chimneys in **chimneys to resist the heat.**
a hot solution of soda will prevent them from breaking
even when exposed to the strongest flames.

Town of Whitby.
Ruins of Abbey of Saint Hilda.

Not far from the light-house is the town of Whitby, interesting on account of the ruin of the once handsome Abbey of Saint Hilda, (founded A. D. 657,) the extensive commerce in jet, mined from the cliffs near by, and as being the port from which Captain Cook sailed in his voyage of discovery.

SOUTER POINT.

The great electric light at Souter Point, which I visited on our return voyage from the north, and a general view of which is given in Fig. 10, is three miles below the mouth of the river Tyne, and I reached it by carriage from South Shields, after a hurried inspection of Sir William Armstrong's great ordnance-works at Newcastle, with Admiral Collinson, to whom I am indebted for the permission which he had thoughtfully obtained from Sir William before leaving London.

Ordnance-works of Sir William Armstrong.

Fig. 10.

View of Souter Point Light-house.

Manufactories, and effect of smoke therefrom.

On both banks of the river, from the mouth to New Castle and beyond, there is an immense number of manufactories of all kinds, and their smoke hangs over the river like a cloud.

Sea-approaches obscured by smoke.

When the wind is from the westward this smoke is driven over the sea-approaches to the river, obscuring, much to the annoyance of the great number of vessels of all classes continually entering or leaving the river, not only the pier-lights at its mouth, but the sea-coast light at Souter Point which indicates the general position of the harbor.

SOUTER POINT LIGHTHOUSE.
ELECTRIC LIGHT.

PLATE IX.

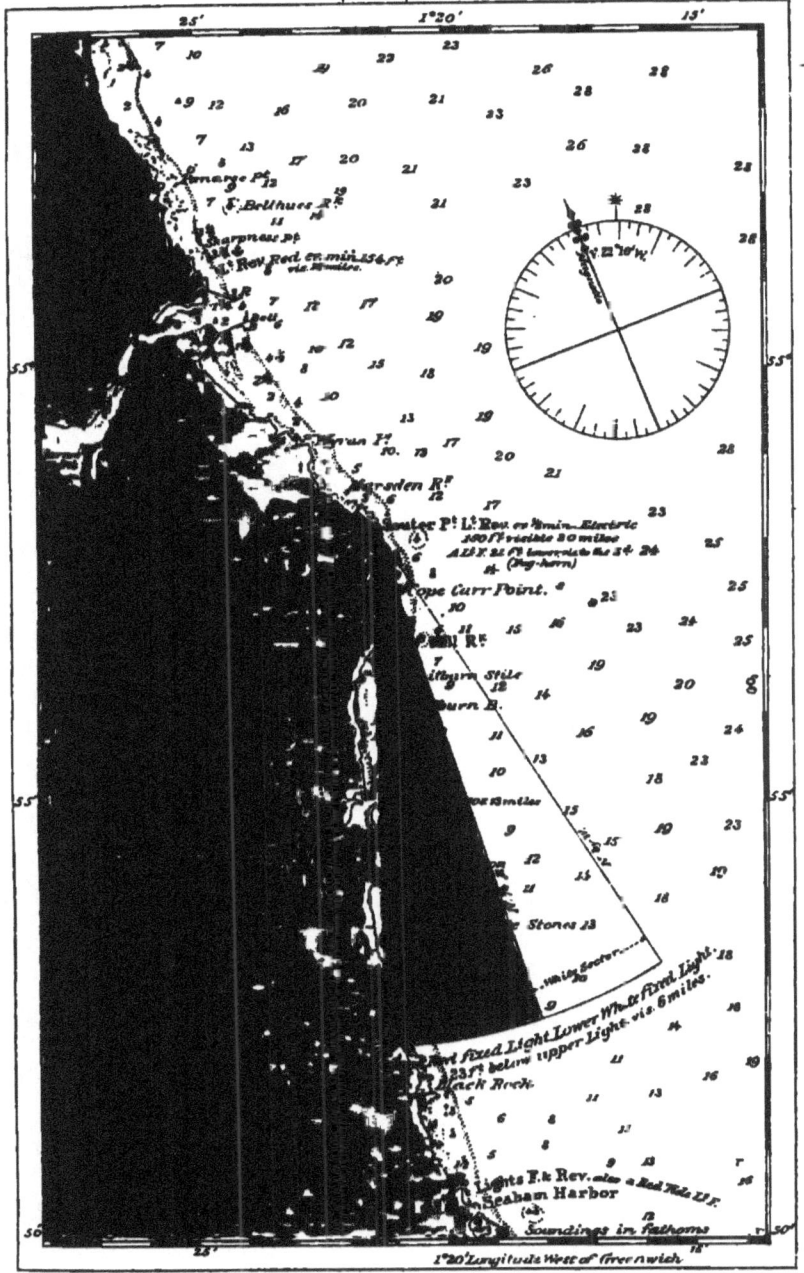

Fogs on this part of the east coast of England are also frequent, and as they mix with the smoke, the problem of light-house illumination of the sea near the mouth of the Tyne is one of great difficulty. *Frequency of fogs.*

To meet it, an electric light was constructed at Souter Point a few years ago, and it is without doubt one of the most powerful lights in the world. *Light at Souter Point one of the most powerful in the world.*

The condensed beams from the most powerful fixed and flashing first-order sea-coast lights of England (with the Douglass four-wick burner, the illuminant being oil) are respectively equal to 9,000 and to 111,000 candles, (ours are much less,) while the condensed beam of the flashing electric light at Souter Point (assuming the power of the lens, as calculated by Mr. Chance, to be 196 times as great as the power of the unassisted light) is equal in power to 800,000 candles! *Powers of first-order lights. (English.) Power of light at Souter Point.*

That even this intense light fails to penetrate a dense fog I know from my own experience, which I will relate further on, and it is not surprising, since the sun itself cannot do so; but that it meets the requirements of those intermediate conditions of "thick weather," between fair weather and impenetrable fog, as well as can be asked for any possible sea-coast light, cannot be doubted, although, considering "flexibility"* as well as power, I believe it yet a question which of the two, the gas or the electric light, may be the better. *Inability of any light to penetrate fog. Question of superiority of gas or electric lights.*

The lens at Souter Point is of the size of those of the third order, or 39.38 inches in diameter, and consists of a fixed-light apparatus covering the sea-horizon, i. e., 180°, and is surrounded by eight panels of vertical condensing-prisms, which in their revolution give flashes at intervals of one minute. *Lens, size and characteristics.*

The lenticular apparatus is a very beautiful piece of workmanship, and much credit has been obtained by the manufacturers, Chance, Brothers & Co., for their skill and ingenuity in disposing the rays of the electric spark to the best advantage to this locality. *Lens made by Chance, Brothers & Co.*

The electric spark used for purposes of light-house illumination differs from other sources of light in the smallness of its dimensions, and, within certain limits, its variability in position. *The electric spark.*

The former feature, although requiring that the apparatus shall be specially designed for its use, is, in the hands of a competent optical engineer, a most valuable characteristic. It is possible to direct and distribute light from a point almost as one wishes and without waste, but the conse- *Apparatus for use in electric lights.*

---

* See Haisborough gas-light.

quences of any error in designing or executing the optical apparatus, are even more serious and far more apparent than when a larger source of light is used.

Accidents to revolving apparatus provided for. It is a fact important to be noted that at this station, as at all others in the English service, the contingency of accident to the clock-work carrying the revolving lens is provided for. A crank for turning it by hand can be attached, and a dial placed before the keeper indicates the velocity of revolution, so that he has no difficulty in preserving the proper intervals between the flashes.

Utilizing the rear light. A part of the light thrown to the rear (toward the land) is taken up by an annular refracting lens surrounded by catadioptric prisms, the whole being about 15 inches in diameter and forming a holophote of the sixth order, and the rays after passing, being formed into a cylindrical beam of parallel rays, impinge against a set of totally reflecting straight prisms, which in turn cast them at right angles, in a beam of parallel rays, down through a vertical wooden tube, passing through a circular aperture in the floor, upon another set of totally reflecting prisms in the room below, and they again, turn the rays at right angles and out through a large *Dangers marked.* plate-glass window upon some dangers southward of the point, which are called "The Mill Rock," "Hendon Rock," and "The White Stones." (See Plate IX, which is a chart showing Souter Point and its vicinity.)

The window through which this borrowed light passes is divided vertically into two parts, the one on the western or land-side being red and the other white; the line of division being produced over the sea gives a "red cut," the *Red cut.* utility of which will be understood from the following sailing-directions, taken from the British light-house list:

Sailing directions. "SOUTER POINT.—*The main light is electric and flashes every minute. A fixed light, also electric, is shown 21 feet below the flashing light, and shows white between the bearings of N. by W. and N. ¼ E., and red between N. ¼ E. and N. by E. ¾ E. When the fixed white light is seen, vessels will be in line of Mill Rock and Cape Carr Point, and when it changes to red, in that of Whitburn Stile, Hendon Rock, and White Stones.*"

Plate X. Plate X shows a plan of the lower light-room and details of the window through which the "red cut" is made, *a* in the plan showing the position of the lower refracting prisms.

Plate XI. The engine-room, a plan of which is shown in Plate XI, with its accessories, (including fuel-rooms for the storing of coke, of which about 100 tons are used annually,) and the dwellings of the keepers form a large quadrangle, (see Plates

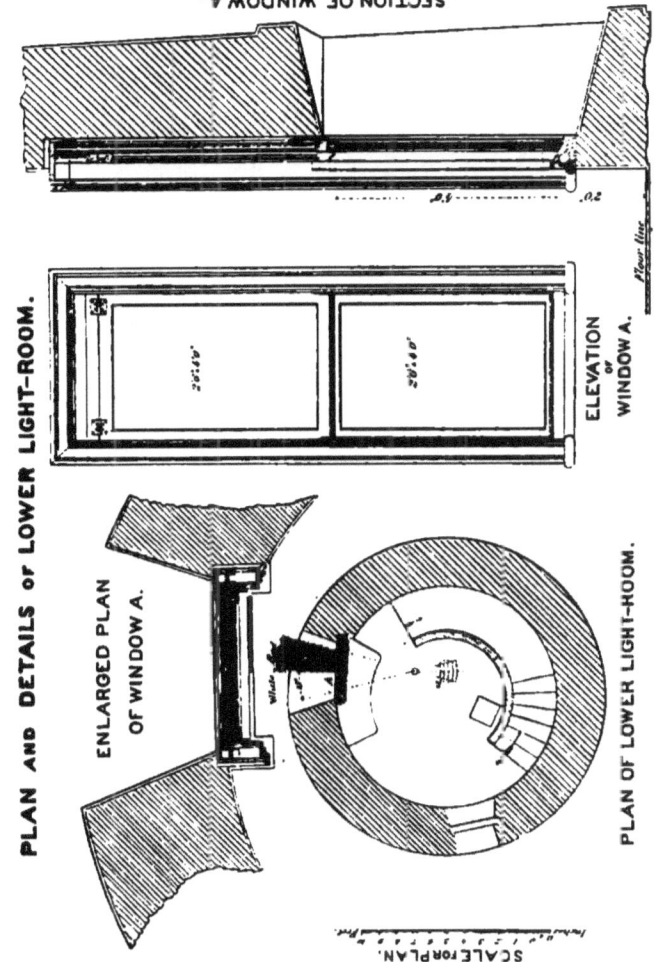

SOUTER POINT LIGHTHOUSE.
ELECTRIC LIGHT.
PLAN OF MACHINE-ROOM.

PLATE XI.

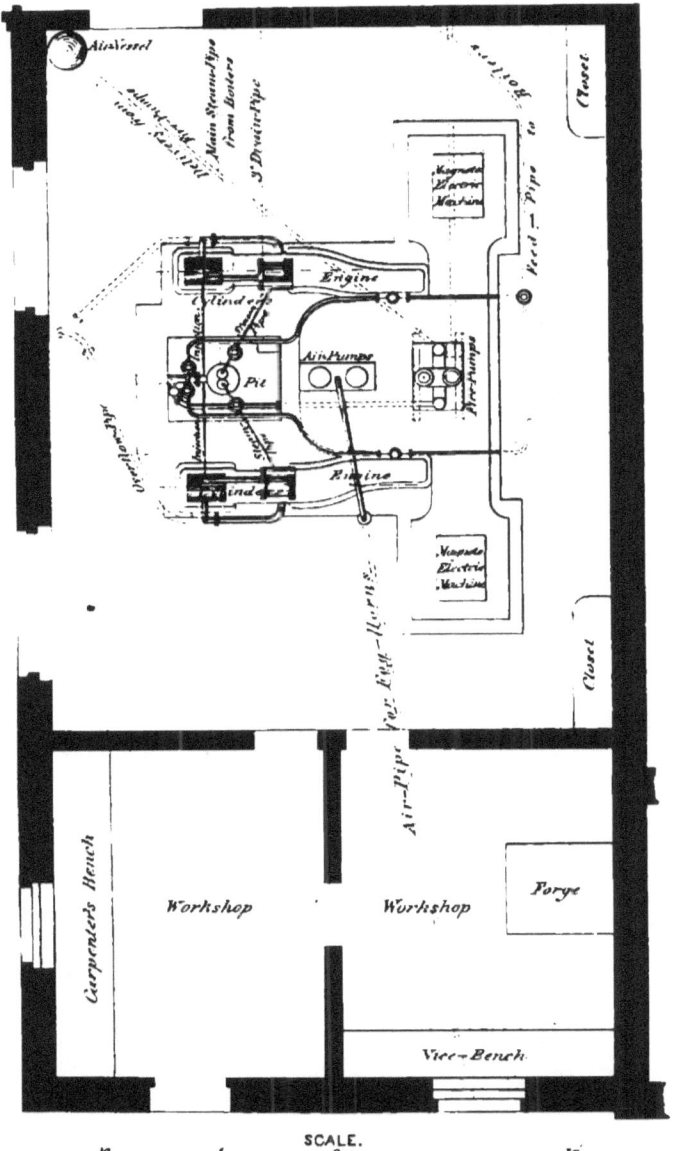

XII and XIII,) the former being on the land-side, while, toward the east or seaward, is the tower, which is detached from the rest of the establishment, except that communication is had by a covered way one story high. *Plates XII and XIII.*

The machinery for generating the electric current is similar to that at South Foreland. *Machinery.*

Two rotary magneto-electric machines of Professor Holmes' patent are driven by two 3 horse-power engines which can be worked up to 6 horse-power each. *Magneto-electric machines and engine.*

Each machine consists of eight radial frames, to each of which are attached 36 magnets, making 288 in all, and the poles are alternately pointed toward and from the axis of the machine. *Construction of the machines.*

A shaft driven by the engine, revolves a series of cylinders composed of helices of wire, past the magnets, which produce the alternately positive and negative currents. These currents are collected by the wires passing up the tower to the electric lamp within the lens. *The magnets. Currents produced.*

The number of revolutions made by each machine per minute is 400, and as 16 sparks are produced by each magnet at each revolution, the number of sparks at the carbon points of the lamp is 6,400 per minute, when one machine only is in operation, as is the case in fair weather, and 12,800 per minute when both machines are at work. These sparks are formed so rapidly that the eye does not separate them, and the result is a continuous beam of light, so dazzling, that the eye of a person within the lantern cannot rest upon it for an instant, without intense pain. *Number of revolutions per minute and consequent number of sparks.*

To insulate the shaft of the machine which conducts the electric current to the wires, it is encased in ebony-wood journals, and where the wires pass through the wall of the engine-room there is a coupling-box so arranged that, by a single motion, they can be connected or disconnected, and the current from one machine or two can be turned on to the lamp at pleasure. *Insulation of shaft.*

The electric lamp, as at South Foreland, consists mainly of two carbon points, each about ten inches long by one-half an inch square in section, placed end to end in a vertical position, and the automatic machines called regulators, feed the points toward each other as fast as they are consumed, which is at the rate of one inch per hour each. *Electric lamp.*

An oil-lamp is placed under the electric lamp, and is always filled and ready to be substituted in case of accident to the latter, or to the machinery; but I did not learn that a necessity for its use had ever arisen. *Oil-lamp for use in case of accident.*

Plate XIV and Fig. 11 illustrate the disposition of the different parts of the lenticular apparatus at Souter Point.

In the former a is the focus; a' the electric lamp; $a''\ a''$ the carbon pencils; b the holophote; c the upper totally reflecting prisms; $d\ d$ the fixed dioptric apparatus; $e\ e$ the revolving frame of flash-panels; $f\ f$ revolving gearing; g the removable bed-plate; h the burner of the oil-lamp; i telescopic tubes for the supply of oil and the overflow, for use if the oil-lamp should be substituted at any time for the electric lamp; l the oil-reservoir; $m\ m$ the oil-supply pipe; n cylindrical shaft for transmission of the beam of reflected light to the lower light-room; o the lower reflecting prisms; $p\ p$ the window of the lower light-room; g a gallery used when cleaning the sash of the window.

Fig. 11.

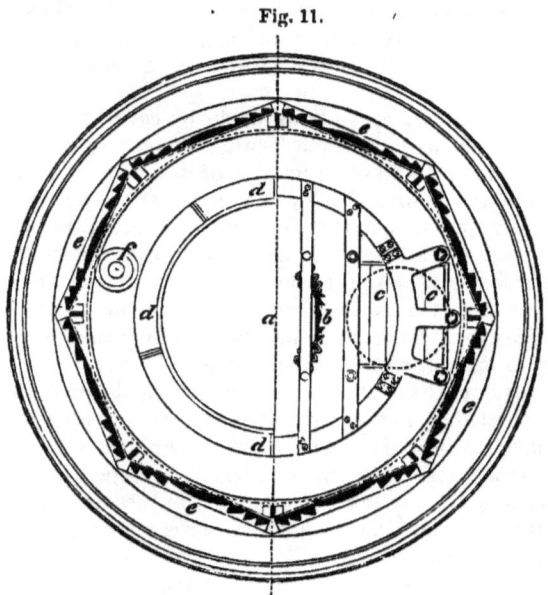

Plan of lens and lantern, Souter Point.

In the figure a is the focus; b the holophote; $c\ c$ the upper reflecting prisms; $d\ d\ d$ the fixed dioptric apparatus; $e\ e\ e$ the flash-panels. The dotted circle under $e\ e$ shows in plan the shaft for transmitting the beam of reflected light to the lower light-room.

The tower at Souter Point, shown in Plate XV, is built like most of the towers I saw in England, being a shell of brick-work into which the steps are let at the outer end only, and with landings at the windows.

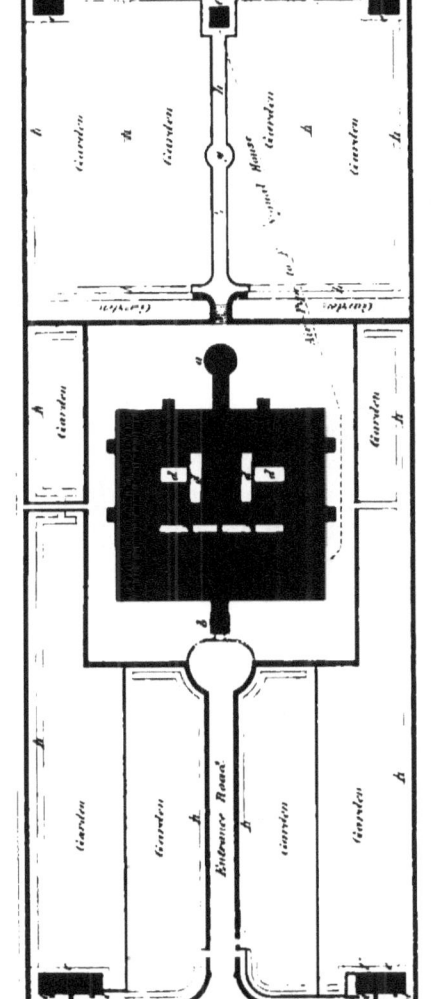

SOUTER POINT LIGHTHOUSE.
ELECTRIC LIGHT.

GENERAL PLAN.

PLATE XII.

a. Tower.
b. Hoister-Shaft.
c. Fog Trumpet House.
d. Yards.
e.e. Dust and Pig sties.
f. Earth-Closet.
g. Flag-Staff.
h.h. Paths.

SOUTER POINT LIGHTHOUSE.
ELECTRIC LIGHT.

PLATE XIII.

GROUND PLAN.

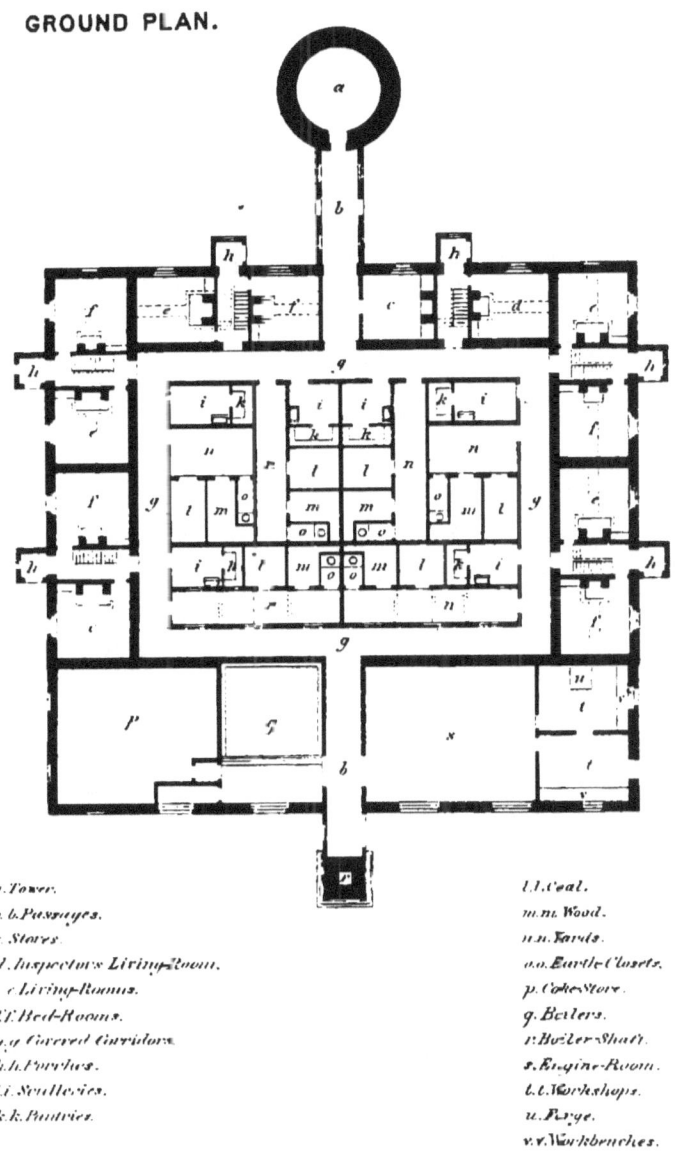

a. Tower.
b.b. Passages.
c. Stores.
d. Inspector's Living-Room.
e.e. Living-Rooms.
f.f. Bed-Rooms.
g.g. Covered Corridors.
h.h. Porches.
i.i. Sculleries.
k.k. Pantries.

l.l. Coal.
m.m. Wood.
n.n. Yards.
o.o. Earth-Closets.
p. Coke Store.
q. Boilers.
r. Boiler Shaft.
s. Engine-Room.
t.t. Workshops.
u. Forge.
v.v. Workbenches.

SCALE.

THE GRAPHIC CO. PHOTO-LITH. 39 & 41 PARK PLACE, N.Y.

The watch-room floor is of iron and supported upon a system of radiating and concentric beams. The watch-room is fitted up with supplies for use in the lantern, viz, oil, burners, and chimneys, and skins and cloths for cleaning the metal-work and glass of the lantern and lens. *Watch-room.*

The lantern is of the size heretofore described, viz, 14 feet in diameter, and it has diagonal sash-bars of steel. *Lantern.*

On the cliff in front of the tower is a Holmes fog-horn, sounded in foggy weather by means of the engine for driving the magneto-electric machines. An ingenious contrivance of the inventor makes the down-strokes of the plungers of the air-pump slow and the up-strokes quick. This is done by means of three eccentric cog-wheels, the middle one (the driving-wheel) of which gives motion to the two others, to which are connected the shafts of the pumps. *Fog-horn.* *Invention for operating the air pump.*

The cost of the station is given as follows: *Cost of station with revolving electric light.*

Building-works	£7,150	$35,750
Lantern, dioptric apparatus, &c	3,436	17,180
Electric apparatus, machinery, &c	4,100	20,500
Miscellaneous	462	2,310
	15,148	75,740

Deducting about £750 ($3,750) on account of difference in cost of revolving and fixed dioptric apparatus, and also the cost of revolving machinery, the above sum would indicate the approximate cost in England of a *fixed* magneto-electric light.

Electric lights, being considered more important than others, receive the preference in appointment of keepers, and the most competent are appointed for these stations, their salaries exceeding that of keepers of their grade at other lights 10 per cent. *The best keepers appointed for electric lights.*

Each electric light-station is in the immediate charge of a principal keeper, who is called an *engineer*. *Engineer the principal keeper.*

At South Foreland, where there are two lights, six assistants are allowed; at Dungeness, five; and at Souter Point, four. *Number of keepers at electric lights.*

At these lights the engineer has sole charge, and is responsible for the premises, property, and stores, as well as for the proper service and efficiency of the light. When he is absent the senior assistant takes his place. He is not required to keep watch, but must visit the lantern and engine-room at various times during the night, besides the regular visits at the end of each four-hour watch; and must always be present in the engine-room when preparing for lighting. *Station in charge of the engineer.*

<small>Watches of keepers at electric lights.</small>

The assistants take equal watches of four hours each, one in the engine-room and one in the lantern.

<small>Engines and boilers.</small>

The engines and boilers are worked alternately, one each week. Steam is to be up in one boiler (the other boiler being filled and the fire ready for lighting) and the magneto-electric machines ready for starting five minutes before sunset. The lamp is lighted at sunset and extinguished at sunrise.

<small>Time of lighting and extinguishing.</small>

In case of accident to any part of the electric apparatus, the oil-lamps must be immediately substituted for the electric lamps, and to keep them in perfect order, it is required that they be lighted and kept perfectly in focus for one hour (during the day) once a week.

<small>Observation of the light at night.</small>

After leaving the Tyne at night we stood off from Souter Point to observe the light from the sea, and it certainly surpassed in brilliancy any I have ever seen, being so bright that at a distance of several miles well-defined shadows were cast upon the deck of the Vestal.

<small>Observing the effect of the red cut.</small>

We afterward took the pinnace of the Vestal and steamed into the white and the red lights from the low light, and across the "red cut" several times and in different directions. We found it quite well defined, so that no vessel in a clear night when observing the sailing-directions could get into the dangers which the low lights are designed to point out.

Admiral Collinson had given directions to have the fog-trumpet sounded when the keepers should observe the Vestal, but we were probably too far off while observing the light from the sea, for we did not hear it.

<small>Light obscured by the fog and smoke.</small>

As before observed, we visited this light on our return voyage from the north, but it had happened that, on going to the north, it being thick and rainy, a dense cloud of fog and smoke shut down over the sea before we arrived off Souter Point, and we ran in toward the land, passing the light as we supposed within three miles, but did not see it.

Finding our position after reaching the mouth of the Tyne, we ran back toward Souter, and in, as far as was thought safe on account of the dangerous rocks in the vicin-

<small>Impossibility of penetrating fog by any light.</small>

ity, but still could not see the light. This confirmed the opinion of the Elder Brethren as well as of ourselves, that there is no light which will penetrate a fog, and all that is possible in light-house illumination is to make light sufficiently powerful to be depended upon in all sorts of thick weather *up to the impenetrable limit.*

<small>Utility of low lights and red cuts.</small>

There can be no doubt of the utility of the low light and the "red cut" in pointing out dangers within range of a

SOUTER POINT LIGHTHOUSE.
ELECTRIC LIGHT.

PLATE XIV

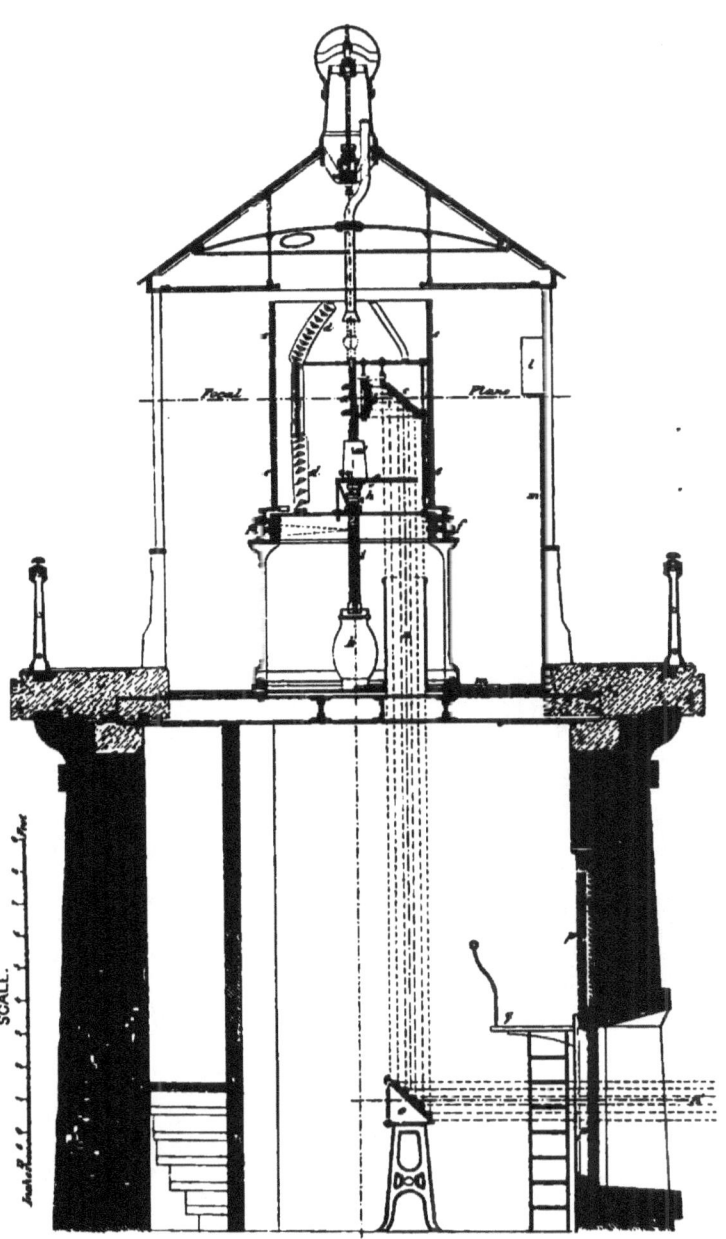

SOUTER POINT LIGHTHOUSE.
ELECTRIC LIGHT.
PLATE XV.
EAST ELEVATION.

light-house. I will describe a more marked case of their application at Coquet Island.

Another application of the use of borrowed light has suggested itself to me. It is this: Our sea-coast lights are often from 150 to 200 feet above the sea, and it frequently happens, particularly on our Pacific coast, that a fog will hang over the sea and shore low enough to envelop a light at this elevation, and yet it remains clear below and at the level of the sea. *Suggestion as to the use of low lights on our own coasts.*

Where the tower is not surrounded by water and there is a land-side, as is almost always the case, a part of the light thrown to the rear or landward can be taken up, as at Souter, and thrown by means of totally reflecting prisms through a tube passing down the tower to a lens placed in a salient lantern at the proper distance, say 15 or 20 feet above the foot of the tower. *Method to be employed.*

In my description of lights on the coast of Wales will be found an account of a separate low light for foggy weather in actual use.

COQUET.

This interesting light-house is on Coquet Island, off the coast of Northumberland, and is of the first order, dioptric, the lens covering about 270° of the horizon. *Position.*

A vertical pane of red glass attached to the lantern covers an area to the northward with red light, and the narrow Coquet Roads inside the island are illuminated by the lamp alone, unassisted by lenticular apparatus. *Red light.*

Areas both to the northward and southward are purposely shut off from all light from the lens, by means of opaque panels in the lantern, the object being to warn vessels of their proximity to danger when the light is lost sight of. *Areas shut off from light.*

The height of the focal plane of this light above the sea is 83 feet. *Height of focal plane.*

Below the watch-room is a lower light-room, with a large plate-glass window looking to the southward, and, on the opposite side of the room, near the middle of the wall, 28 feet below the focal plane of the main light, is a catoptric apparatus, consisting of three reflectors provided with lamps having red chimneys. *Lower light-room.*

The jambs of the window limit the red rays emanating from these lamps, and mariners are warned to use great caution in approaching the shore after they get into the red light. *Limitation of red rays.*

For the further purpose of marking the position of an important buoy, a red "cut" is produced by means of a fourth *Red cut produced.*

reflector placed near the northwest corner of the room, which throws a beam of *white* light through the *red*, the cut being made at the edge of the ruby-glass, with which a part of the window is glazed. This red cut intensifies the cut produced by the interception by the east jamb of the window of the rays from the three red-light reflectors.

Fig. 12.

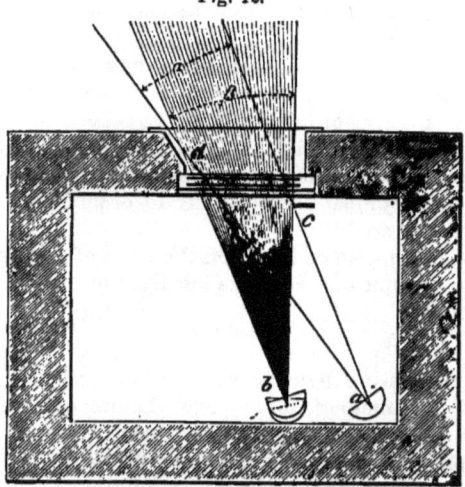

Low-light room, Coquet island.

Figure 12 shows the arrangement of light in the lower light-room; *a* is the single reflector for white light; *b*, the three reflectors for red light, placed one over the other, but with their axes slightly divergent in plan; *c*, a vertical strip of red glass, the western edge of which divides the light from *a* into red and white lights; *d*, the eastern window-jamb limiting the red light from *b*; *a*, the sector of white light, and *β* the sector of red light; *a c* and *b d* are parallel.

The purpose of the ingenious arrangement of lights in the lower light-room will be better understood by an inspection of the chart in Plate XVI, and from the following sailing-directions:

Sailing-directions.

COQUET ISLAND.—*The upper light, white, is visible seaward between the bearings from the sea of S. by W. ¾ W., and N. ¾ E.; and red from S. by W. ¾ W. to S. by W. ½ W., to cover the Bulmer Bush and Bulmer Stile rocks.*

"*A dim white light from the naked lamp of the apparatus is shown between the bearings S. and N. E. ¾ N. to cover the anchorage inside the island.*

PLATE XVI.

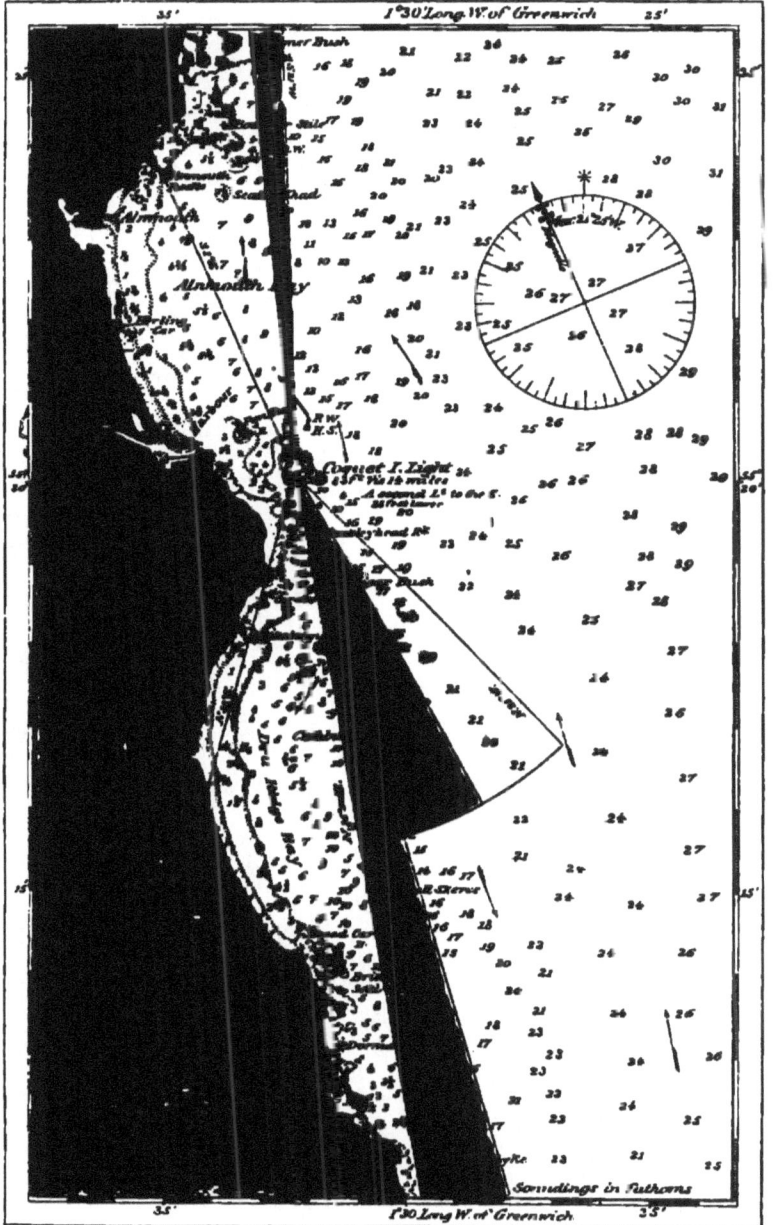

"A second light, twenty-eight feet below the main light, is shown from the same tower. It is white between the bearings from the sea of N. by W. ¾ W. and N. ½ W., to guard the shoal off Hauxley Point, and red from N. ½ W. to N. by E. ½ E., to cover the Bondicar Bush shoal.

"When the upper light is lost sight of westward of N. ¾ E., the line of Hauxley Point and Bondicar Bush will be passed, and while in the low red light great caution is necessary in approaching the shore."

Division of the sea-area around the light into eight sections.

It will be observed that the sea around Coquet Island is divided into eight distinct areas, each of which is easily recognizable by an inspection of the lights; that is to say, in going from the northward round by the eastward and southward, the following changes are seen:

Bearings of the light-house.	Appearances of the lights.
From S. 21° W. to N. 21° W.	Main light (only) white.
From N. 21° W. to N. 6° W.	Both the high and low lights, white.
From N. 6° W. to N. 7° E.	The high light white, the low light red.
From N. 7° E. to N. 15½° E.	The lower light (only) red.
From N. 15½° E. to N. 28° E.	No light.
From N. 38° E. to S. 1° E.	Dim high light unassisted by lens, white.
From S. 1° E. to S. 17° W.	No light.
From S. 17° W. to S. 21° W.	The high light (only) red.

Arrangement of lights the best that could be had.

The main light could not have been arranged to warn the mariner of the near dangers, as the low light so successfully does, since the former must throw its light uniformly over the sea and many miles farther to the southward beyond the dangerous rocks and reefs.

Tower and dwelling.

The tower and dwelling at Coquet are built of stone and are very picturesque, the station having been built upon the remains of a Norman monastery.

Buoy-depot.

Adjoining the light-house is a substantial buoy-depot, and a well-built wharf is provided with a traveling-crane for loading and unloading.

Acknowledgments to Mr. Morton.

The lights on this part of the coast are under the superintendence of Mr. Morton, whose polite attentions I desire to acknowledge.

S. Ex. 54——9

While lying in Coquet Roads I went at night with Admiral Collinson in the steam-pinnace to observe the red cut of the low light; its utility, which had been explained to me in theory, I found fully confirmed in practice.

Observing "red cut" from the sea at night.

I am also under obligations to the admiral for an opportunity of visiting with him one of the most interesting places in England, the ruins of the once magnificent Castle of Warkworth, the home of the Percys, which is a few miles from the landing-place opposite Coquet Island.

Visit to Castle of Warkworth.

INNER FARNE ISLAND.

There are two lights on the Inner Farne, which is off the coast of Northumberland, and they form a range or lead. The main light is catoptric and revolving, there being seven reflectors, (one on each face.) The low light is about 200 yards northwest from the main light, and contains a single fixed reflector.

Lights form a range or lead.
Characteristic of main light. Of low light.

The keeper of the main light watches the other by means of a small reflector, which catches a little of the ligh tof the latter and throws it back toward the lantern of the main light through an aperture in the smaller tower.

Keeper of main light watches the low light; method.

The only peculiarity to be observed at this station was the chimneys of the reflector-lamps, which had no shoulders, but were conical from top to bottom, with a large flare.

Chimneys.

THE LONGSTONE.

The Outer Farne or Longstone light-house is, with the exception of a small pier-light at Berwick, at the mouth of the Tweed, the most northern of the North Sea lights of England, and is in view of the light on St. Abb's Head, the first of the Scottish lights.

The most northern of North Sea lights.

It is a rock light-house of the peculiar construction shown in Plate XVII, the tower and dwellings being surrounded by high walls to protect them from the sea, which frequently rolls with great violence over the rock, which is long and narrow.

Construction.

A new first-order revolving lens, made by Chance Brothers & Co., has recently been placed in the tower, and the entire station has been repaired and refitted, the mechanics being still at work at the time of my visit.

Lens.

While at Coquet Island I saw this light very distinctly from the deck of the Vestal, say 10 feet above the sea, at a distance of twenty miles, which was remarkable, as the focal plane is but 85 feet above the sea.

Light seen twenty miles away at an elevation of but ten feet.

Plate XVII.

THE LONGSTONE LIGHT-HOUSE.

There is only the ordinary number of keepers (two) at this station, but they are supplied with provisions from the mainland and but rarely leave the rock. *Keepers.*

This light-house is interesting as having been the home of Grace Darling, the daughter of a former keeper, and to whom owed their rescue the nine out of the sixty-three who were on the Forfarshire when she struck the "Hawker Rock," near the Longstone, on the 5th of September, 1838. *The home of Grace Darling.*

The keepers had much pride in showing us the bed-closet occupied by the heroine and the window through which she first saw the wreck. A beautiful tomb is erected in her memory, at Bamborough Castle, near by on the mainland. *Tomb of Grace Darling.*

The foregoing comprises the notes of my journey among the lights of the North Sea, and I must, in concluding my account of it, express my thanks to Admiral Collinson and Captain Weller for their unremitting efforts to make it for me a journey of pleasure as well as of profit. *Thanks to Admiral Collinson and Captain Weller.*

The Vestal, in which this cruise was made, is a handsome sea-going side-wheel steamer, about the size of our supply-steamer Fern, and is used for conveying oil and other supplies to the light-houses, and for purposes of inspection. *Description of the Vestal.*

A yearly inspection of the light-stations is made by some of the Elder Brethren; at other times the superintendent in charge of each district inspects the stations and causes the necessary repairs to be made. *Yearly inspection made by the Elder Brethren.*

The Trinity House has several of these steam tenders, or "yachts," as they are called, each of them carrying a steam pinnace or launch outboard.

The weight being too heavy for the davits, two swinging brackets were placed under the bottom of the launch and stepped upon the rail of the steamer. These supports can be raised or lowered by means of a screw-thread and a stationary nut in the rail.

These launches are of much utility in landing stores and towing the other boats, (of which the Vestal carries four,) and they steam from six to eight knots an hour. In going to beaches or rocks the launch is always accompanied by a dingey; this, as does each of the other boats, carries on its thwarts a gang-plank, which is made with a triangular cut at one end, to be hung over and against the stem of the boat. These gang-planks, which are battened on the upper surface, and are about 10 feet long, are found to be very useful. *Use of the pinnace.* *Dingey carried by the pinnace.* *Gang-planks.*

The Vestal has no house on deck except a small one over *House on deck.*

the main companion-way to the saloon, and another over the galley between the paddle-boxes.

Dinner at the Trinity House in honor of the Younger Brethren.

Having landed from the Vestal at Harwich, on the 19th of June, we returned to London by rail, and I had the pleasure of attending an annual dinner given at the Trinity House, according to custom, in honor of the Younger Brethren.

Allusions to the Light-House Board of the United States.

There were about one hundred persons present, and the very complimentary allusions made by Sir Frederick Arrow, Deputy Master of the Trinity House, in the course of the after-dinner speeches, in reference to the Light-House Establishment of the United States, and to our Chairman, Professor Henry, in particular, were received with enthusiasm.

Naval review in honor of the Shah of Persia.

On the 22d I went down to Portsmouth by rail, with Sir Frederick, to witness the grand naval review at Spithead in honor of the Shah of Persia, which was to take place on the 23d, after which I was to embark on the Vestal for an inspection of the lights on the south and southwest coasts of England.

Sir Frederick was kind enough to invite me to accompany him on board the Galatea, a fine large steam-yacht belonging to the Trinity House, which with the Royal Yacht Alberta formed the escort to the Victoria and Albert, which carried the Shah, his Royal Highness the Prince of Wales, and other members of the royal family through the lines, so that I had an excellent opportunity of seeing this magnificent pageant, which comprised nearly all the celebrated iron-clads of Great Britain, including the Agincourt, the Hercules, the Devastation, and eight others, and surpassed even the review of Sir Charles Napier's Baltic fleet during the Crimean war, the only one in history comparable with it.

Joining the Vestal.

On the 24th I left the Galatea and joined the Vestal to accompany Captain Webb, of the Elder Brethren, on the proposed cruise of inspection.

On this journey we had the pleasure of the society of an agreeable guest of the captain, Colonel Sim, of the Royal Engineers, who remained with us till we left the yacht at Saint Ives, on the west coast of Cornwall.

Detention at Cowes.

We steamed over to Cowes, on the Isle of Wight, where we were detained a day by a severe storm, but the detention afforded me an opportunity of inspecting the Trinity House depot in the harbor, the light-house of Saint Catherine on the outer side of the island, and a drive to Osborne House.

Trinity House depot at Cowes.

The depot is very complete, and is fitted up with every convenience for the supply of the lights of the district.

The buildings and landing-place are unusually fine. the latter being used by the Queen during her residence on the island.

ST. CATHERINE.

This station is on the extreme point of the southern or sea-face of the Isle of Wight, and we reached it by carriage from Cowes, crossing the entire breadth of the island, the topographical features of which are everywhere beautifully diversified. Position.

The tower, which is 122 feet high, and carries a fixed light of the first order, is octagonal in form, and is, as are all the buildings of the station, of a pleasing design, being built of stone in the crenelated style, and in the most thorough manner. Tower.

Unfortunately there is occurring here what I have observed before only on the coast of California, i. e., the entire surface of the side of the great cliff which rises behind the light-house and the plateau on which it stands are gradually moving toward the sea, sliding upon some stratum below, carrying the fine tower, which is already slightly out of the vertical, with it; and unless some means are taken to prevent further motion of the tower much trouble is anticipated. Gradual moving of the upper surface of the land.

Below the main floor of this tower is the oil-cellar, a circular room, having the one hundred gallon oil-cans (or cisterns as they are termed in England) ranged around the wall. Oil-cellar.

There is an opening in the middle of the floor above this cellar through which, when oil is being delivered at the station, is passed a tube about 4 inches in diameter, which is bent below the arch, so that it can be turned into any of the cans, while at the upper end and above the floor is fitted a large funnel, into which the oil is poured, so that it is not necessary to carry it below the main floor of the tower. (See Fig. 13.) Filling oil-butts.

This is a fog-signal station, and on the point in front of the light-house is the engine-house, containing an Ericsson hot-air engine of 24-inch cylinder, 4 horse-power, and running under 12 pounds of pressure. Engine for fog-signal.

The trumpet rises through the iron roof, and its face, which is vertical, turns through the sea-arc of the horizon 215°. Fog-signal.

As an instance of the effect of projecting points which produce *sound-shadows* that very often interfere with the utility of fog-signals, I will mention that on the night be- Effect of projecting points producing sound-shadows.

134   EUROPEAN LIGHT-HOUSE SYSTEMS.

fore our visit to Saint Catherine a large vessel had gone ashore in a bay three miles to the westward, the night being foggy and an intervening point preventing the sound of the signal from covering that part of the shore. There are three keepers at this station.

Fig. 13.

Filling oil-butts.

THE NEEDLES.

Location and characteristics.   This light-house is situated on an island at the western extremity of the Isle of Wight, and has a first-order fixed dioptric apparatus, producing two red and two white sectors of illumination.

Red cuts.   The following, from the British light-house list, shows another application of the system of "red cuts," of which I have before made mention, and of which there are, as will be observed, many examples in the English light-houses:

"NEEDLES LIGHT.—*Red when bearing from N. W. ¼ N. (round northerly) to E. White from E. to E. S. E.; red from E. S. E. (round southerly) to S. W. by W.; and white from S. W. by W. to S. W. by W. ½ W. The white light shows in the direction of the Needles Channel, its southern limit bearing E. passes one and a half miles S. of Durlston Head, and about a* cable S. of outer part of Bridge Reef. Its northern limit, bearing E. S. E., passes two cables south of Dolphin Bank, and the S. W. buoy of the Shingles.

"The ray of white light between the bearings of S. W. by W. and S. W. by W. ½ W. clears Warden ledge."

We did not visit this light, but as we passed it early in the morning it shone brilliantly and gave us a good illustration of the system of pointing out dangers by "red cuts."

THE BILL OF PORTLAND.

We passed in view of these two fine light-houses on the coast of Dorsetshire without stopping.

They were built in 1716, and in 1788 the coal-fires which burned on their summits were extinguished and oil first came into use. Date of building.

They are about five hundred yards apart, and form a range or lead between the Race and Shambles.

THE START.

The Start light-house is a bold headland on the coast of Devonshire, and when I visited it there were extensive renovations going on, including the placing of one of the latest first-order lanterns, of Mr. Douglass's design—13 feet inside diameter—in lieu of the old one. Repairs.

The light is revolving, and the difficulty of making this change without extinguishing the light was overcome by the use of a ship's revolving catoptric apparatus of the same interval, building the new lantern up around it. Change of lantern without extinguishing the light.

This lantern has diagonal bars of steel formed of two thicknesses of ½ inch each, so the bar, when finished, is 3 inches by 1 inch. Diagonal sash-bars.

The glass (½ inch thick) is not set in a rebate, but the lozenge-shaped panes abut against each other outside the bars, and the joints are covered by strips of brass fastened by screw-bolts, through the glass and into the steel bars. The cost of these lanterns is about £1,700, ($8,500.) Setting of the glass.

Cost of lantern.

There is a lower light-room, from which a single reflector throws a *fixed* light over a danger called "The Skerries." Lower light-room.

There is a fog-bell at this station, weighing thirty-five hundred pounds, and struck on the inside by machinery. It strikes five hours without winding up the weights, an operation requiring fifteen minutes, during which the striking is interrupted. Fog-bell.

THE EDDYSTONE.

Resemblance to Point Bonita, Cal. — The resemblance of Start Point to Point Bonita, at the entrance of the Golden Gate of San Francisco, is remarkable.

View of the tower. — On the morning of the 26th of June we came in sight of the light-house which, more than any other extant, is known throughout the world as a splendid proof of the ability of man to overcome the force of the sea—the famous Eddystone, which lies off the coast of Devonshire.

Swell of the sea prevents landing. — There was a heavy swell from the Atlantic, and as the Vestal neared the light-house we were disappointed to see the waves running high up the tower, and the keepers' signal from the gallery at its summit that a landing was impracticable; but I was much gratified at having even such an opportunity of seeing this historical work.

Appearance. — Neither in height nor in appearance is it the equal of either of the modern light-houses The Wolf or The Longships, off Land's End, which I afterward visited, yet I could not but feel a thrill of admiration as I gazed at this grand old tower which has so successfully battled with the sea for one hundred and fifteen years.

Date of building the first light-house on the Eddystone. — The first light-house on the Eddystone was commenced in 1696, finished in 1698, and was destroyed in a terrible storm in 1703. Not a vestige of the building remained, and neither its keepers nor Henry Winstanley, its builder, who had wished to be in his light-house "in the greatest storm that ever blew under the face of heaven," were heard of afterward.

Date of building second. — The second light-house was built by John Rudyerd. It was lighted in 1709, and destroyed by fire in 1755.

Date of building the present one. — The builder of the present light-house was John Smeaton, who commenced the work in 1756, and finished it in 1759.

Account of building. — Of the construction he has given a most interesting narrative, or, as he styles it in the dedication to his King, "*a plain account of a plain and simple building that has nevertheless been acknowledged to be in itself curious, difficult, and useful,*" which clearly exhibits the industry, perseverance, and genius of one of the most remarkable of men. His plan was entirely different from those of his predecessors, he having conceived the idea that a light-house in a position like this, in order to withstand the sea, must depend upon its weight.

Method of joining the stones. — He therefore built it of stone, dovetailing the joints, as shown in Plate XXI, so that no stone can be moved without displacing the others, and his work has furnished a model upon which all rock light-houses built from Smeaton's

Plate XVIII.

THE EDDYSTONE LIGHT-HOUSE.

time to the present have been constructed, except as regards some of the details, which have been modified in some degree by their respective engineers.

Illuminating-power first used. The science of illumination, as applied to the Eddystone, was far behind the science of construction, and while Smeaton sprang at once from the prejudice of his time to a full conception of the true principles which should govern the construction of a work of this character, it remained lighted for many years as at first, by "*twenty-four candles burning at once, five whereof weighed two pounds.*"*

Reflectors used. Reflectors were not introduced until early in the present century, and in 1845 these in turn gave way to a second-order Fresnel lens, (fixed,) the beam from which, with its *Fresnel lens.* Douglass burner, is equal to 4,650 candles. This was the first catadioptric apparatus ever constructed.

SAINT ANTHONY.

Position. This is an old station on the coast of Cornwall, standing on a rugged promontory projecting into the bay which leads into the harbor of Falmouth.

Tower. The tower is square, and contains the dwelling for the principal keeper and his family, the assistant occupying a cottage connected with the tower by a covered way.

Apparatus. The apparatus is catoptric, and is composed of eight reflectors, one on each face of the revolving frame.

Fog-bell. There is a bell, struck by machinery, but quite unlike the others we had seen, and, indeed, there appears to be no uniformity in the bell-machinery of the English service, as many being struck by the hammer on the inner as on the outer side.

Low light. A single reflector is placed in the living-room of the principal keeper, and shows through a square window a fixed white light to guide clear of some dangerous rocks, called "The Manacles."

PLYMOUTH.

This light-house, which corresponds to the general character of "rock" light-houses, is placed on the end of the fine breakwater which protects the harbor of Plymouth, and is one of the most elaborate pieces of stone-work I have ever seen. The material of which it is built is a beautiful granite from Penrhyn, in Wales, and on the interior *Material of tower.* the immense expense of the construction is shown by the exquisite finish, almost polish, of the surfaces. The floors, *Beauty of finish.* ceilings, partitions, and walls are all of granite, and no lining of any kind is used.

---

* Smeaton's Narrative of the building of the Eddystone Light-house.

Strength of the structure. This tower is exposed to the heavy seas which roll over the breakwater, and for a distance of 10 feet from the bottom it is a solid mass of masonry. Its entrance is through a heavy gun-metal door, sliding upon rollers at top and bottom.

Arrangement of rooms. The lower floor contains the store-rooms; the second floor the oil-rooms; the third floor the kitchen and living-rooms; the fourth the bed-room, and the fifth the service-room. All the doors and window-frames are of gun-metal.

Characteristics of the light. A small segment of a fifth-order dioptric apparatus in the watch-room throws a beam of leading white light through a small window upon a buoy and the fairway, the beam being limited by placing in front of the lens a metallic case in which there is a narrow slit about $2\frac{1}{2}$ inches wide.

The main light shows *white* within the anchorage and *red* to seaward.

Red light, how produced. The red light is produced by surrounding the lamp, (one of the second order,) except for the small arc covering the anchorage, with a red cylindrical glass about 9 inches in diameter.

Method adopted to more sharply define the "cuts." To more sharply define the "cut" between the red and white light, narrow vertical strips of red glass are placed opposite the edges of the segments of this cylindrical glass and outside the lens.

This mode of producing red light by means of a cylinder around the lamp is different from any I saw elsewhere, and is shown in Fig. 14, in which a, a', represent the red shade, c the lamp, and a the sector of white light.

Fig. 14.

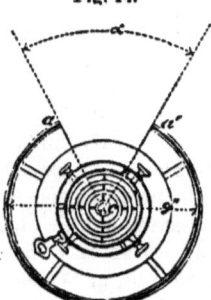

Red Cut, Plymouth Breakwater.

Libraries As before observed, the corporation of Trinity House furnishes its keepers with books by means of circulating libraries, and there are also fixed libraries at the station. The latter contains bibles, prayer-books, dictionaries, religious

Plate XIX.

THE IRISH GAS LIGHT FOR LIGHT-HOUSES—108 Jets.

works, and others of permanent interest or value. The circulating libraries are carried in neat, strong boxes, about 12 inches square by 18 inches long, and contain bound volumes of Punch, illustrated periodicals, novels, and other light reading. There are from eight to ten books in each box; these boxes, which are numbered and charged to the principal keeper of the station where left, are exchanged on the visits of the supply-vessel.

The keepers are uniformed by the Trinity House, one suit of clothing being supplied annually. At rock light-houses they are also supplied with great-coats. *Uniform of keepers.*

They are required to wear their uniforms on Sundays and holidays, when they go to the villages or to church; also whenever any of the officers of the Corporation visit the station.

The keepers are a bright and intelligent class of men, who seem well instructed in their duties. They are neat, trim in their appearance, and manifest pride in the stations. *Character of keepers.*

Flag-staffs at light-stations are universal, and whenever the light-house has a land-side the mast is stepped on the lantern-gallery. *Flag-staffs*

The Trinity House flag is displayed on Sundays and holidays, and when national vessels are passing, or the Trinity House yachts are at the stations. *Trinity House flag displayed.*

THE LIZARD.

This is a large establishment consisting of two towers, connected by a long building occupied by the keepers and their families, and they also contain the oil-cellars and store-rooms of the station. *Establishment.*

The towers, the focal planes of which are 229 and 232 feet respectively above the sea, were first illuminated by coal-fires on the tops in 1752. Oil was substituted in 1812, and the original lanterns and first-order catoptric apparatus are still in use. *Height of towers. Coal-fires first used. Oil substituted.*

The lanterns are excessively heavy, but both they and the reflectors are good examples of the light-house engineering of the beginning of this century. Although they are sixty-two years old and have been subjected to the thorough burnishing which the English light-keepers certainly have not failed to perform faithfully every day of that long period, the reflectors are to all appearances as bright and serviceable as when new. Originally, reflectors were made by beating the silver into parabolic form, and they were no doubt better than can be obtained at the present day. *Lanterns and reflectors. Original method of constructing reflectors.*

Characteristics of lights. The lights are fixed, and the apparatus consists of nineteen 21-inch reflectors in each tower, each lighted by an Argand burner.

Importance of this light. Lizard Point is a bold headland on the coast of Cornwall, projecting far beyond the general trend of the southwest coast of England, and Captain Webb informed me that as it is the first land-fall of most of the oversea commerce which enters the English Channel, it is one of the most useful light-stations of the kingdom; also that the brilliancy of its lights *Satisfactory character.* is often praised by mariners, and no desire for a change has been expressed, for which reason the Trinity House does not propose to make any, certainly at present, though it is a rule that when extensive repairs are necessary at large stations, catoptric apparatus, of which there are but few examples remaining, is changed for dioptric.

Superiority of dioptric apparatus. The English have no doubt of the great superiority of the latter, and its very great economy in consumption of oil, but when small areas are to be lighted, as in case of range or *Catoptric apparatus, when used.* leading lights, reflectors are in many instances used instead of more expensive lenses of the smaller orders.

Excellent condition of station. The station was in excellent condition, and maintained by three keepers, the principal of whom had been forty years in the service, being a large portion of the time at the light-house at Gibraltar, which belongs to the Corporation of Trinity House.

Fog-gun to guard against outlying rocks. There are dangerous reefs and rocks far out in front of the Lizard, and it is the intention to place a fog-gun here; the necessity of some powerful signal is very apparent.

THE WOLF.

Position. Ten miles to the southwest of Land's End is one of the latest achievements of modern light-house engineering, The Wolf Rock light-house, a view of which is shown in Frontispiece.

Date of building. It was commenced in 1862, under the direction of the father of Mr. James N. Douglass, the present engineer of the Trinity House, and finished by the latter in 1869.

Exposed position of the rock. The rock, which is 17 feet above low tide, (the tide rising 19 feet,) has twenty fathoms of water around it; is exposed to the full force of the waves of the Atlantic, and was for centuries the dread of the mariner; now its very distance from the shore adds to its value as a site for a light-house to guide into the English Channel; but there is probably no position whose occupation has required more skill and perseverance, or more courage in overcoming difficulties and dangers.

The position of this light-house is shown in Plate XX, which is a chart taken from Mr. Douglass's interesting paper, concerning the history and the peculiarities of construction of this interesting work; to the same paper I am indebted for many of the data concerning it which I have embodied herein. *Chart showing position.*

In the year 1795 a day-beacon was erected here, and at subsequent times, others; but they were all carried away, sometimes by the force of the waves, at others by the *débris* of wrecks striking against them. *Day-beacons placed on the site.*

Solid wrought-iron shafts of different and increasing diameters were, from time to time, sunk in the rock, and the difficulties of the site were such that, during the construction of the last one, which occupied five years, but 302½ hours of work could be obtained on the rock, and the cost was more than £11,000, ($55,000.) *Iron shafts sunk in the rock.*

The rock is submerged at high water, and is but little larger than the base of the tower, which is 41 feet 8 inches in diameter, 116 feet high, and solid from base to a height of 39 feet or to the door of the light-house. The thickness of the walls at the doorway is 7 feet 9½ inches, and at the top, which is 17 feet in diameter, it is 2 feet. The shaft is a concave elliptic frustum, the generating curve of which has a major axis of 236 feet and a minor axis of 40 feet. *Size of rock, and diameter of the base of the tower.* *Walls.*

The stones are laid in offsets to the level of 40 feet above the rock, with a view of breaking the sea, and above that height the surface is smoothly cut. *Manner of laying the stones.*

Each face-stone is dovetailed vertically and horizontally into the adjoining stones, and every stone is bolted to the course below it by two 2-inch bolts, of yellow metal for the exterior, and galvanized steel for the interior stones. *Dovetailing face-stones.*

The dovetailing was adopted, not only for increase of strength, but to prevent displacement by the sea during construction, before the superincumbent weight of the additional courses could be obtained, and to protect the cement mortar of the joints from being washed out before it could be set. *Reasons for dovetailing.*

Mr. Douglass stated that the additional cost of the dovetailing was not more than 1 per cent., and that during the construction there were lost but thirty-four stones, they belonging to an incomplete course which it was impossible to finish at the end of the working-season before the winter of 1865. *Additional cost of dovetailing.*

On Plate XXI will be found horizontal sections through the masonry of The Wolf and five other rock light-houses of the same general character, viz: Eddystone, Skerryvore, Inch Cape or Bell Rock, Minot's Ledge, and Spectacle Reef.

These sections are taken uniformly at 10 feet above high water, and are interesting as exhibiting the different methods of arranging the dovetail joints of the stones to prevent displacement by the sea.

Number of hours' work during the first year. Considering the exposure of the rock, which may be estimated from the fact that but twenty-two landings and thirty-eight hours' work could be had upon it during the first year of the construction; that the depot where all the stones were cut was at Penzance, seventeen miles from the rock; that the light-house contains 44,506 cubic feet of granite, and weighs 3,296 tons, it is not surprising that it cost £62,726, or more than $300,000, a cost which compares favorably with that of other structures of similar character.

Cubic feet of granite.
Weight.
Cost.

I find in Mr. Douglass's printed narrative of the work the following interesting table:

Comparative table of costs of seven rock light-houses.

	Total cost.			Cubic feet.	Cost per cubic foot.		
	£	s.	d.		£	s.	d.
Eddystone	40,000	0	0	13,343	2	19	11
Bell Rock	55,619	12	1	28,530	1	19	0
Skerryvore	72,290	11	6	58,580	1	4	7¼
Bishop	34,559	16	9	35,209		19	7¼
Smalls	50,124	11	8	46,326	1	1	7¼
Hanois	25,296	0	0	24,542	1	0	7¼
Wolf	62,726	0	0	59,070	1	1	3

Store-room. The first room above the solid part of the tower, (to which access is obtained by a strong ladder reaching from the rock and bolted to the tower,) as well as the next above, is used for stores.

Hoisting-derrick. The latter room has an opening through the wall of the tower, through which a derrick can be run out; by means of this and a winch inside are hoisted the oil and other supplies of the light-house.

Oil-room. The next room above is the oil-room; then come successively the kitchen, the bed-room, with five recesses in the walls for beds, and lastly the service or watch-room.

Apparatus. The dioptric apparatus is of the first order, manufactured by Chance Brothers & Co., of Birmingham, and shows alternate red and white flashes at intervals of thirty seconds.

Investigation by Prof. Tyndall to determine loss of light caused by intervention of red glass. Previously to the construction of the lens an investigation was entered into by Professor Tyndall, the scientific adviser of the Trinity House, to determine the loss of light caused by the rays passing through the red glass, and it was found that for an equal range of the red and white light from the same lamp it was necessary to make the arcs

PLATE XX.

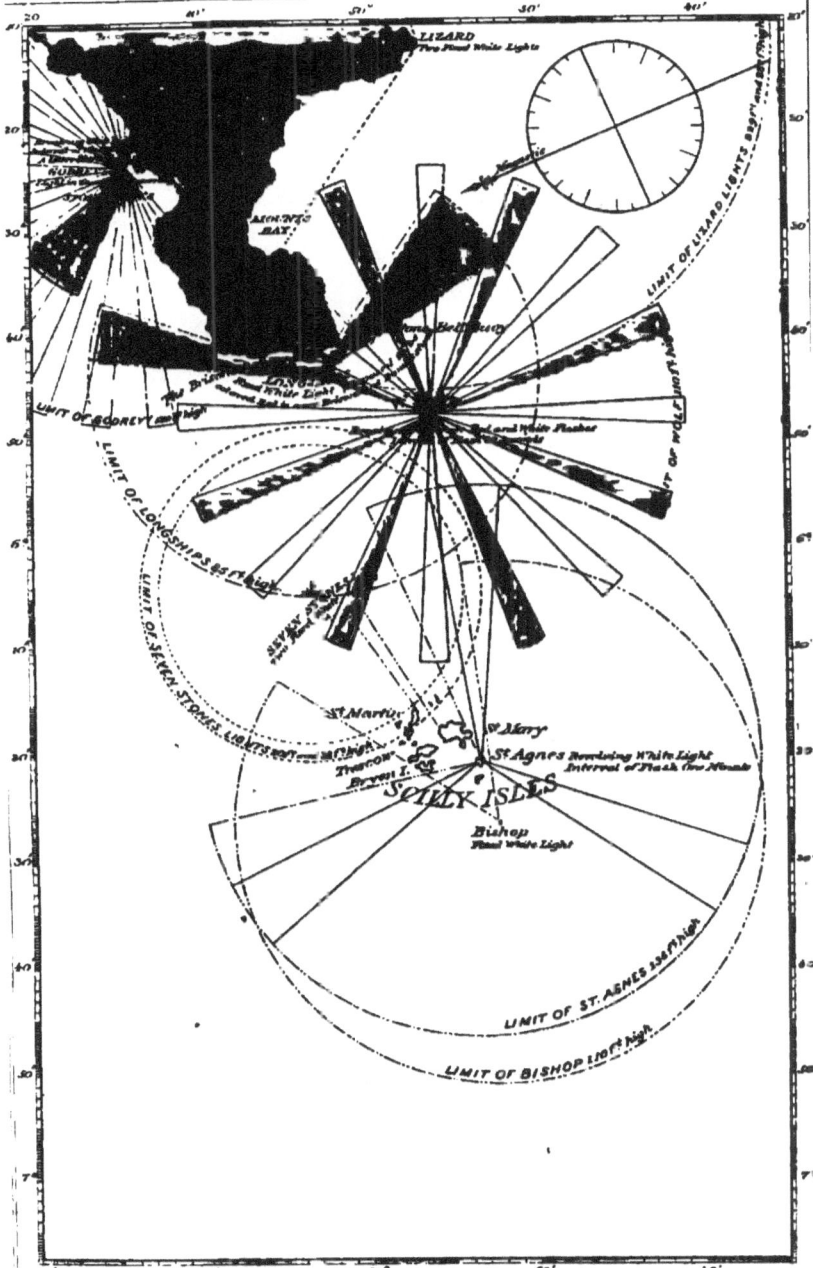

of the red and the white sectors in the ratio of 21 to 9 nearly; and this rule was followed at The Wolf, so that its beams of red and white light have the same value. Ratio of areas of red and white sectors.

The lantern is of the cylindrical helically framed kind, and upon the lantern-gallery is placed a fog-bell, struck by means of machinery placed within the pedestal of the lens. Lantern. Fog-bell.

There is but little wood used in the construction of the interior of the light-house, and all of the doors, the window-frames, and storm-shutters are of gun-metal. I was told that bronze was habitually used for window-frames and sash-bars at rock-stations, and oak for those of shore-stations. Gun-metal used for window-frames, &c.

The windows of the watch-room are arranged, as shown in Plate VIII, for admitting air to support combustion in the lamp, by means of a valve in the upper part, the current passing over the heads of the keepers and through the grating which forms the lantern-floor. Watch-room windows.

As at other stations, I observed that the floor of the exterior lantern-gallery, unlike ours, is made of stone with raised joints, *i. e.*, the surfaces for about one-half an inch on each side of the radial joints do not partake of the inclination of the general surface, but are quite level. Floor of lantern-gallery.

There are four keepers belonging to the station, and three of them are constantly in the tower, while the fourth is on shore with his family. Number of keepers.

The stated term of service on duty on the rock is one month, but it sometimes happens that eight weeks or more elapse before a sea can be found sufficiently quiet to make a landing practicable. Term of service before relieved.

As before stated, the rock upon which the light-house stands is submerged at high water, and the winch, mast, and boom of the derrick used for landing the keepers, visitors, and provisions are, when not in use, laid into deep troughs or recesses in the stone and strongly fastened down to protect them from the sea.

It was a comparatively calm day when I went to The Wolf, and I was fortunate in being able to land upon the rock; but it is an undertaking attended with a good deal of danger, and many trials and much delay were experienced before we were successful. Method of landing upon the rock.

The landing-boat, which is well adapted for the purpose, is built diagonally of two thicknesses of elm-plank, without timbers or floor, and is provided in the bow with a landing-deck and stake. Landing-boat.

This deck and the forward part of the gunwale are covered with rough rope-matting to prevent slipping in jumping into or from the boat, which is warped in by means of a line made fast to a buoy astern and two lines from the bows, the latter of which are managed by the men on the rock.

The person who is to land is provided with a cork life-belt, and stands on the landing-deck forward, holding the stout mast or stake with both hands, and when the proper instant arrives, of which he is warned by the coxswain, who watches the waves and manages the line astern, he seizes the rope which is lowered from the end of the derrick-boom, places one foot in the loop at the end, and is quickly hauled up by the men at the winch on the rock.

Danger attending landings. Landing by the mode I have described is comparatively safe, but is often impracticable, and sometimes when the keepers are relieved they are pulled *through* the surf into the boat when it cannot get near enough to the rock to permit of their being *dropped* into it.

This light-house is one of the most striking examples of rock light-house engineering for which Smeaton's Eddystone has furnished the model.

More of this class of light-houses in Great Britain than elsewhere. There are now several of this type in the various countries of the globe, but Great Britain possesses more than any other.

Notable instances in the United States. In the United States we have notably two; one built by General Alexander, of the United States Engineers, on Minot's Ledge, off the coast of Massachusetts, where the rock is exposed to the full force of the Atlantic, and is only uncovered at extreme low water; the other proposed by General Raynolds and built by General Poe, both of the United States Engineers, and the latter now a member of the Light-House Board, on Spectacle Reef, in Lake Huron, the site of which is 10 feet below the surface.

Spectacle Reef, difficulties met in building. The latter, however, was quite a different problem from any of the others in that the structure was to withstand the immense fields of moving ice by which it is assailed in the spring.

I regret that I could not visit the other rock light-houses of England, or the Skerryvore and Bell Rock, the latter of which have given so enviable a reputation to the Stevensons, the distinguished family of Scottish light-house engineers.

THE RUNDLESTONE BELL-BUOY.

The bell-buoy which marks the Rundlestone, off the point of Land's End, is 10 feet in diameter at the water-line, moored *Moorings.* with 45 fathoms of 1½-inch chain and a 30-cwt. sinker, which is backed with 30 fathoms of 1½-inch chain, and a second sinker of the same weight.

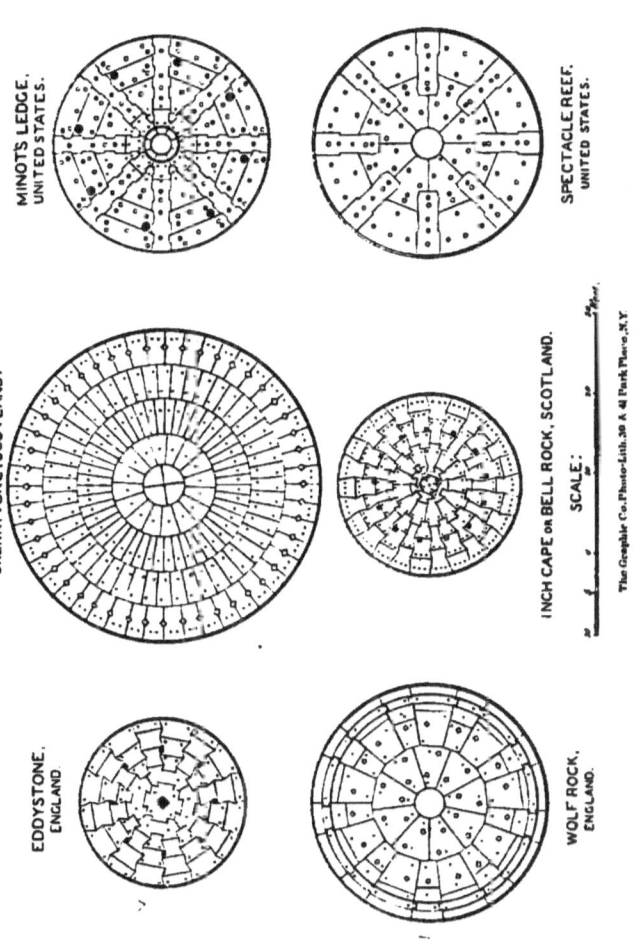

The bell weighs 3 cwt. This is a water-ballast buoy, and is shown in Fig. 15, in which *a* is the outer water-tight compartment; *b*, the inner water-tight compartment; *d, d*, the India-rubber springs.

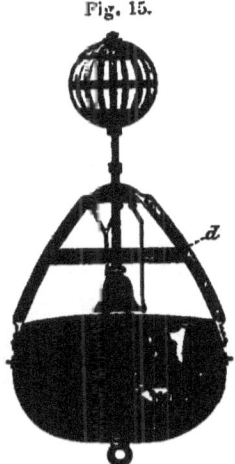

Fig. 15.

Rundlestone Bell-buoy.

THE SEVEN STONES LIGHT-SHIP.

Position. This vessel, which marks the dangerous shoal of rocks indicated by its name, lies about twenty miles to the westward of Land's End, or about midway between it and Saint Agnes (Scilly Islands) light, in what is probably one of the most exposed positions in the world for a light-ship.

Age. It is one of the latest constructions, and has been on the station about two years.

Masts and apparatus. It has two masts, besides the mast for a jigger or mizzen, which is stepped on the taffrail. The former carry two fixed catoptric lights, 28 and 20 feet above the sea.

Fog-trumpet. A fog-trumpet, of Professor Holmes's patent, shown in Plate XXII, is operated by a hot-air engine, the cylinder of which is 24 inches in diameter.

Engine and horn, position of. The engine is placed between-decks, amidships, and the horn between the masts on deck.

Length of time required to generate steam impairs utility of the signal. I should think the utility of the signal much impaired by the length of time required to generate sufficient pressure to operate it, which I was informed is an hour and three-quarters.

This is a serious objection in any fog-signal, since the fires are ordinarily not started till the fog comes on, and in the long interval before the warning is given serious accidents are liable to occur.

"Heaters" used at our steam fog-signals. For our signals operated by steam, we frequently provide "heaters," by which, with a very small expenditure of fuel, the water in the boiler is kept hot, so that when the signal is required, a quick fire will raise the pressure to the required point in a few minutes.

Fog-horn can be operated by hand. The fog-signal on the Seven Stones can be operated by hand, by means of a pair of air-pumps on deck, in case of accident to the hot-air engine, or when waiting for the required pressure, but it is extremely hard work for the seamen, and I fancy the signal may not be satisfactorily sounded at such times.

Signal-gun fired when vessels are seen standing into danger. There is also a signal-gun on deck which is fired when a vessel is seen standing into danger.

Measurements and moorings. The measurement of the Seven Stones is 188 tons. It is moored in 41 fathoms of water by 200 fathoms of 1¼-inch chain to a mushroom anchor weighing 40 cwt. Three hundred and fifteen fathoms of chain can be run out when necessary.

Crew. This light-ship carries a crew of fifteen men besides the master and mate; one of the latter and five of the seamen are constantly on shore with their families. They are all uniformed, and the name of the vessel is marked upon their hats and shirts. This rule applies to the crews of all the light-ships of England, and to those of the yachts or tenders belonging to the Trinity House.

Uniforms of crew.

THE LONGSHIPS.

Location. The Longships is the name of some rocks about a mile and a quarter to the westward of Land's End, on the largest of which is the new first-order light-house of that name, which, at the time of my visit, was on the point of completion.

Rough seas prevalent. As at The Wolf, the sea is generally rough at The Longships, and when we approached it on our return from the Seven Stones light-ship to Penzance, we found it impossible to land. The following day, however, we made another attempt, and were successful.

Form of rock. The rock on which the tower is built is conical in form, and rises about 60 feet above the sea; and the tower, which is precisely like that at The Wolf, is placed on a ledge in front, about 20 feet above the sea. The rock in the rear was being blasted down to the level of the top of the solid part of the tower, at which is the door or entrance.

HOLMES'S FOG-HORN APPARATUS. PLATE XXII.

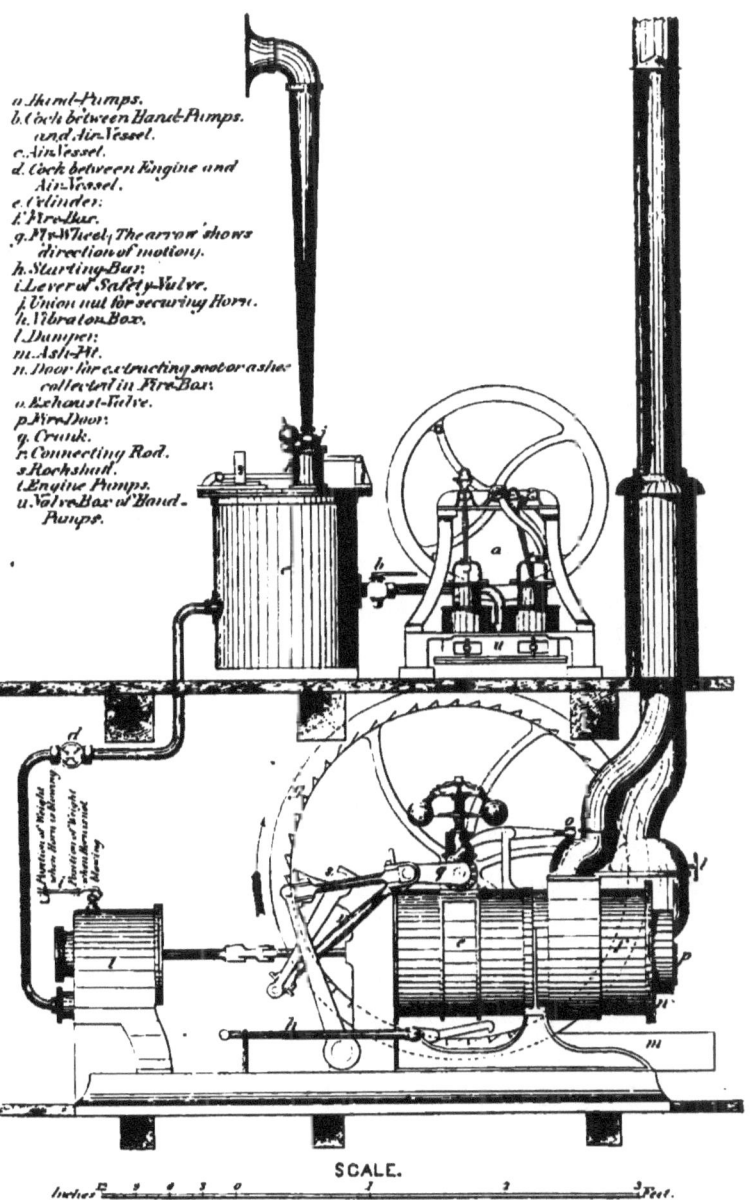

a. Hand-Pumps.
b. Cock between Hand-Pumps and Air-Vessel.
c. Air Vessel.
d. Cock between Engine and Air-Vessel.
e. Cylinder.
f. Fire-Bar.
g. Fly-Wheel, The arrow shows direction of motion.
h. Starting-Bar.
i. Lever of Safety-Valve.
j. Union nut for securing Horn.
k. Vibraton Box.
l. Damper.
m. Ash-Pit.
n. Door for extracting soot or ashes collected in Fire Box.
o. Exhaust-Valve.
p. Fire Door.
q. Crank.
r. Connecting Rod.
s. Rockshaft.
t. Engine Pumps.
u. Valve Box of Hand-Pumps.

Mr. Michael Beazeley, the resident engineer in charge of the construction of The Longships, was also in charge of The Wolf light-house, and I am much indebted to him for kind attentions on the occasion of my visits to these stations, and for information in regard to the works. *Mr. Beazeley, engineer.*

The courses of stone at Longships were dovetailed, as at The Wolf, but as the rock which was blasted away for the site for the tower, was an extremely hard slate, a "core" of it was left in the interior to a considerable height. *Method of building.*

These courses, to the height of the solid part of the tower, were 2 feet in thickness, and above that height, 1 foot 6 inches. *Thickness of the stone-courses.*

Quick-setting cement was used for setting the lower courses, and strong muriatic acid for removing the sea-weed from the rock, though Mr. Beazeley stated that lime is better for this purpose, if at least two hours can be had before it is covered by the tide. *Cement used. Sea-weed removed by acid.*

The focal plane of Longships light-house is 110 feet above the sea. *Focal plane of tower.*

The handsome new lantern was in place, but the lens was not set, and nineteen reflectors were ranged temporarily around the inside of the lantern to cover the proper arc of the horizon. *Reflectors in use.*

As at nearly every light-house which I visited on the coast of England, there are outlying dangers marked by "red cuts." (See Plate XX.) *Red cuts.*

At Longships these dangers are on one hand "The Brisons" rocks, and on the other "The Rundlestone," each about six miles from the light-house, and being well out from the land, they were for many years the terror of navigators. *Dangers marked.*

By the system of marking out these dangers by means of red light, they can, however, be avoided with absolute certainty in all weathers, except when the light is obscured by fog.

As an evidence of the force of the waves off Land's End, it may be mentioned that before the commencement of the new tower at The Longships, a granite light-house, the focal plane of which was 79 feet above high water, occupied the site; but owing to the terrific seas to which it was exposed, the lantern was so often under water in stormy weather that the character of the light could not with certainty be determined by mariners, and the erection of a higher tower became necessary. *Height of waves off Land's End.*

GODREVY.

This light-house is on a rock off the west coast of Cornwall, and about twenty miles to the northward of The Longships.

Manner of landing. Between this rock and the mainland is a smaller rock, on which the landing is effected. A wire rope is stretched from one rock to the other, and a basket suspended from it, is run over the rough water intervening, and by this means the keepers and stores are carried to the station.

When the Vestal approached the rock the keepers signaled that a landing was impracticable, and we steamed away without visiting this interesting station, much to my regret.

"The Stones." A mile and a quarter outside of Godrevy are some submerged rocks called "The Stones," a source of great danger to the mariner. They are unmarked except by a buoy, but are covered at night by a beam of red light from Godrevy light-house. (See Plate XX.)

Intention of Trinity House to place the light on "The Stones." As an evidence of the intention of the Trinity House to serve the best interests of commerce, Captain Webb mentioned that before the light-house was built, the Corporation proposed to place the light upon one of "The Stones," but was overruled by the Board of Trade. The expense of the proposed structure was but a few thousand pounds more than the cost of the actual light-house, and the latter has the same expense of maintenance and the same dangers in landing supplies of oil, &c., which the former would have had, while the very important advantage of marking "The Stones" is not gained.

"THE STONES" BUOY, OFF GODREVY.

This buoy is an "egg-bottom," water-ballasting, 13 feet *Moorings.* long, moored in 60 feet of water with 45 fathoms of 1½-inch chain and a 30-cwt. sinker, which is "backed" by 30 fathoms of 1½-inch chain and another sinker of 30 cwt.

Although this is probably one of the most exposed places around Great Britain, the moorings of this buoy are so excellent that it has not been adrift in three years.

Return to London. After leaving Godrevy we landed at Saint Ives, on the west coast of Cornwall, and went up to London by rail, arriving there on the 1st of July. I cannot conclude the notes on my journey on the southwest coast of England without expressing my sincere thanks to Captain Webb for his kindness and his constant efforts to make the cruise instructive and interesting for me, and for the many polite attentions which I received from him, from the time I first arrived in London till my departure for America.

On the 2d of July I attended the annual dinner at the Trinity House, which takes place on Trinity Monday, at which were present His Royal Highness the Prince of Wales, the Czarowitch (eldest son of the Czar) of Russia, the Dukes of Argyll and Richmond, and others of the nobility, the Queen's Ministers, and other distinguished personages. Annual dinner at the Trinity House.

His Royal Highness the Duke of Edinburgh, Master of the Trinity House, presided, and during the evening I had the distinguished honor of being presented by him to His Royal Highness the Prince of Wales. They both kindly referred, as the former had done before (as I have mentioned) at the dinner of the Lord Mayor, to the satisfaction which Sir Frederick Arrow and Captain Webb, of the Elder Brethren, had expressed in regard to their visit to the United States and inspection of some of our light-houses.

Before proceeding to Ireland and Scotland I visited the Continent, but I will here continue my account of the British lights in the order in which I visited them after my return.

HOLYHEAD.

Leaving London by train on the 14th of August, I arrived at Holyhead, on the Island of Anglesea, the following day, where I met, by appointment, Mr. Douglass, the engineer of Trinity House, and had the pleasure of inspecting with him the Trinity House light-stations on the island.

The great breakwater for the protection of the harbor of refuge at Holyhead, together with the light-house at its outer end, had just been completed; the latter, although quite finished, was not to be lighted until a few days later, when there was to be a grand demonstration to celebrate the completion of the harbor works, at which the Prince of Wales and several members of the government were to be present. The light-house, which was built from designs approved by the Trinity House, by Mr. Hawkshaw, the distinguished engineer in charge of the breakwater, at the expense of the fund appropriated for the latter, is a handsome tower of granite surmounted by a flashing lens of the third order, made by Chance Brothers. Outside the parapet of the tower is placed a fog-bell, rung by machinery, which, together with the usual revolving machinery, is contained within the pedestal of the lens. The latter revolves upon conical wheels, the outer bearing-surfaces of which are flat and not beveled as in the French system and our own. Breakwater to protect the harbor of refuge.

Light-house.

Lens.

Fog-bell.

Flat wheels used for revolving machinery.

150   EUROPEAN LIGHT-HOUSE SYSTEMS.

Mr. Douglass stated that wheels of the latter kind wear into the rail rough channels, which increase the friction and interfere with the regularity of motion, which is essential in revolving machinery of light-houses. The lamp, a "moderator" of Chance's manufacture, is like our own, except that the weights are hung underneath the cylinder or reservoir, and are connected with the plunger by means of exterior rods.

<small>Weights of the lamp.</small>

The plain pulleys and rods formerly used for revolving and fog-bell machinery have been replaced by chain-wheels and chains which are much more durable and reliable.

Fig. 16.

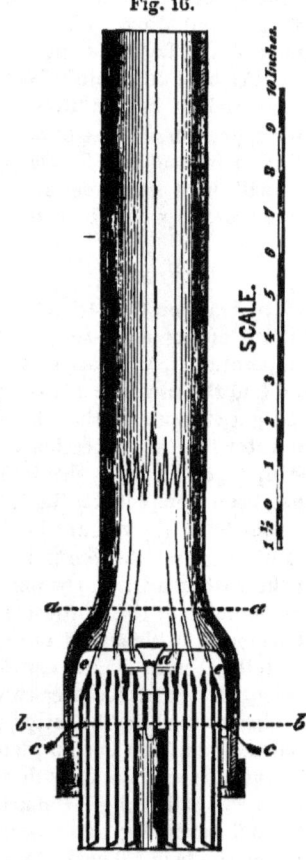

Lamp of Single and Double Power.

<small>Lamp of single and double power.</small>

The burner, shown in Fig. 16, contains all the recent improvements, one of which gives to the lamp its name of the "lamp of single and double power." In the figure $a\,a$ is the focal plane; $b\,b$, the level of the mineral-oil; $c\,c$, the overflow-holes, eight in number, and three-sixteenths of an inch in diameter; d, the interior, and e the exterior deflector. The adjustable chimney, the perforated button, and the conical tips to the burners are also shown. Explanation of figure.

Between the exterior deflector or jacket, and the chimney, there is an air-space, open at the bottom, through which a draught of cold air passes and keeps the base of the chimney so cool that it can be removed by the naked hand, as I found by experiment, without the use of the tongs usually required. These burners have three wicks, the exterior $2\tfrac{5}{8}$, the middle $1\tfrac{3}{4}$, and the interior $\tfrac{7}{8}$ of an inch in diameter.

By a simple movement the inner wick can be lowered into the burner and extinguished, or raised and lighted. This and the change of the perforated buttons require but an instant, and the result is, that the flame and the amount of oil consumed can be exactly adjusted to the quantity of light required. Power of adjustment of light and consumption of oil.

In fair weather only the outer and middle wicks are burned, but when fog or thick weather comes on, the perforated button is changed, and the inner wick is raised and lighted.

This is of great importance in the economy of light-house illumination, and the English have been, as far as I am aware, the first to attempt to vary the power of oil-lights; a result which has been effected, as I have before stated, in both the gas and the electric lights. Economy of the burner.

Photometric experiments have shown that the flame of the outer two wicks is equal to 146 candles, and the lighting of the small interior wick raises the power to 208 candles, showing that, though the inner wick has but 19 per cent. of the total burning-surface, it yields 30 per cent. of the entire amount of light, an amount very much in excess of its power when burning alone; but it must be considered that the circumstances of combustion are much more favorable when the wick burns inside the larger flames. Power of the inner-wick flame.

The lantern at Holyhead is of the latest kind. The sash-bars run diagonally from top to bottom, and the panes of glass are of a lozenge shape, cast in cylindrical form, corresponding to the diameter of the lantern. With regard to the diagonal sash-bars, Mr. Douglass stated that Faraday, the predecessor of Professor Tyndall in the office of scientific adviser of Trinity House, found that the vertical bars, which were at one time used, obstructed at least 48 per cent. of the light in certain directions, whereas with diag- Lantern with diagonal sash-bars and cylindrical glass. Obstruction of light by vertical bars.

onal bars the shadow was lost at a distance of less than a hundred feet. In discussing this matter Faraday considered that the rays of light issuing from the lens in any direction, form at the exterior surface, a beam, a vertical section of which is a column of light of the height of the lens; that with diagonal bars the beam from this column of light intersects them in *points*, the shadow of which, the beam being broader than the sash-bars, is terminated not far outside the lantern, while with vertical bars the best part of the beam is partially obstructed, and a vertical shadow is thrown upon the sea.

<small>Extract from report of Prof. Tyndall.</small> The following extract from a report by Professor Tyndall of experiments with gas and oil burners at Howth Baily light-house, Dublin Bay, confirms this view:

"I did not, however, think it safe to limit myself to this particular point of observation, and to meet my wishes, Captain Roberts was good enough to engage a steamer, which enabled me to proceed down the river and to pass across it from side to side between the North Wall and the South Wall, thus varying the points of observation.

"It soon became manifest that the oil-flame at Howth Baily varied in intensity with our position, and that the direction of *minimum* intensity corresponded almost exactly with that in which our observations had been made on the previous evening from the North Wall.

"No safe conclusion, therefore, could be drawn from the observations made on the evening of the 7th, for it was manifest that some cause existed which prevented the oil-lamp from displaying its full power in the direction of the North Wall, thus giving the gas a relative superiority, which in reality it did not possess.

"It is to be borne in mind that the oil-lamp in these experiments was placed in the first-order dioptric apparatus of the light-house, the apparatus being, as usual, surrounded by its glazed lantern. The gas-flame, on the contrary, was placed in a temporary hut at some distance below the lantern. A refracting-panel, similar in all respects to those of the dioptric apparatus, was placed at the proper distance in front of the gas-flame, and to make the conditions alike, the upper and lower reflecting-prisms of the apparatus were shut off. It was, therefore, the lights transmitted through the two lenticular belts that were compared together.

"In company with Captain Roberts, I again visited Howth Baily on Wednesday, the 9th.

"We were preceded by Captain Hawes and Mr. Wigham, who were instructed to examine with all care whether any obstruction was offered by the lantern to the passage of the light toward our station on the North Wall.

"*It was soon found that one of the vertical sashes of the lantern was directly interposed between the light and our point of observation, and that to this obstruction the enormous apparent superiority of the gas flame over the oil, manifested on the evening of the 7th, was to be ascribed.*

"Since my return to London, Captain Roberts, at my request, placed the gas-flame in the dioptric apparatus and the oil-flame in the hut below. From the North Wall the oil-flame was the brightest, thus affording additional evidence, if any were needed, as to the influence of the obstruction offered by the sash of the lantern."

M. Quinette de Rochemont, engineer *des Ponts et Chaussées*, who is charged with the supervision of the French lights north of the Lower Seine, in his "*Note sur les Phares Électriques de la Hève*," referring to experimental comparisons between the electric lights at La Hève and the oil-lights at Honfleur, Fatouville, and Ver, remarks as follows: {Extract from report of M. de Rochemont.}

"The observations at Fatouville show an anomaly, but this is easily explained, as it was in consequence of one of the uprights of the lantern of the northern (oil) light-house being placed in the direction of Honfleur, and thus obscuring a considerable part of the light from the apparatus."

This matter is still further illustrated by Fig. 17, in which a section of the beam is shown, partially eclipsed in one case by a vertical sash-bar of the lantern *a*, constructed on the American and French plan, and in the other crossed by the diagonal sea-bars of an English lantern, *b*, the width of the beams being drawn the same in both cases. {English and American Lanterns.}

In 1873, the engineer of our seventh light-house district reported to me that grave complaints were made of the light at Key West, it having been reported by mariners to have been extinguished on several occasions. {Complaints made of the light at Key West due to shadows cast by broad sash-bars.}

On investigation it was found that such large shadows were cast by the great sash-bars of the old-fashioned lantern of the light-house that vessels were sometimes in shadow for a long time, and to this cause, and not to the want of vigilance on the part of the keeper, were the above reports due. This old lantern has since been replaced by one of the new model, in which the width of the sash-bars is very much less than in the old.

I have been thus particular in referring to the subject of obstruction by vertical sash-bars in light-house lanterns, since in our service vertical-barred lanterns have always been {Error of the American system of using vertical-barred lanterns.}

154 EUROPEAN LIGHT-HOUSE SYSTEMS.

used, and I believe it to be a very serious error into which we have been led by an improper consideration of the matter.

Cylindrical lantern-glass stronger than plate. Mr. Douglass stated that he had found by experiment that cylindrical lantern-glass is 68 per cent. stronger than plate-glass, an important consideration to us, particularly with regard to our southern coast, where our lantern-glass, *Lantern-glass broken by sea-fowl.* although one-half an inch in thickness, is frequently broken by sea-fowl, which, blinded by the light, fly against the lantern, damaging not only that, but the lenticular apparatus, to such a degree that we have been obliged to cover the entire face of lanterns with netting, necessarily of such strength and size of wire as to impair the value of the light.

Fig. 17.

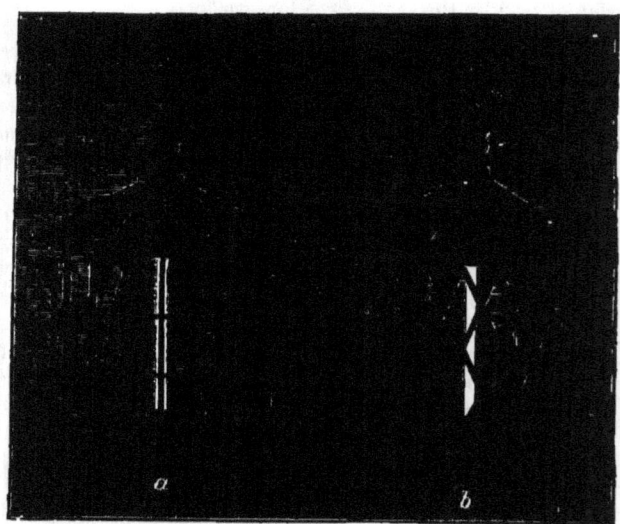

American and English Lanterns.

Low parapet used on English towers. With further reference to the English lantern, I will remark that the substitution of the low parapet in place of the high one used by the French and ourselves does not seem to me to be an improvement. The former necessitates the use of a step-ladder when cleaning the lenticular apparatus, and the door for reaching the outer gallery is necessarily low and inconvenient. *Ventilation.* The English provision for egress of heated air and smoke is not as good as our own, and the arrangement for turning the cowl with the wind I should think liable to become obstructed by soot and dust. I have the same opinion with regard to the manner of providing fresh air to support combustion in the lamp. The English admit this into the watch-room and thence through the grating of the lantern-floor. (See Plates VII and VIII.) It is true that the tops of the ventilating windows in the watch-room are above the heads of keepers on watch, but in our cold latitudes this would be quite inadmissible, and in no case do I perceive the English mode to be in any way superior to ours of admitting the exterior air directly into the lower part of the lantern.

The excess of 2 feet in the diameter of their lanterns over our own I think unnecessary, as we have found 12 feet to be quite sufficient, affording abundance of space between the lens and the lantern-sides. On our eastern coast, with our necessarily high towers, the expense of 2 feet additional to the diameter of lanterns and towers is much greater than in England, where the towers are ordinarily on elevated sites, requiring but small elevations of the lights.

Superior size of lantern used by the English no advantage to compensate for the consequent increased cost of towers.

NORTH STACK.

This fog-signal station is about four miles south of Holy-

Fig. 1ᵈ.

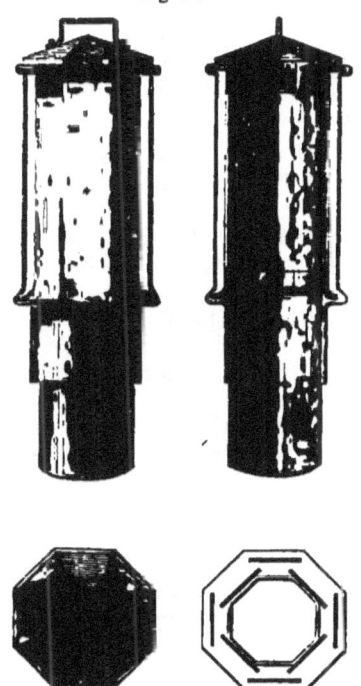

Wind-guard.

head Harbor, on the east side of Saint George's Channel. The signals, which are in charge of two keepers, whose sole duty is to attend to them, are a pair of 18-pounder guns, placed in a masonry building and fired through embrasures, in thick and foggy weather, at intervals of fifteen minutes, the charges being three pounds of powder.

As most of the steamships and other trade to Liverpool pass quite near this point, and fogs are common, this is a most important signal; but I am convinced that the long intervals between the discharges make it less valuable than a powerful steam-signal, either a whistle, siren, or Daboll trumpet, would be.

It is to be remembered that an increase in the number of discharges would materially add to the already heavy expense.

I observed on the dwelling of the keeper a wind-guard, which Mr. Douglass stated was designed by Faraday, and is in common use in the light-house service as well as throughout the country. In the most trying positions, such as North Stack, which is under high land rising immediately in the rear, it effectually prevents the annoyance, so frequent in such localities, of the smoke being driven down the chimney during high winds. A sketch of this chimney-top is shown in Fig. 18.

SOUTH STACK.

This light-house, at the extreme westerly point of Holyhead, the extremity of the Island of Anglesea, is about five miles south of the harbor, in a remarkably picturesque locality.

From the elevated land in the rear, one descends by a flight of 450 steps to a handsome suspension-bridge thrown over the chasm separating the South Stack from the Island of Anglesea.

This light is catoptric, revolving, and of the first order, having eleven reflectors on each face.

At this station the fog-bell, weighing two and one-half tons, is placed with its mouth uppermost. A counterpoise-weight is hung underneath the axle, which is turned by machinery placed under cover. The hammer is within the bell. The axle of the cog-wheel which moves the bell and its counterpoise, passes through a slot in the side of the machine-house, and has a slight vertical motion. Great power is not required for ringing this bell, which has been in use for many years, giving entire satisfaction to mariners.

OR OCCASIONAL LIGHT AT SOUTH STACK,
ST. GEORGE'S CHANNEL.

The only fog or "occasional" light for use at light-stations in thick weather of which I have any knowledge, is at South Stack.

Fog or "occasional" light.

It is well known that while lights in high towers and on considerable elevations of land, can be discerned in clear weather at the maximum distance, (in our service, as in others, about twenty miles,) an object of the first importance, great elevation of site is a disadvantage in foggy weather, since the fog-clouds frequently maintain themselves at a considerable height above the sea, and envelope the light when it is clear below.

High lights obscured by fogs.

At many points in our Pacific States, the coast rises so abruptly from the sea that no sites can be found at a sufficiently low elevation to avoid this difficulty, even when low towers are erected, and the arrangement I saw at South Stack provides a remedy for the obscuration of light during foggy weather.

Cases where low lights are needed in foggy weather.

An inclined plane has been excavated in the rock, and a tramway laid thereon. The fog-light is contained in a carriage, which in clear weather is kept at the summit of this plane near the main light-house, but during a fog is lowered by means of a windlass to a position where, safe from the waves, it is still as near the sea as possible. The forward part of this movable light-house is glazed, and contains a catoptric apparatus of three reflectors in the same plane, on a revolving frame which has a reciprocal motion through the exact arc to be illuminated, thus giving the characteristics of the main light.

Manner of using the low light.

A weight for driving the machinery in this case being, as is evident, out of question, the motion is produced by means of a powerful spring.

Means of producing the motions necessary to give the characteristics of flashing lights.

This ingenious light has been in operation for many years, (I believe since 1832,) and has proved to be of great advantage. Captain Moodie, of the Cunard line, with whom I returned to America, told me that frequent occasions had been afforded him for testing its value, the light at Holyhead being of the first importance to the immense traffic through Saint George's Channel, where, at some seasons of the year, habitual fogs prevail. Plate XXIII will give an idea of this construction.

Before closing my notes in regard to the English lighthouse service, I take pleasure in recording my obligations to Sir Frederick Arrow, the Deputy Master, to the Elder Brethren, and to Mr. Douglass, the engineer of the Trinity House, for the attentions and kindnesses of which I was constantly the recipient, from the time I arrived in England

Acknowledgment of attentions received from the members of Trinity House.

till my departure for America. There was no facility for acquiring information in regard to the object of my journey, no act of hospitality which could suggest itself, which was not proffered with that hearty generosity for which their countrymen are distinguished.

<small>Drawings, descriptions, &c., received from Trinity House.</small> While in London I was furnished with drawings, descriptions, &c., of many of their light-houses and accessories, and since my return I have received others which have been of much assistance to me in the preparation of this report, and I am happy to believe that such relations are now established as will lead in the future to that interchange of information which is desirable in this most interesting and important service.

Although I should have been glad to avail myself of the kind invitation of Admiral Schomberg, Harbor-Master at Holyhead, to whom I am indebted for polite attentions, to remain to witness the demonstration which was to celebrate the completion of the great breakwater, the limited time at my disposal would not permit, and on my return from South Stack, the last English light-station that I visited, I embarked in the Irish mail-steamer for Dublin.

IRISH LIGHTS.

<small>Letter from Commissioners of Irish lights.</small> Soon after my arrival in England I received the following letter from the Commissioners of Irish lights:

"IRISH LIGHTS OFFICE,
"*Westmoreland Street, Dublin, May* 24, 1873.

"SIR: The Commissioners of Irish lights having been informed by the inspector of lights to this department that you propose visiting Ireland shortly, and that you were desirous of availing yourself of such opportunity to inspect one or more of the gas light-house establishments under the management of this Board, I have the pleasure to acquaint you that the Commissioners will be most happy to afford you every facility to carry out your wishes.

"Will you be so kind as to let me know a day or two previously, as to the probable period of your arrival in Ireland?

"I have the honor to be, sir, your obedient servant,

"W. LEES,
"*Secretary.*"

"Major GEORGE H. ELLIOT,
"*Corps of Engineers, United States Army.*"

I reached Dublin on the 16th of August, and on my arrival proceeded to the office of the Commissioners, where I was politely received by Mr. Lees, who stated that the Com-

Plate XXIV.

GAS LIGHT-HOUSE AT HOWTH BAILEY, IRELAND.

missioners were at that moment, according to previous arrangement, embarking for Holyhead to take part in the celebration at that place, and wished him to convey to me their regrets that they were unable to meet me, and to inform me that Captain Hawes, the inspector of Irish lights, (whom I had the pleasure of meeting at South Foreland on the Straits of Dover at the commencement of the fog-signal experiments of which I have given an account,) and Mr. Wigham, the inventor of the Irish gas-light for light-houses, would take me to such of their establishments as I should desire to visit. Inspection of Irish gas-lights, with Captain Hawes and Mr. Wigham.

We soon set out by rail for Howth Baily, the northern head of Dublin Bay, where is a first-order fixed gas-light; after describing this station and that at Wicklow Head, where is a first-order *intermittent* gas-light, I shall give a general description of the Irish gas-lights for light-houses, derived from information received from Captain Hawes and Mr. Wigham as well as from observations made on my visits to the above-named stations, and to the English gas-light at Haisborough.

HOWTH BAILY.

This station, (see Plate XXIV,) on a bold promontory at the outer northern limit of Dublin Bay, is of interest as being the first station at which the Irish gas-light patented by Mr. Wigham was established; this was in 1865. First light where Wigham's gas-light was established.

An inspection of Plate XXV will show the compactness of the gas-works, comprising retort-house, gas-holder, &c. Few light-house sites are too limited to contain buildings necessary for the apparatus. The gas-holder at this station contains 800 cubic feet, but is considered too small, a disadvantage not found in later establishments of this kind. Little space required for gas-works.

Uniformity of pressure of gas at the burners is obtained by means of a regulator, also of Mr. Wigham's design; and I will here mention that this burner is but one of many curious and ingenious inventions led to by, and necessary for, the proper development of the system. Regulator to produce uniformity of pressure of gas.

As at Haisborough, I saw here from actual trial that the ordinary first-order oil-lamp, which is always ready, can be substituted for the gas in less than two minutes, and the Irish regulations require the keepers, for the purpose of keeping in practice, to change the gas for the oil light once a month. Substitution of oil for gas-light when necessary. Change of lights required monthly.

The changes of the burners and the mica oxydizers to meet the varying atmospheric conditions, require but an instant, the change of the burners being accomplished Manner of changing burners.

simply by setting into, or lifting out of, the mercury-cups, the short pieces of supply-pipe to which are attached the semi-cylindrical rings of jets, and by opening or changing the cocks.

Diameter of burner and power of flame. The largest burner, having a diameter of 10½ inches, contains 108 jets, and the immense body of flame carried to the mouth of the mica oxydizer (see Plate 19) is most dazzling.

Heat. The heat, as may be conceived, was very great, but the keeper said it gave him no inconvenience, nor did it injure the lenticular apparatus, thus corroborating the information given me at Haisborough. The lens is in no respect different from the ordinary fixed lens of the first order.

Keepers. Employed at this station are two keepers, one apprentice under instruction, and a gas-maker, the latter receiving 2s. 6d. (62½ cents) a day. Captain Hawes considered the employment of the gas-maker quite unnecessary, the labor at any gas light-house being easily performed by two keepers.

Fog-bell. There is at Howth Baily a fog-bell, operated at present by an Ericsson engine, but for some years a gas-engine was used, which was, I believe, discontinued in consequence of the insufficient size of the gas-holder.

Superiority of light at Howth Baily. Mr. Wigham, who constructed the works at both Howth Baily and Haisborough, stated that at the former place the light was much the purer and whiter, owing to the burner being supplied under much greater pressure. Captain Hawes is of the opinion that 28 jets are here quite enough in clear weather, while 48 are habitually used at Haisborough.

WICKLOW HEAD.

This station, on the western side of Saint George's Channel, south of Dublin, we reached partly by rail and partly by "jaunting-car," a two-wheeled vehicle, which is the common conveyance of the country.

Manner of disposing gas-works. The gas-house, containing the furnaces and retorts, the gas-holder, and other buildings, are ingeniously disposed on the face of a high cliff, and occupy but little space, as shown by Plate XXVI.

Characteristics of the light, and manner of producing flashes. This is a first-order intermittent light, the lens-apparatus being that of an ordinary fixed light. The gas is let on and shut off by an automatic arrangement which allows ten seconds of light and three seconds of darkness. This arrangement does not cut the gas entirely off, and each jet during the three seconds' eclipse shows a tiny blue flame, which, while it produces no illumination of the lenticular

GAS LIGHT-HOUSE AT HOWTH BAILEY, IRELAND.

apparatus and can scarcely be detected in the daylight, is still sufficient to light the main body of gas when the supply is turned on.

To guard against all danger of total extinguishment of the light by gusts of wind through the ventilators or door of the lantern, each jet is surrounded by a small pipe called the "by-pass," the top of which, being at a level with the tip of the jet, is pierced by several minute holes through which gas is supplied from a pipe quite independent of the automatic cut-off; thus protected, these little jets of flame burn from the moment of lighting at sunset to that of extinguishing at sunrise, and it is impossible even if the "cut-off" or a gust of air should completely extinguish the main flame that the burner should not be lighted at regular intervals of thirteen seconds. *"By-pass," to guard against extinguishment of light.*

The two keepers at Wicklow Head agreed with those at Howth Baily in saying that the gas gives very much less trouble than the oil light. *Gas-light less trouble than oil-light.*

DESCRIPTION OF WIGHAM'S GAS-APPARATUS FOR LIGHT-HOUSES.

FIXED GAS-LIGHTS.

The light is produced by a burner shown in the Plates XXVII and XXVIII. It is capable of producing five different degrees of power according to the state of the atmosphere, the first power being produced from 28 jets of gas. Each jet consists of a hollow tube with an ordinary fish-tail burner of lava at the top. At the bottom of each tube is placed a small lava cone bored with fine holes, (one on each side,) the effect of which is to act as a regulator and to allow the gas to enter the air at such a rate of speed as is found most conducive to its combustion in connection with the overhanging oxydator, which is formed of talc, and which is fitted with terminal pieces to suit the respective sizes or the various powers of the burner. The second power is produced from 48 jets as above, except that some of the orifices in the lower cones are slightly larger than in the smaller burner. The third power is from 68 jets, with slight alterations in the orifices of the lower cones and also of the upper jets. The fourth is from 88 jets, and the fifth is from 108 jets as above. The flame of the latter is shown in the frontispiece. *Capacity for producing different powers of light. First power. Regulator. Second power. Third power. Fourth and fifth powers.*

The rings which contain the jets for the several powers, except the first, are removable and replaceable at pleasure. Mercurial cups, each fitted with a ground-in valve, are the *Removable rings.*

means employed to facilitate this application. The whole of the gas supplied to the burner passes through a chamber of cast brass, at the bottom of which there is fixed a mercurial lute which enables the whole gas-burner to be removed in a moment, and the ordinary oil-lamp used in dioptric apparatus to be substituted for it. The talc chimneys are also removable, and arrangements are made by which the ordinary oil-lamp condenser can be fixed for the use of the oil-lamp. This arrangement for rapidly substituting the oil-lamp in place of the gas-burner was instituted when gas was first lighted at Howth Baily lighthouse, in the year 1865, as a precaution against any accident occurring to the gas-light, but during the time that has since elapsed (about eight and a half years) no occasion has ever arisen for putting this precautionary plan into operation.

Talc chimneys, and substitution of oil-lamp.

By the use of gas in a fixed-light apparatus of any size there is no occasion to alter the existing lenses, but in some lanterns it may be necessary to provide for additional ventilation. It will be seen by the table to be found further on that the photometric values of the flames of the respective powers of the gas-light are largely superior to any photometric results obtained from the oil-lamp, either for paraffine or colza oil. The cost of gas-light in Great Britain is said to be less than that of oil, comparing only the ordinary power of the gas-light. Of course, when the fog-powers of the gas-light are turned on, the cost per hour is greater, but taking the average of a year's consumption (including consumption for fogs) at several light-houses in Ireland which are lighted by gas, it appears from a return made by the engineer and accountant of the Board of Irish Lights that there is a saving of about £65 ($325) per annum at each lighthouse by the use of gas instead of oil.

Lenses used.

Cost of gas-light.

INTERMITTENT GAS-LIGHT.

The burner used in this case is precisely the same as that in the fixed lights. The intermission in the light is caused by the opening and shutting of a gas-valve, which cuts off the supply of gas to the burner for any required period, and the re-exhibition of the light takes place as soon as the valve is opened, a small by-pass being provided for keeping a supply of gas in the burner, so small as to render its flame invisible, but sufficient for the re-exhibition of the light immediately on the opening of the valve. The opening and shutting of this valve is accomplished by means of a small piece of clock-work fixed in the room under the lantern, and which requires to be wound up every four, six, or eight hours, according to the height of the tower.

Burner. Flashes.

GAS LIGHT-HOUSE AT WICKLOW HEAD, IRELAND.

Professor Tyndall very clearly illustrates the application of the intermittent light to revolving lenses in a report to the Board of Trade dated the 7th of February, 1871.

Statement of Professor Tyndall.

After stating that a gas-burner of 28 jets is almost identical in size and sensibly equal in illuminating power to the Trinity four-wick burner, that it is quite as applicable as the latter to a revolving light, and that a saving can be secured by a periodic extinction of the gas-flame, he says:

"The central octagon figure (see Fig. 19) from which the rays issue is intended to represent the eight-paneled revolving apparatus. The points of the stars are to be regarded as the centers of the beams issuing from the respective panels. The blackness of the disk underneath the star is intended to denote the darkness of night, while the circle round the disk represents the horizon.

Fig. 19.

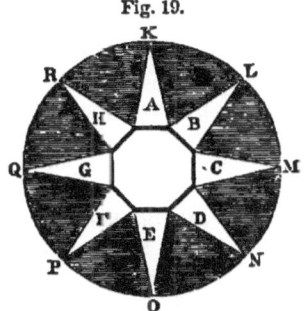

Diagram illustrating revolving intermittent gas-light.

"Every part of the horizon receives a flash every second minute, the gas being lighted one minute, during which the lens is moved one-eighth round, and each part receives a flash, then the gas is extinguished one minute, and so on in succession. Thus half of the consumption of gas is saved.

"Let the star be turned so that the beam A shall point to and illuminate the station K on the horizon. The ray B would at the same moment illuminate L; C would illuminate M; D, N; E, O; F, P; G, Q; while the ray H from the eighth panel would illuminate R, the eighth station.

Description of Fig. 19.

"Let the star be turned, or in other words let the apparatus be supposed to revolve, until A reaches L. During this time B travels from L to M; C from M to N; D from N to O; E from O to P; F from P to Q; G from Q to R; and H from R to K. Thus, through the passage of A from K to L, in other words during the rotation of the apparatus through one-eighth of a revolution, *every point on the horizon is once illuminated.*

"If the flame continue burning, the same effect is produced during every succeeding eighth of a revolution. Every point of the horizon is once illuminated. At Rockabill the time of a complete revolution is ninety-six seconds; hence, the number of panels being eight, as in the model, each point of the horizon receives flashes which succeed each other in intervals of twelve seconds.

"But suppose at the moment the beam A points toward L the supply of gas to be cut off, and the apparatus permitted to revolve in darkness until A reaches M. During this eighth of a revolution no gas is consumed, and no flash is received by any point on the horizon.

"When A reaches M let the gas be relighted. During the succeeding eighth of a revolution every point of the horizon would be once illuminated as before. It is quite manifest that this process may be continued indefinitely; the gas being lighted and every point of the horizon once illuminated during every alternate eighth of a revolution.

"It is also plain that the intervals of darkness between the flashes instead of being, as they now are, twelve seconds, would be twenty-four seconds. This reasoning, which as far as it goes is of the character of a mathematical demonstration, has, as stated in my report of the 8th of October, been verified by actual observations on the Rockabill light.

"There is, as far as I can see, but one drawback to the perfection of this scheme; and this I will now try to point out with distinctness. Let the beam A point, as at the outset, toward K; at a certain moment we start from K toward L with the gas lighted. According to the new scheme K ought to be in absolute darkness for twenty-four seconds. But just as A reaches L, and before the gas is cut off, a glimmer is seen at K, this glimmer being only twelve seconds distant from the preceding flash. The same is true of the other seven points of the horizon faced by the panels the moment before the flame is extinguished. In all the remaining sweep of the horizon this secondary glimmer is absent."

Professor Tyndall further says in regard to this glimmer that he noticed it in his experiments at Rockabill lighthouse, but attached no practical importance to it; and that on the isolated points where it is seen it is so masked by the superior brilliancy of the true flash that no mariner could be deceived by it.

Substitution of oil-lamp for gas-burner. The same method of substituting an oil-lamp for the gas-burner is arranged for intermittent as for fixed lights, the only difference being that if the oil-lamp were used the machine which actuates the cutting off of the gas is made

THE IRISH GAS LIGHT.

108 JET BURNER.
SECTION.

PLATE XXVII.

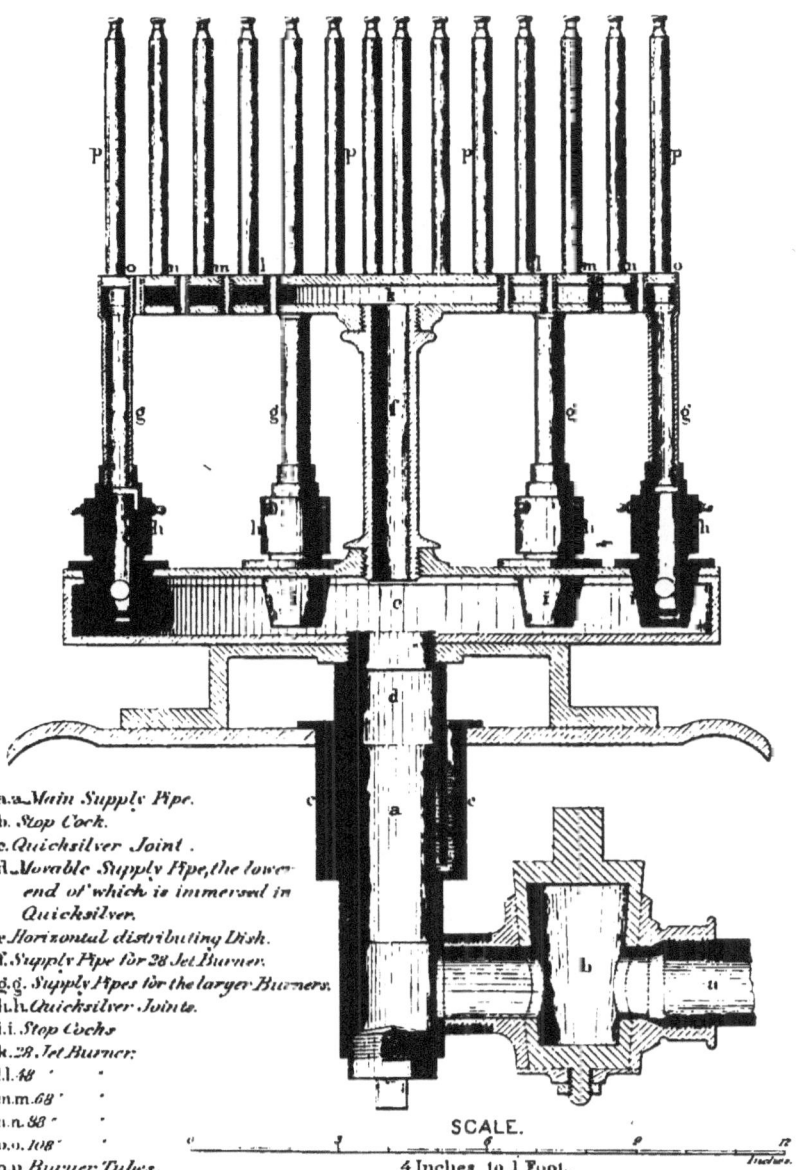

a.a. Main Supply Pipe.
b. Stop Cock.
c. Quicksilver Joint.
d. Movable Supply Pipe, the lower end of which is immersed in Quicksilver.
e. Horizontal distributing Dish.
f. Supply Pipe for 28 Jet Burner.
g.g. Supply Pipes for the larger Burners.
h.h. Quicksilver Joints.
i.i. Stop Cocks.
k. 28 Jet Burner.
l.l. 48 "
m.m. 68 "
n.n. 88 "
o.o. 108 "
p.p. Burner Tubes.

SCALE.
4 Inches to 1 Foot.

to open and shut semi-cylindrical shades for eclipsing the oil-lamp for the required period of darkness. As in the case of Howth Baily, no occasion has ever arisen at Wicklow for falling back upon the oil-lamp since gas was lighted at that station, about seven years ago. The advantage of the use of intermittent lights as applied to gas light-houses is obvious, for during the period of darkness the consumption of gas is saved, whereas in the case of oil-lights, the flames of which are merely eclipsed by shade, the consumption of oil continues during the intervals of darkness as well as during the intervals of light. This peculiarity of gas renders its use for intermittent light still more economical than for fixed lights. All the four *fog*-powers of the gas-burner may be intermitted precisely as is the case with the first or ordinary power. *[Advantage of use of intermittent light.]*

By the use of gas in an intermittent-light apparatus there is no occasion to alter the existing lenses.

REVOLVING GAS-LIGHTS.

The burner used in this case is precisely the same as that in the fixed and intermittent lights. The lenses by which the flashes are directed to the observer revolve around the gas-light just as they revolve around the oil-light. An arrangement similar to that in fixed lights for substituting the oil-lamps for the gas-burner is also established. The remarks as to the superiority of gas over oil in fixed lights apply equally to revolving lights, with this additional advantage, that by the use of gas-burners which have large diameters each flash is of longer duration than in the case of the oil-lamp, and this is stated to have been highly appreciated by mariners. *[Burners. Revolution of lenses.]*

GROUP-FLASHING GAS-LIGHTS.

The burner used in this case is precisely the same as that in fixed lights. The groups of flashes are caused by the continual extinction and re-ignition of the gas-flame. This is accomplished by similar means to that which causes the intermission of a fixed light, except that the machinery which causes the lenses to revolve is availed of for actuating the cutting-off gas-valves instead of a special machine being provided for that purpose. By means of this system of group-flashing the same economy may be attained which the use of gas presents in the case of intermittent lights, with this difference, that the economy is always at a fixed ratio, viz, about one-half the consumption of the gas. The lenses are continually revolving and the gas-light is continually intermitting. By reducing the speed of the revolutions of the *[Burner. Flashes.]*

lenses (which themselves require no alteration for this system) a group of seven or eight distinct and powerful flashes recurring at regular intervals may be obtained, and the effect of this kind of light is said to be exceedingly arresting to the eye of the mariner.

TRIFORM FIXED GAS-LIGHT.

Three burners used. This light is produced by three burners similar to that above described, placed vertically over each other as shown *Air-chambers for upper burners.* in Plate XXIX, but the two upper burners are surrounded by two air-chambers for the supply of pure air to the flame and the carrying off of the foul air from the flame of the burner below. Tubes for the introduction of pure air are placed obliquely in these chambers, and the effect of the arrangement is that not only are these three burners of equal power when in a light-house lantern in place of one as in an ordinary case, but the light is much intensified by the manner in which the burners are supplied with heated air.

Description of plate. In the plate, a is the main supply-pipe, b, b', b'' are the gas-burners, c, c' are the distributing-pipes, d, d', d'' are the mica chimneys, e, e', e'' are the sheet-iron chimneys, f, f' are the outer air-chambers, h, h' are flues leading to the inner air-chambers g, g', which at their lower ends have the form of inverted cones; i is the main escape-pipe, and k, k, k are the refracting belts.

Removal of top and bottom prisms. The top and bottom prisms of the dioptric apparatus are removed, and instead of them segments of refracting belts are placed above and below the central belt of the dioptric apparatus. (See Fig. 1, Plate XXX.)

Amount of gas consumed. The consumption of gas in this form of apparatus is three times that of the single lamp, but the quantity of light is more than three times as great. We must, however, deduct the light which would be transmitted from the top and bottom prisms of the single lamp apparatus, which is valued at about one-third of the whole apparatus, and it may be said that the light of a light-house illuminated by the triform *Ratio of illuminations produced by triform light and single lamp.* apparatus is to that of a light-house illuminated by a single lamp and an ordinary lenticular apparatus as 2 to 1. It will be seen by the table of photometric values that the power of the largest size of the gas-burner is stated by Mr. Wigham to be equal to 2,577 candles. Assuming this to be correct, there is therefore transmitted by means of triform apparatus three times that amount, viz, light equal to 7,731 candles, and if this is increased more than eighteen times by the agency of the cylindric refractors, as is the case where the ordinary oil-burner is used in a first-order fixed lens, the immense power of more than 139,158 candles is transmitted to the observer. This power, great as it is, is of course exceeded by the triform light in revolving apparatus.

TRIFORM INTERMITTENT GAS-LIGHT.

The apparatus in this case is precisely similar to that described for intermittent lights with one lamp, except that refracting belts are substituted for top and bottom prisms as described in the case of the triform gas-apparatus for fixed lights.

TRIFORM REVOLVING GAS-LIGHTS.

The arrangements of burners, lenses, &c., are nearly similar to those for fixed lights. The lenses are placed vertically (see Fig. 2, Plate XXX) in order to give the most powerful beam that can be transmitted to the horizon, but in case it were desired that the ordinary flash should recur more frequently than is possible with an apparatus containing only one light, the lenses may be placed eccentrically, as shown in Fig. 3, Plate XXX, and this in a very striking manner attains that object. *Arrangement of burners, lenses, &c.*

TRIFORM GROUP-FLASHING LIGHT.

The remarks made under the head of group-flashing light for a single lamp apply equally in this case, but the groups of flashes proceed from three lenses in place of one, and are consequently of much greater power.

EXPERIMENTS WITH THE TRIFORM LIGHT.

On the evening of our return from Howth Baily to Dublin we proceeded to a point near Kingstown, on the south side of Dublin Bay, distant six miles from the Howth Baily fixed gas-light, to observe some experiments to be made with a triform light, (previously placed temporarily, for purposes of experiments made under direction of Professor Tyndall, in a small cabin near Howth Baily light,) arrangements for which had been made by Captain Hawes and Mr. Wigham while we were at that station. *Experimental comparison between fixed and triform gas-lights.*

The three Wigham burners, each capable of burning from 28 to 108 jets, were placed vertically over each other in the foci of three panels of the refracting belt of a first-order lens, as already described, stop-cocks being provided, so that any of the rings of either burner could be lighted or extinguished *Arrangements of burners in the triform light. Manner of increasing or decreasing light.*

at pleasure. The principal keeper at the station had received instructions to cover and uncover the catadioptric prisms of the first-order lens in the tower, while the assistant was to vary the powers of the triform lights, both acting according to a memorandum given to each, a copy of which we were provided with.

Programme of experiments with a triform light at Howth Baily, Dublin Bay, August 6, 1873.

Tower lights.	Triform lights.
P. M.	P. M.
9.30.—28 jets; cover the catadioptric prisms....	9.30.—28 jets; upper belt only.
9.35.—28 jets; uncover catadioptric prisms.....	9.35.—28 jets; the three belts.
9.40.—28 jets; catadioptric prisms uncovered ..	9.40.—48 jets; the three belts.
9.45.—28 jets; catadioptric prisms uncovered ..	9.45.—68 jets; the three belts.
9.50.—28 jets; catadioptric prisms uncovered ..	9.50.—88 jets; the three belts.
9.55.—108 jets; catadioptric prisms uncovered .	9.55.—108 jets; the three belts.
10.00.—28 jets; catadioptric prisms uncovered ..	10.00.—108 jets; the three belts.
10.05.—End of experiments.	

Appearance of the lights. From our point of observation at Kingstown the 28-jet light, shown only in the upper belt of the triform arrangement, appeared slightly inferior to the 28-jet light in the tower when the catadioptric prisms were covered. They should have appeared equal, but the difference is accounted *Reason for inferiority of gas-light burning in the upper belt of the triform apparatus.* for by the facts that the ventilation of the cabin and the old refracting belts used for the experiments were imperfect, and neither the latter nor the glazing of the window were so clear as in the lantern of the tower-light.

When the catadioptric prisms of the tower-light were uncovered and the entire beam from the lens thus coming to our view was compared with the triform lights, changed, *Superiority of the triform-light.* as shown in the table, from 84 to 324 jets, the superiority of the latter was very marked, it having the appearance of a great globe shining with a pure and dazzling white light. The effect of the changes in the triform light was peculiar, and one could hardly believe that he was observing two lights equally distant from him, for, as the power of this light was gradually increased, the one in the tower appeared rapidly to recede and to be thrown further and further to the rear, so that when 108 jets burned simultaneously behind each of the refracting belts, the tower-light, burning 28 jets and equal in power to an ordinary first-order oil-light, appeared to have receded many miles, and to have dwindled until it presented the appearance of a star.

When 108 jets were burning in the tower the comparison with the triform light of 324 jets was still very much in favor of the latter, though the superiority was not so marked.

THE IRISH GAS-LIGHT.   PLATE XXVIII.

28-JET BURNER.

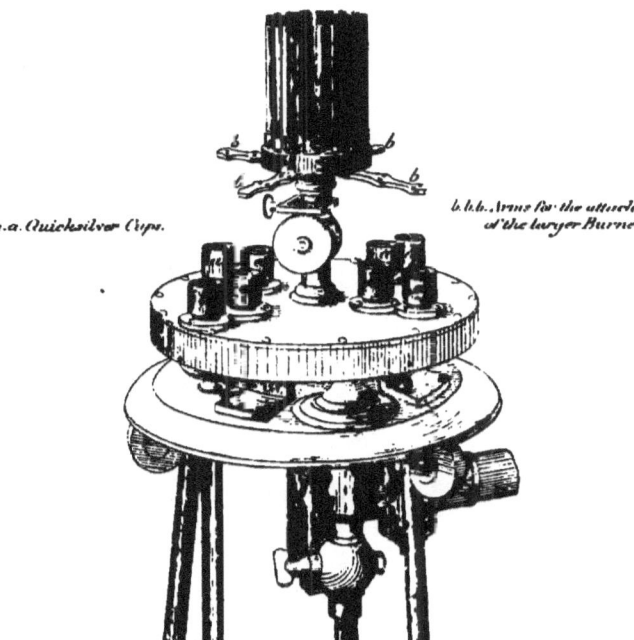

a.a.a. Quicksilver Cups.   b.b.b. Arms for the attachment of the larger Burners.

It may be well to state that the columnar form which the triform light might be supposed to assume is not observable, as from Kingstown it appeared to us of precisely the same *shape* as the single light in the tower.

<small>Form of the triform light like the other in appearance.</small>

ILLUMINATING POWERS OF GAS-LIGHTS.

The following are the results of observations recently made with a photometer, as stated to me by Mr. Wigham:

Gas-burners—No. of jets.	Consumption in cubic feet per hour.	Illuminating power, the unit being a sperm candle, consuming 120 grains per hour.
23	50	330.69
48	100	668.29
68	150	1,002.09
88	220	1,667.49
108	290	2,577.3

In this connection it should be observed that Trinity House four-wick oil-lamp, as improved by Mr. Douglass, consuming 34.13 ounces per hour, (160 ounces to the imperial gallon,) gives an illuminating power equal to 328.18 sperm-candles. <small>Illuminating power of four-wick lamp.</small>

According to actual experiments, the first-order fixed dioptric apparatus transmits to the mariner 27.39 times, and the first-order revolving apparatus of eight panels 338.74 times, more light than the unassisted first-order oil-lamp. These ratios would give for single lenses illuminated by Wigham's gas-burners of 108 jets powers, equal respectively to 70,592 and to 873,034 candles. It is stated (see *M. Reynaud's Mémoire sur l'Éclairage des Côtes de France*) that the luminous beams from the refracting belt of a dioptric apparatus for a fixed light of the first order is .7 of the beam from the entire lens, and that the beam from the refracting portion of one of the panels of a revolving octagonal lens of the first order is .647 of the entire beam; also that in the first case the power of the refracting portion of the lens is 19.13, and in the last case 218.1 times more powerful than the unassisted first-order oil-lamp. The same ratios for triform lights, supposing each lamp to burn 108 jets, would give for the fixed light 147,914 candles, and for the revolving light 1,686,228 candles! I am not aware that any experiments have been made of the actual powers of lenses illuminated by gas-lights, but as a considerable

<small>Additional power furnished by fixed-lens apparatus. By revolving apparatus.</small>

<small>Power of refracting belt of first-order lens, fixed.</small>

<small>Revolving.</small>

<small>Ratios applied to triform lights.</small>

<small>The above results not good in practice, owing to the divergence of the rays.</small>

portion of the light is exfocal, the divergence is so much increased that the above results would be very far from holding good in practice.

Results arrived at by Professor Tyndall on experiment. I should further add in regard to photometric experimental comparison between oil and Wigham's gas-burners for light-houses that Professor Tyndall arrived at the following results:

(The four-wick lamp being taken as the unit, the illuminating power of the gas-flame is expressed in terms of that unit.)

Four-wick lamp.	Number of jets.	Gas-flame.
1	28	2½
1	48	4½
1	68	6
1	88	9¾
1	108	13

That is to say, the photometer showed the 28-jet flame to have two and one-half times, the 48-jet flame four and one-half times, the 68-jet flame six times, the 88-jet flame nine and three-fourths times, and the 108-jet flame thirteen times the illuminating power of a four-wick flame of a first-order sea-coast light-house lamp.

The above experiments anterior to recent improvements in lamp-burners. It should be mentioned in this connection that these experiments were probably made anterior to the improvement in oil-lamps made by Mr. Douglass, the present power of his four-wick lamp as compared with the lamp formerly used being as 328 to 269.

COST OF WIGHAM'S GAS-LIGHT APPARATUS FOR LIGHT-HOUSES.

Cost of apparatus packed for shipment and delivered at Liverpool. Mr. Wigham informed me that the cost of gas-making apparatus of a size similar to that used at the Irish stations, Rockabill, Wicklow Head, Hook Tower, and Minehead, packed and delivered at Liverpool ready for shipment, would be, for a single tower, about £1,000, ($5,000,) and for

Resulting cost of placing gas-lights at Cape Ann, or Navesink. a station with two towers, like that at Cape Ann or Highlands of Navesink, about £1,600, ($8,000.) These estimates include furnaces, retorts, gas-holders, and tanks for the same, together with the burners and every other item of expense except the freight from Liverpool to America, the cost of erection at the site, and the additional buildings required. The expense of the latter item is, in Ireland, about £250, ($1,250,) and in the United States would be somewhat larger in consequence of the higher price of labor. Expense of buildings.

Mr. Wigham further stated that he would personally inspect any site at which the United States Government might desire to use his apparatus, and give the Board all the information and assistance in his power, on condition that his traveling and other expenses be paid; also that he would bring with him for the purpose of assisting in and superintending the erection of the entire work, one of his most competent foremen, the charge for whose time would be 10s. 6d. ($2.62) per day in addition to his expenses for board, lodging, and traveling. Conditions on which Mr. Wigham would consent to erect a gas-light apparatus in the United States.

In regard to the use of his gas-light in the United States Mr. Wigham stated that although his patent did not extend to this country he would have no hesitation in building a gas-light for our use, having no fear but that, should others be required, his labors for the improvement of light-house illumination would be recognized and rewarded.

COST OF THE TRIFORM-LIGHT APPARATUS.

The cost of this gas-apparatus is greater by about £250 ($1,250) than for the ordinary single-light apparatus. Cost of triform gas-apparatus.

The expense of removing the upper and lower catadioptric prisms of the lens and substituting for them two refracting belts is in Ireland not great, as Chance, Brothers & Co., who supply most of the lenticular apparatus for the British lights, offer to take the prisms of the first-order lens in exchange for refractors on payment of a difference of £150, ($750,) provided the prisms are of their own manufacture. Cost of changing ordinary apparatus for the triform.

If a new light-house were building, or if a new dioptric apparatus were to be placed in an existing light-house, it would be rather less expensive to have it arranged on the triform system than in the ordinary way, with a central belt and upper and lower catadioptric prisms. Cost of placing triform lens-apparatus less than that of the ordinary kind.

COST OF MAINTENANCE OF GAS-LIGHTS.

The use of tar for fuel in Great Britain within the last few years has so lessened the cost of production of gas that it may be calculated, it is stated, not to exceed, even at outlying stations, 10s. ($2.50) per thousand feet, including 4 per cent. interest on the original outlay for apparatus. Use of tar for fuel.

If gas could be supplied to our light-houses at this rate, the cost per annum to a first-order seacoast-light, burning habitually 28 jets during the average period of illumination, Approximate annual cost of gas.

(4,311 hours,) would be $550; adding for an extreme case (as West Quoddy Head on the coast of Maine for example) 20 per cent. for additional consumption in foggy and thick weather, we have $660 as the approximate annual cost of the gas for such light.

Cost of annual consumption of oil. The annual consumption of oil in our light-houses of the first order is about 760 gallons, amounting at the last average contract-price of 89 cents per gallon to $676.40.

Less cost for gas per thousand feet at stations having two lights than where there is but one. At stations with two towers the cost per thousand feet is considerably less than at single tower-stations, since the cost of the labor and other items is proportionately less the greater the quantity manufactured.

ILLUMINATION OF BEACONS BY GAS.

Beacons on outlying rocks. Mr. Wigham has devised an arrangement for lighting by gas, beacons on detached rocks which are inaccessible during heavy weather. Gas for the illumination of such positions cannot ordinarily be carried in submarine pipes, on account of the condensation of moisture within the pipe, the lowest part thus becoming filled with water and the flow of gas being consequently obstructed.

Drying gas with chloride of calcium. Mr. Wigham's plan is to dry the gas by chloride of calcium, and he proposes to light and practically extinguish the beacon by means of variations of the pressure of gas in the supply-pipe; that is to say, a high pressure of gas, say of six inches of water, closes a stop-cock at the beacon and keeps it closed during the day; at the time of lighting, this pressure is decreased to the ordinary working-pressure of, say, three inches of water, and the cock opens. The burner is lighted by means of a little flame supported by a small "by-pass," such as preserves the light from extinguishment during eclipses at the Wicklow Head intermittent light heretofore described. The full power of the light can be kept up till sunrise, when the increased pressure of gas closes the cock and extinguishes the beacon.

The invention not yet fully tested. Experiments sufficient for determining the utility of this invention have not yet been made; but it seems a step in the right direction, and affords another indication of the ingenuity of Mr. Wigham.

Opinions of the Irish Board as to the economy of gas-lights. In concluding my remarks on the subject of gas as an illuminant for light-houses, I will only say that the Irish Board and its officers state most positively that the actual use of gas at five of its sea-coast stations proves it to be more economical than oil, and specifically, at Howth Baily the saving is £50 ($250) annually, taking into account all

Annual saving at Howth Baily.

THE IRISH GAS LIGHT.
TRIFORM BURNERS.

PLATE XXIX

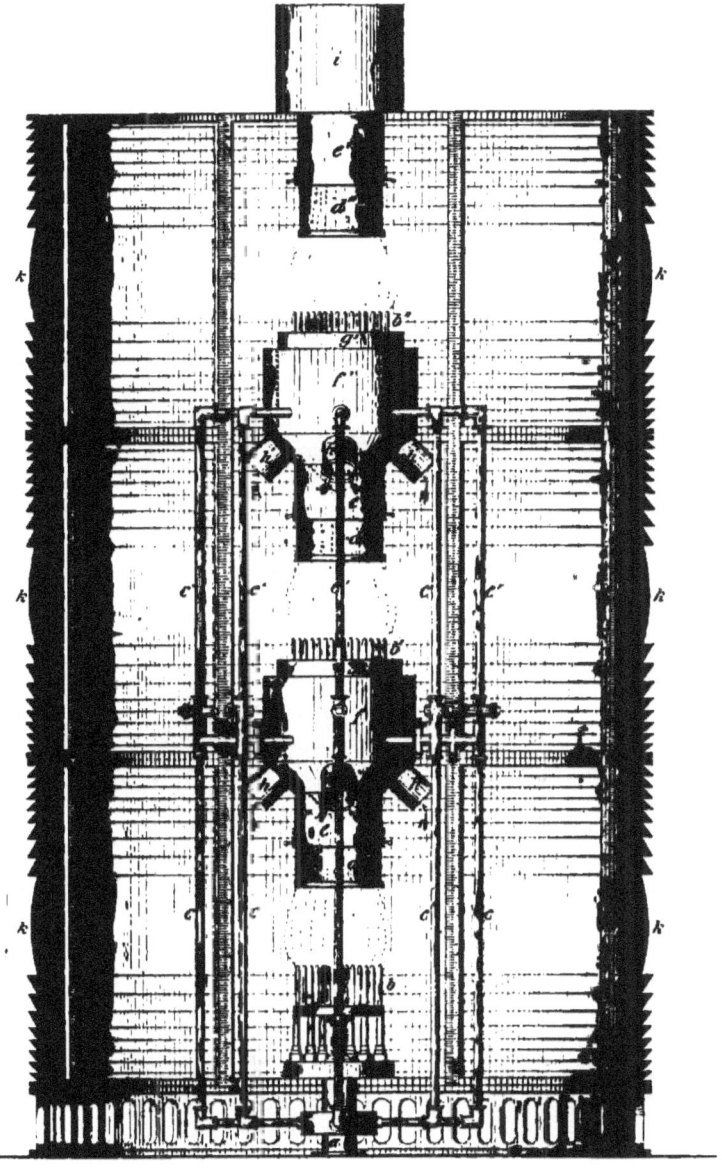

expenses of making gas and the interest on the first cost of apparatus.

Professor Tyndall states that the 28-jet burner, the lowest power of the gas-lamp, gives a light in no degree inferior to the first-order four-wick oil-lamp used in light-houses. *Opinion of Professor Tyndall as to the comparative power of gas and oil lights.*

The oil-lamp is susceptible of few variations in regard to power, (our lamps have none; the Douglass six-wick lamp used by the Trinity House has three, and the power of the light can be increased from 324 to 722 candles;) while the gas-light of the triform system can be carried through many more gradations reckoning from the lowest,* so that a skillful keeper can suit his light to any condition of the atmosphere. *Variability in regard to power not possessed by oil-lamps, except those of Mr. Douglass's recent patent.*

Assuming the facts to be as stated by Professor Tyndall, either the four-wick oil-lamp or the Wigham 28-jet gas-lamp are sufficient in clear weather to send their rays from the lanterns of sea-coast towers to the sea-horizon a distance of twenty miles. *Either the four-wick lamp or the 28-jet a sufficient illuminator for fair weather.*

As has been stated in speaking of Haisborough, no light which has been or ever can be invented can be seen through a dense fog, which obscures even the sun itself.

It will be seen, therefore, that the required improvement in sea-coast lights is that of a varying capacity of power that can be suited to all stages of the atmosphere, and the Irish gas-light certainly appears to me to meet this requirement more fully than any other known, with this additional advantage: during the eclipses of revolving and intermittent lights the consumption of the illuminant may be entirely suspended, and when, as is often the case, the total amount of eclipses is six hours or more out of twelve, the economy is evident. *Improvement to be desired, the power of increasing light in thick weather. Additional advantage of economy in flashing gas-lights.*

Professor Tyndall, in one of his reports to the Board of Trade, thus sums up his conclusions: *Conclusion of Professor Tyndall.*

"The results assure me that with gas as a source of illumination an amount of variableness and consequent distinctiveness is attainable which is not attainable with any kind of oil. It would, I think, be easy to give to every light-house supplied with gas so marked a character that a mariner on nearing the light should know with infallible certainty its name.

"As stated in a former report, I look in great part to the flexibility with which gas lends itself to the purposes of a signal-light for its future usefulness.

---

* It has been calculated that the actual number of possible gradations is 155, although in practice not more than fifteen would probably be made.

"It may be beaten in point of cheapness by the mineral-oil now coming into use, (that is to be proved;) but in point of handiness, distinctiveness, and power of variability to meet the changes of the weather, it will maintain its superiority over all oils."

WIGHAM'S GAS-GUN FOR FOG-SIGNALS.

Gas-gun for use as a fog-signal. Mr. Wigham has also invented a gas-gun, to be used as a fog-signal at stations illuminated by gas; and I had an opportunity of testing it, both at the manufactory in Dublin and at Howth Baily light-station. Captain Hawes kindly directed that the gun should be fired during our observation of the triform-light from Kingstown, so that, at a distance of six miles, I could judge of its efficiency as a signal.

Construction of the gun. The gun is simply a tube of iron connected with the gas-holder by a half-inch pipe; in fact, in these experiments the guns were nothing more than pieces of ordinary gas or water-pipe of different diameters. The charge of the gun is a mixture of oxygen, coal-gas, and common air, one-fourth of the mixture being common air and the remainder composed of equal volumes of oxygen and ordinary illuminating gas.

Charge.

Manner of filling the gun. The proper quantities of the gases are allowed to flow from their respective reservoirs into a holder, and the mixture is thence transferred to the closed end of the pipe or breech of the "gun," the flow being regulated by a stop-cock. The mixture is lighter than common air, and when it fills the feed-pipe and gun, the latter being lower than the source of supply, it will remain charged or full until fired, which may be done by touching a match to an orifice at any point of the connecting-pipe desired, taking care that communication with the holder is closed by the stop-cock.

Product of the explosion. The product of the explosion is carbonic acid gas and water, and, as the latter would rapidly fill any part of the feed-pipe which might be lower than the gun, it would probably be a fatal objection to the use of the invention which immediately suggests itself, viz, its application as a fog-signal on outlying rocks difficult to approach or nearly submerged. The defect is all the more to be regretted, as it is at precisely these points that fog-signals are most needed and the erection of other kinds is impracticable.

Reasons for not applying this signal to outlying rocks.

At Mr. Wigham's extensive works at Dublin the feed-pipe was several hundred feet long.

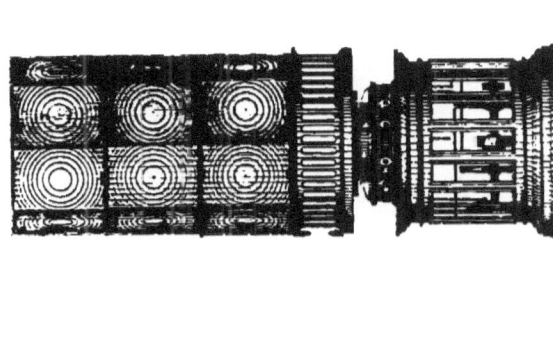

The use of the gun at any gas-light station would be extremely simple, and the keeper need not go to the gun itself, but could easily fire it from his watch-room at the required intervals. I do not know that the experiment has even been tried, but it will readily be seen that by using the electric spark the service of the gun might be made still easier, for a system of clock-work connected with a battery could be easily devised by which an electric circuit could be formed and a spark produced at any desired interval, and thus the gun be fired without any attention on the part of the keeper except what might be required to keep the apparatus in order. *Simplicity of the signal. Suggestion that the gun might be fired by electric clock-work.*

At Howth Baily the guns were twelve inches in diameter and from six to nine feet long. The latter were duplicated, and consisted of two connected pipes, fired simultaneously. Near at hand the reports seemed loud and clear, but when heard from Kingstown a high wind prevailed over Dublin Bay, and I was disappointed in the results. It is true that the distance was six miles, and a comparison with other signals would have been more satisfactory, but I fancied that the 18-pounder fog-signal gun at North Stack, on the other side of the channel, would have been more distinctly heard under the same circumstances. *Description of guns at Howth Baily. Near report loud. Superiority of the 18-pounder gun at North Stack.*

The flash from this gun is said to illuminate fog much better than that from a gunpowder-gun. *Illumination produced by the flash.*

I have no doubt of the utility of the invention for fog-signals at stations illuminated by gas, if the very great expense attending the manufacture of oxygen can be overcome; and, as Professor Tyndall is now charged by the Board of Trade with the conduct of a complete series of experiments with the gas-gun, it is to be hoped that the inventive genius of Mr. Wigham will overcome all objections to which it may now be subject.

In concluding my observations on Irish lights I must express sincere thanks to the commissioners and to Captain Hawes, the very intelligent inspector of lights, as well as to Mr. Lees, the secretary of the board, and to Mr. Wigham, for the pleasure and instruction I derived during my limited sojourn in Ireland.

SCOTTISH LIGHTS.

From Dublin I proceeded to Belfast by rail; thence by steamer to Glasgow, and by rail again to Edinburgh. Immediately on my arrival I called at the office of the Commissioners of Northern Lights, and introduced myself to Mr. *Visit to Commissioners of Northern Lights*

Alexander Cunningham, for many years the secretary of the Commission, who received me with great cordiality, and with whom I had an interesting conversation concerning the Scottish system of light-house administration, and especially in regard to the appointment, payment, and regulations affecting the keepers of the northern lights. The

Regulations. regulations are quite severe, and for any neglect of duty or other misconduct the keeper is peremptorily dismissed or otherwise punished, and a printed circular, advising keepers of the facts in the case, is at once sent to all the stations in the service. The warnings thus received tend greatly to promote the efficiency and good management of the lights.

The following extracts from the regulations of the Scottish light-house service will give an idea of the great care that is taken to promote the interests of the keepers and to secure efficient lights.

Appointment of keepers. All light-keepers are appointed by the Board, after an examination in reading, writing, and arithmetic, and a probation of three months' instruction at light-houses, (viz, six weeks in a dioptric or lens light-house, and six weeks in a catoptric or reflector light-house.)

Instruction of "expectant" keepers. While on probation the " expectant " is carefully instructed by the principal keeper of the light-house where he is assigned, cautioned as to the responsibility he is undertaking, and the *invariable rule* of the board, that if he goes to sleep at his post he cannot be admitted into the service.

He is specially instructed in the management of the lamp, cleaning the lenses and mirrors, and in taking apart and re-adjusting the various machinery. He makes the monthly returns, and keeps the books of the station while there. At the expiration of his term of instruction the principal keeper certifies to his competency, or gives reasons for not doing so. If found competent, he is appointed when a vacancy occurs.

The following are the ordinary rates of pay allowed to keepers:

Term of service.	Principal keeper.	Assistant keeper.
	Per annum.	Per annum.
Under 5 years' service	£56, ($280)	£44, ($220)
Above 5 and under 10 years' service	£58, ($290)	£46, ($230)
Above 10 years' service	£62, ($310)	£48, ($240)

Additional pay is given for rock and other detached stations, in some cases as much as £20 ($100) per year.

Each keeper has a furnished house, with annual supplies of coal and oil, and where no land is attached to the station an allowance of £10 ($50) per annum is made. They have also an allowance for washing and for expenses when traveling on public service. They are uniformed at public expense. Three pounds ($15) per annum is retained from the salaries of each and applied toward effecting an insurance on their lives. *Dwelling and land for keepers.*

Retiring pensions are allowed, and gratuities if they are constrained to quit the service before being entitled to a pension by reason of injury sustained in the discharge of duty or from other infirmity of mind or body. *Pensions.*

The ground attached to light-houses is carefully cultivated and turned to the best account, and the growing crops are transferred when one keeper relieves another.

Light-keepers at rock-stations are allowed daily rations, as follows:

One pound of butcher-meat.
One pound of bread.
Two ounces of oatmeal.
Two ounces of barley.
Two ounces of butter.
One quart of beer.
Vegetables when procurable.
For tea, sugar, salt, and other table necessaries, 4d. per day.

The light-houses are arranged in groups, and each group is supplied with a moderate amount of current literature and periodicals, which circulate in the group, remaining a specified time at each station, and afterward are bound and form part of the library of the last station. Each light-house is in turn the last of its group, so as to give each station its fair share of books. The Weekly Scotsman and the Illustrated London News are sent to each light-house. *Libraries.*

An ordained clergyman of the Church of Scotland is appointed to visit annually those remote stations where keepers and their families cannot attend divine worship. He remains about two weeks at each station conducting divine service and instructing the children at the station in ordinary branches of education as well as in their religious duties. *Attendance of clergyman.*

It is recommended by the Commissioners that each light-keeper or his wife spend some time daily teaching their children after the clergyman leaves, and when he returns the following year he examines the children as to their progress.

Medical attendants are also appointed for remote stations, *Medical attendants.*

and are allowed a fixed sum per annum, exclusive of the fees paid by the keepers.

They are to attend on the keepers, who pay them a fixed fee for each visit. Medicines and medical instructions are furnished each station.

Report or quality of stores received.
Test of oil.
Keepers are required to report annually the quality of the stores received after a trial of them in detail. A special trial is made of the oil for ten nights from December 1 to 10; the result of each night's trial is noted on a form prepared for that purpose and finally reported to the Board.

Precautions with mineral-oil.
Special precautions are taken with mineral-oil. The tanks have tight-fitting covers, and the oil is tested in the presence of the keepers to ascertain that the flashing-point is not below 120° Fahrenheit. In addition to this the keeper is required to test it before commencing to use out of a new tank.

Appointment of "occasional" light-keepers.
For each station a person resident near the light-house is appointed an "occasional" light-keeper, and is required to attend the station whenever required by the regular keepers. They are regularly trained, are under the supervision of the commissioners, and are allowed regular rates for each day's attendance at the station. They are obliged to attend the light at least twenty nights per annum in order to keep in practice.

Boatmen for island-stations.
At each island-station a boatman is appointed and paid either a fixed salary per annum or a certain rate per trip, and when he has no boat of his own, the Commissioners furnish one. He is obliged to make at least four trips to the light-house every month, and to visit it whenever signaled.

Sketch of the "board of commissioners."
The "Board of Commissioners for Northern Lights" was established in 1798. Up to that time the Trinity House exercised direct control over the Scottish lights, and it does so now in some small degree. The Commissioners receive no salary. They are all *ex-officio* members, viz, the Lord-Advocate and Solicitor-General of Scotland, the chief municipal authority (whether Lord Provost or Senior Bailie) of Edinburgh, Glasgow, Aberdeen, Inverness, Campbell-town, Dundee, Leith, and Greenock, and the sheriffs of the maritime counties of Scotland. The committees of the Board meet twice a month, but the entire executive functions are exercised by the secretary and engineers.

The latter are Messrs. David and Thomas Stevenson, whose published writings on light-houses and their illumination have not only given them a world-wide fame, but have established the reputation of the light-house system of Scotland as second to none but that of France, which is acknowledged to be the model for all others.

Both were unfortunately absent the first morning I called, and I took the opportunity of seeing somewhat of Edinburgh, which, I think, is justly called the most picturesque city of Europe.

As I found I would have sufficient time, I made a quick journey to Stirling Castle, Loch Lomond, Loch Katrine, and the Trossachs. On my return, I had the pleasure of meeting Mr. Thomas Stevenson, and had a prolonged and interesting conversation with him, gathering much information on subjects connected with the object of my visit. He showed me a reflector for light-houses, which was made after designs of his grandfather nearly a century ago. The interior reflecting-surface is composed of little facets of mirror-glass set into a paraboloidal form, and it is apparently as bright and useful to-day as when it was new, showing that such reflectors, which suffer no wearing of the surface by polishing, are very durable. In the opinion of Mr. Stevenson the silvered copper reflectors, which depend for their efficiency on the polish given them by the keepers, are really no improvement, they having no advantage over those previously used. Mr. Stevenson has invented a new form for harbor and ships' lights, which he calls the *differential reflector*, in which the vertical sections are parabolic and the horizontal elliptical; and he showed me a model. None of this kind, however, had been made for service. *Reflector made of mirror-glass. Differential reflector.*

He also showed me models illustrating the use of dioptric lights in light-ships; also his holophote, hemispherical dioptric mirror of total reflection, and holophone or sound reflector. The latter is shown in Figs. 20 and 21, and Mr. Stevenson kindly promised to send me a model of the latter as soon as the mechanics employed by the Board could find time to make it, (I have since learned that it is *en route*, having been sent according to promise,) so that from it a fog-signal reflector can be made for actual trial in our experiments. *Stevenson's sound-reflector.*

In regard to reflectors for fog-signals, Mr. Stevenson confirmed the opinion entertained by Professor Henry, that if they are of metal they should be covered with plaster or some other substance to prevent vibration; also, that wooden surfaces would be as efficient reflectors as any others. This would enable us to construct a holophone cheaply and expeditiously, if it should be desired to use one in our experiments.

Mr. Stevenson is the inventor of several important modifications in the form of dioptric apparatus for light-houses, and at the time of my visit, the Northern Lights Board occupied a large space in the industrial exhibition at Edinburgh, having an exceedingly full and interesting display of illuminating-apparatus, and of models of some of Scotland's famous light-houses, including Skerryvore and the Bell, or Inch Cape Rock. Among other interesting objects I noticed a fixed azimuthal condensing-apparatus, designed for "leading," or, as we say, " range" lights, for the river Tay. It collects the rays of the lamp and distributes them equally over an angular space of 45°, and combines for this purpose five optical agents, viz, Fresnel's fixed light-apparatus, Stevenson's condensing prisms, a half holophote, right-angled conoidal prisms, and a hemispherical mirror of totally reflecting prisms.

Apparatus for range-lights.

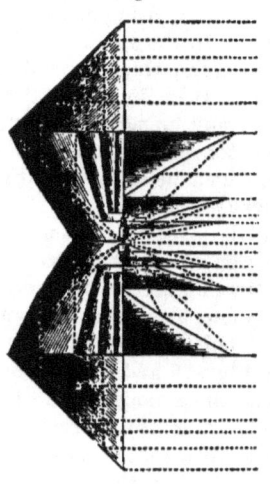

Fig. 20.   Fig. 21.

Vertical section of holophone.   Front elevation of holophone.

Diagonal sash-bars an invention of Mr.Stevenson. There was also on exhibition a model of a light-house lantern with diagonal sash-bars, the placard of which stated that the first lantern of that description was constructed by Mr. Stevenson in 1836.

Fog-signals. Mr. Stevenson remarked, in regard to fog-signals, that the Board have none operated by steam, but that bells are placed wherever they are useful. The only instance in Scotland of a steam or hot-air fog-signal is in the river Clyde, and it is maintained by a steamship company; yet the coasts of Scotland are habitually foggy at some seasons of the year, being, I infer, as unfortunate in this respect as our northern Atlantic or our Pacific coasts.

Neither have the Scottish Board any light-ships, nor indeed is there any occasion for them, the coast being everywhere abrupt with no outlying shoals or reefs. Mr. Stevenson gave as his opinion, derived from observations, that revolving lights should be used as much as practicable for light-ships, for the reason that they have a greater range of visibility than fixed lights. *No light-ships in Scotland.*

Messrs. Stevenson and Cunningham confirmed the statement made to me by Captain Doty in London, viz., that the mineral-oil lamp invented by the latter had been adopted by the Scottish Board, subject to the approval of the Board of Trade, and with the understanding that he should receive as remuneration for his patent the saving effected in one year at each light-house where it might be introduced. At the time of my visit, however, it had been actually introduced into but five light-houses, as the Board of Trade hesitated at the terms of Doty's offer, which would give him for the use of his patent about £75 ($375) for each first-order light-house, and proportional amounts for smaller lights. The Board of Trade, acting under the advice of Trinity House, directed that the substitution of mineral for colza oil in Scotch light-houses should be deferred until the experiments on the relative values of the Trinity House (Douglass) and the Doty lamps should be completed; in order, I presume, to avoid paying a royalty to Doty if the Douglass lamp should be found superior or equal to his.* *Adoption of the Doty lamp, subject to the orders of the Board of Trade.*

In this connection I quote, from a parliamentary paper in my possession, the following extracts from a report made in 1870, before the adoption of mineral-oil in British light-houses, by the Messrs. Stevenson, engineers of Scottish lights: *Extracts from the report of Messrs. Stevenson.*

"1st. The paraffine-flame produced by the four-wick mechanical lamp is 2½ inches in height, and of great purity and intensity. *Flame produced by paraffine.*

"2d. There is no difficulty, or even trouble, in maintaining the flame. *Care necessary.*

"3d. According to those in charge, the light is, on the whole, more easily attended to than that from colza-oil.

"4th. The same wicks have been used throughout the whole thirty days' experiments, and are still quite fit for use. *Wicks.*

---
* The Board of Trade have since given its authority for the substitution of mineral-oil in all of the Scottish light-houses.

Lamp-glass.	"5th. The lamp-glass used for the experiments has stood during the month without breakage.
Ventilation.	"6th. The ordinary ventilation of the light-room has been found quite sufficient.
Absence from smoke.	"7th. No inconvenience has occurred from smoking of the wick or smell of the paraffine.
No rise in temperature of the room.	"8th. No undue rise of temperature of the light-room or apparatus has occurred.
Temperature of the oil.	"9th. The temperature of the paraffine in the cistern of lamp did not, after twelve hours' burning, rise above from about 55° to 63°.
Flashing-point.	"10th. The safe vaporizing temperature, or that to which Young's paraffine may be heated without giving inflammable vapor, as tested by us with Mr. Rowat's patent instrument, is about 140°.
Quantity consumed.	"11th. The quantity of paraffine consumed in the first-order light was at the rate of 718 gallons per annum. The consumption of colza-oil is about 800 gallons per annum.

* * * * * * *

Cost.	"Taking the cost of colza-oil at 34s. per cwt., (2s. 9d. a gallon, 68¾ cents,) which was the price in 1869, adopted in our recent reports on illumination by gas, and paraffine at its present price of 1s. 4d. (33⅓ cents) per gallon, we find that the cost of maintaining a first-class light with colza and paraffine will be £110 ($550) and £47 17s. 4d. ($239.33⅓) respectively, thus giving a yearly saving on each first-class light of £62 2s. 8d., ($310.66⅔;) but if we take the present contract rate of colza of 38s. 6d. per cwt., (3s. 1d. per gallon, 77 cents,) the saving would amount to £75 9s. 4d., ($377.33⅓.) On the supposition of paraffine being used for all the lights under the charge of the commissioners, the
Annual saving.	saving, calculated on the same basis, would amount to about £2,874 ($14,370) per annum, but at the present contract rate of colza the saving would amount to £3,478 15s. 7d. ($17,393.89½) per annum.

* * * * * * *

"We have perfect confidence in recommending the use of paraffine for light-house illumination. Its introduction would require to be done gradually; the light-keepers would require to receive some instructions in its use, and a slight alteration would in each case require to be made on the level of the burner with reference to the optical axis of the apparatus, and the marks for testing the adjustment of the lamp to be carefully altered. A full set of directions would also require to be drawn up and furnished to all the stations when the change is made."

No more convincing proof of the utility of permanence in the peculiar service of light-house administration can be given than the excellent reputation the Scottish lights bear throughout the world for economy and efficiency, and it is well known that the Commissioners are eager to adopt any improvement which tends to the increase of either. Mr. Cunningham has for many years most ably filled the position of secretary, and for nearly a hundred years the Stevenson family has supplied engineers.

<small>Good results of permanence in light-house service.</small>

The time at my disposal was too limited to allow me to visit any of the Scottish light-houses, and I especially regretted that I could not accept Mr. Stevenson's invitation to visit Bell-Rock light-house. My thanks are due both to Mr. Thomas Stevenson and Mr. Cunningham for their polite attentions while I was at Edinburgh.

THE MANUFACTORY OF DIOPTRIC APPARATUS FOR LIGHT-HOUSES OF CHANCE BROTHERS AND COMPANY, NEAR BIRMINGHAM.

On my return to London from Edinburgh I visited the extensive glass-works of Chance Brothers & Company at Spon Lane, near Birmingham, in compliance with an invitation which I received from Mr. J. T. Chance soon after arriving in England. This establishment is most extensive, and is mainly devoted to the manufacture of plate-glass, which is sent from here to all parts of the world.

A part of the works is, however, devoted exclusively to the manufacture of apparatus for light-houses, a manufacture commenced by this firm in 1855, in competition with the lens-makers of Paris, who until that date monopolized this branch of industry.

Mr. Chance stated that in establishing this part of their trade they had lost more than $100,000, but that their reputation is now established, and they supply not only Great Britain, but many other countries, with lenses, lanterns, lamps, and accessories of all kinds necessary for the service of lights.

<small>Cost of establishing the manufacture of light-house apparatus.</small>

They have in use a great variety of machines for grinding and polishing the prisms, and the establishment appears to be as complete in every particular as any which I saw at Paris.

<small>Machines for polishing prisms.</small>

The scientific branch of this industry is in charge of Dr. Hopkinson, who is responsible not only for the correctness of the forms of the various parts of every optical apparatus, but for their correct assembling, and he personally tests each lens before it leaves the manufactory.

<small>Dr. Hopkinson.</small>

Flashing-lens for Start Point. Among other works in hand, I saw a revolving lens of the first order for the light-house at Start Point, on the coast of Devonshire, remarkable for having flash-panels that cover arcs of 60°, which is larger than any before attempted, as far as I am informed. Heretofore the arcs of first-order lenses have not extended 45°, and as the amount of light in the flashes is nearly in proportion to the size of the panels, it follows that the power of this lens, when compared with those of similar character heretofore made, is nearly as 3 to 2.

Red cut to be produced. A red cut showing the position of outlying rocks near Start Point, will be produced, as is done at the electric lights at Souter Point, by collecting a portion of the rear light and throwing it down a tube to a lower light-room upon a set of totally reflecting prisms, which in turn bends the beam and turns it out upon the sea.

Apparatus for Longships light-house. A new first-order dioptric apparatus for Longships light-house, on the coast of Cornwall, was also in progress, and I had the pleasure of witnessing Dr. Hopkinson's final test of the accuracy of this lens.

Dr. Hopkinson's photometer. The doctor also kindly presented me with a photometer, of his own invention, for comparison of lights at a distance, which is designed to be free from the defects inherent in those depending on absorption.

It is very compact, and consists of two Nicol prisms, which can be moved relatively to each other in azimuth. A little tube carries both the analyzing prism and a second tube containing the polarizing prism. The latter being turned till the light, which is viewed through the axis of both prisms, is eclipsed, the angle through which the polarizer is moved, is read. The other light being then observed in the same way, a comparison of the angles gives the relation of the powers of the lights.

For many years the place now occupied by Dr. Hopkinson was filled by Mr. J. T. Chance himself, and to him the service of light-house illumination is indebted for several treatises on the subject.

THE LIGHT-HOUSES OF FRANCE.

THE COMMISSION DES PHARES.

I arrived at Paris on Saturday, the 5th of July, and on the 7th I went to the offices of the *Commission des Phares*, or Light-House Board of France, situated on the hill Trocadéro, which overlooks the Seine and the Champ de Mars. This **French Light-House Board.** Board is composed as follows: Four engineers, two naval officers, one member of the institute, one inspector-general of marine engineers, one hydrographic engineer.

The executive officers of the establishment at Paris are M. Léonce Reynaud, Inspector-General of the Corps of Engineers *des Ponts et Chaussées*, who is Director of the French light-house administration, and M. Allard, engineer of the same corps, and Engineer-in-Chief and Secretary to the Commission.

Light-house administration.

The entire administration on the seaboard is in the hands of the engineers, who, in addition to other duties, are charged with the work of river and harbor improvements.

At the time of my visit to the *Dépôt des Phares*, M. Reynaud, who is, I believe, much occupied with other duties, especially at the *École des Ponts et Chaussées*, was absent, but I had the pleasure of meeting M. Allard, upon whom the major part of the executive duties devolves.

During my long interview with M. Allard, he kindly showed me the different parts of the establishment at Trocadéro, all of much interest, particularly the grand hall or council-chamber of the Commission, the museum, the experimental rooms, and store-rooms.

Dépôt des phares.

The buildings are placed around a rectangular court-yard in which are models of light-houses, buoys, &c.

Tower for experiments.

The principal building, which contains the offices of the commission, is a handsome structure 150 feet long and two stories in height, built of brick and limestone in alternate courses. It is surmounted by a tower and a first-order lantern, where experiments are made, and from which a magneto-electric light is exhibited on occasions of public display; as, for example, the *fête nocturne* in the Champ de Mars, in honor of the Shah of Persia. This fête M. Allard was kind enough to invite me to witness from the *Dépôt des Phares*, which afforded a most desirable site from which to view the magnificent spectacle.

The grand entrance hall also contains many models, and those of the rock light-houses of France at once arrest the attention of the visitor.

Council-chamber.

The council-chamber of the commission is richly decorated, and upon its walls are painted two large charts, each occupying an entire side, one of the world, on Mercator's projection, showing all the lights now in existence, thus, as M. Allard happily observed, marking the progress of civilization in a most striking manner. The other chart was one of France, showing not only its lights, but the illuminated areas.

A bust of Augustin Fresnel, engineer of *Ponts et Chaussées*, the first Secretary of the Commission, and the inventor of the system of lights which now illuminates the coasts of most countries of the globe, occupies a prominent place, not only here, but, as I found afterward, at all French light-stations, similar ones being placed over the entrance-doors at each station I visited. The museum contains all kinds of illuminating apparatus, both dioptric and catoptric, though the latter is not now used in any French light-house.

<small>Museum.</small>

The collection of dioptric apparatus embraced many articles of historical interest, among which I was shown the first lens made from the designs of Augustin Fresnel, which was placed in the Tour de Cordouan, and various apparatus showing the successive steps by which he arrived at the lens which is now used in all parts of the world.

<small>Hall.</small>

The hall also contains models of buoys and beacons. French bell-buoys have four hammers or clappers, and at the top of the frame are placed small mirrors to catch the eye of the mariner.

<small>System of coloring buoys.</small>

A uniform system is used for the coloring of all the buoys and beacons of the coast of France. All of those marks which should be left by the navigator on the starboard hand when approaching from seaward are painted red;* those which should be kept on the port side are painted black; those which may be left indifferently on either side are painted with horizontal stripes, alternately red and black.

<small>Beacons.</small>

Beacons are colored in this way only above high-water mark; below that level they are painted white.

The red and black are varied as circumstances may require by painting in white, design of checks, vertical bands, &c.

<small>Color of buoys marking anchors.</small>

Buoys marking anchors, &c., are painted white. On each buoy or beacon is painted either the entire or abridged name of the bank or rock that it marks, and buoys and beacons belonging to the same passage are numbered serially, commencing to the seaward. The even numbers are given to buoys and beacons which the navigator leaves on the starboard hand, that is, to the red ones, and the odd numbers are given to the black buoys and beacons. The buoys and beacons painted with red and black stripes bear names, but no numbers.

<small>Numbering of buoys.</small>

---

* The red is soon tarnished by the sea-water, and in order to prevent any error it has been decided that those marks which are painted red shall hereafter have a white crown a little below their summit.

Small heads of rocks in frequented passes may be painted the same as the beacons, except that only the most prominent part is colored when they show a surface larger than is necessary for clear distinction.

No oils are kept at the depot, nor are they tested there, as they are sent by the contractors directly to the light-houses and there thoroughly tested by the engineer of the district. *Oils sent directly to the light-houses.*
Lenses and lamps, however, undergo a thorough trial at the depot. The photometer used is different from Bunsen's, used by us, and instead of the standard light and the one under test being both fixed in position, the former or unit is moved until the beams, passing through a slit or opening in the photometer and falling upon a pane of glass which has a ground surface or is covered with a sheet of paper, are, when viewed on the reverse side, equal in intensity. The distances from the photometer are then measured by a tape-line, and reference to a calculated table shows at once the intensity of the light under test in terms of the standard or unit. This unit in France, both in the practice of the officers of the light-house administration and that of the lens-manufacturers, is always the Carcel burner, consuming 40 grams (61.728 grains) of colza-oil per hour. *Test of lenses and lamps. Photometers used.*

The relation between the French and English unit is not accurately known. I have been informed by Mr. Douglass that the French unit is estimated by the French and English gas-engineers as equal to 9.6 candles. *Relation of English and French units of light.*

M. Lepaute informed me, however, that it has heretofore been considered equal to 11½ candles. M. Lepaute has recently been desired by the French government to ascertain the exact relative value. M. Allard stated that the French engineers prefer the Carcel to the candle unit used in England and the United States on account of greater variability in power to which the latter is subject in consequence of irregularities in the wick.

I found this photometer to be easily operated, and the bringing into contact upon the pane, the two images of the slit, seemed to me to show with more precision when the desired equality of intensities of beams were arrived at than Bunsen's apparatus.

M. Allard showed me the mode adopted by the Commission for testing mineral-oil in apparatus especially designed for the purpose, and shown in Figs. 22 and 23.

A box containing these instruments is supplied to every light where mineral-oil is used. The qualities to be tested are the specific gravity and the flashing-point. The specific gravity is tested by an areometer, (see Fig. 22.)

Test for specific gravity. The test-glass is filled with the oil to be tested to within three-quarters of an inch of the edge; then the areometer is plunged in and the specific gravity is read from it. The standard required is between .810 and .820 at 15° centigrade, (59° Fahrenheit.) This temperature is obtained in winter by heating the oil in a water-bath, and in summer by cooling the vessel containing it by means of fresh water. If these methods are not available the specific gravity and temperature of the oil are taken, and a correction of .00074 is added for every degree below 15° centigrade, or deducted for every degree above.

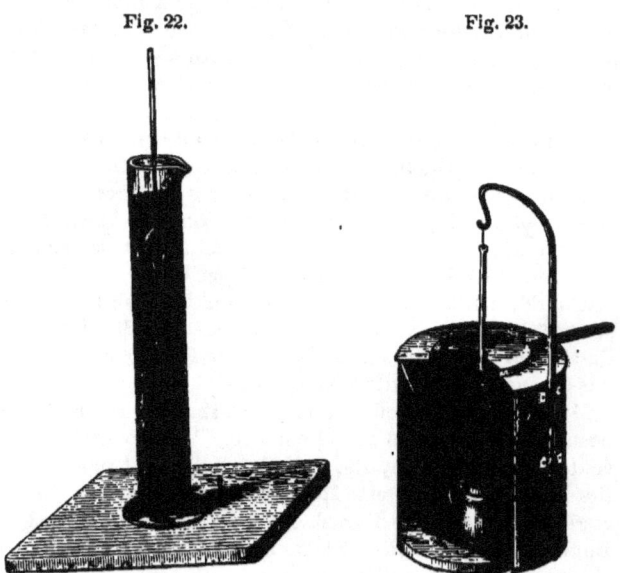

Fig. 22. Areometer.   Fig. 23. Apparatus for testing the flashing-point.

Test for flashing-point. To test the flashing-point the instrument shown in Fig. 23 is used. The oil to be tested is poured into a little copper capsule placed in a water-bath heated by a spirit-lamp; a thermometer so suspended as to dip into the oil gives the temperature. While heating, matches are applied at short intervals near the surface of the liquid. The degree of the thermometer at which ignition takes place is the flashing-point. It is considered that an oil is unsafe for illuminating purposes if it gives out inflammable vapors below 50° centigrade, (122° Fahrenheit,) and the contracts require that the oil actually used shall not have a flashing-point lower than 60° centigrade, (140° Fahrenheit.)

The principal keepers of the lights are required to test *Oils tested by the keepers.* the oil when delivered by the contractors. They heat it in the capsule until the thermometer shows more than 60° centigrade, then extinguish the lamp, and while the mercury slowly descends apply the match.

Before the introduction of mineral-oil the colza-oil for- *Colza-oil tests* merly used was required to be of the first quality, perfectly purified and clarified. The oils delivered by the contractor were first tested by drawing a certain quantity from each cask and burning it in two or three night-lamps with floats. The oil was refused if the lamps, properly arranged, went out of themselves before burning twelve hours. If they did not burn twenty hours the oil was considered, if not bad, at least of doubtful quality. In this case, or any other where there was doubt, the decisive test was to compare the oil *Decisive test of colza-oil.* with colza known to be of the first quality. The oil to be tested was not considered acceptable if, during combustion of sixteen hours, it did not give as good a flame, or if it deposited more carbon on the wick than did the standard oil, other circumstances being equal. In this comparative test either ordinary constant-level lamps or the regular light-house lamps were used.

In regard to lamps, M. Allard stated that the French pre- *Lamps preferred.* fer the mechanical lamp, although the "moderator" is used in some of the light-houses, and the constant-level lamp in others, but in no case is any part of the lamp allowed to obstruct any portion of the light when the entire horizon is required to be illuminated. For small orders (below the fourth) lamps in which the oil is drawn up by the capillary attraction of the wick are used in most cases.

The French are experimenting with a six-wick burner, but in M. Allard's opinion the heat generated would be found to be too intense for the safety of chimneys. As I have stated, the English are using the six-wick burner in the Douglass "lamp of single and double power" for the purpose of producing a powerful flame in thick whether, but, as far as I am informed, the French have not manufactured any lamps with more than five wicks.

M. Allard confirmed Mr. Douglass's opinion concerning *Care to be exercised in purchase of wicks.* the great care which should be exercised in regard to the wicks used in light-house lamps, and stated that in the French service very great precaution is adopted in procuring and inspecting them.

After my visit to the *Dépôt des Phares* I visited the lighthouses at the mouth of the Seine, and on my return to Paris I found the following note from M. Reynaud:

"COMMISSION DES PHARES,
"*Paris, July* 19, 1873.

"M. Reynaud much regrets not having been at the *Dépôt des Phares* when Major Elliot did him the honor to call.

"If Major Elliot wishes to see M. Reynaud on business, he will find him at the *Dépôt des Phares, Avenue de l'Empéreur*, corner of *la place Troeadéro*, to-morrow, Wednesday, from 2 to 4 o'clock. His approaching departure obliges M. Reynaud to indicate an hour instead of asking Major Elliot to name one that would be convenient for him. He hopes to be excused, and presents Major Elliot his best compliments.

"L. REYNAUD."

On my arrival at the *dépôt* at the time appointed, M. Reynaud received me with cordiality and expressed his desire to assist me in the object of my mission by any means in his power.

Question of illuminants and lamps. Our conversation turned mostly upon the subjects of illuminants and burners for light-houses, they being the questions now of greatest interest to the light-house engineers of Europe, and I found the French fully as much interested in this subject as are the English.

Order of French government to substitute mineral for colza oil. M. Reynaud stated that the Commission des Phares had recently given an order for the substitution of mineral for colza oil in all of the French light-houses, and their lamps were being changed for this purpose as fast as possible; that no difficulty whatever is found in its use; that it is perfectly safe when inspected, and received after proper tests; and that, while the consumption by the larger orders of lamps is about the same, it is, for the smaller orders, very much less than the consumption of colza,* while its cost per gallon is about one half, and the light produced is superior and the lamp more easily managed, a change of wicks being required only at long intervals, the full power without trimming being kept up from sunset to sunrise.

Burner in use. The burner, or *bec* as it is called by the French, now in use, is denominated the *bec de l'Administration*. It combines the peculiarities of the Doty and the Fresnel burners, with modifications made by the Commission, especially in regard to the overflow. By an ingeniously devised screw the overflow is regulated to suit the temperatures of different climates, which is highly important.

---

* It will be found by reference to my account of the improvements in lamp-burners made by Mr. Douglass, engineer of Trinity House, that in *consumption* and *illuminating power*, colza has been brought to an equality with mineral oil.

In regard to the claims of Captain Doty, which have caused much controversy in England, M. Reynaud said that Captain Doty presented his burner without conditions to the French government, and it had paid him a certain amount, (not stated to me,) not as payment for an *invention*, but rather as a reward for calling attention to the economy and other advantages of mineral oil for use in light-houses, and pressing its claims for adoption over the vegetable and animal oils. I afterward ascertained that the French government paid to Captain Doty an indemnity of 10,000 francs, ($2,000,) and purchased his burners at 44 francs ($8.80) each, being an excess of 19 francs ($3.80) over the cost of the burners formerly used. Question of claim of Captain Doty as inventor.
Indemnity paid Captain Doty by the French government.

In this connection the following recent letter from the Minister of Public Works of France to the British Embassador in Paris will be found of interest:

"VERSAILLES, *August* 12, 1873.

"M. L'AMBASSADEUR:

"In acknowledging the reception on the 16th of June last of various documents relative to the illumination of light-houses with mineral-oil, your Excellency has done me the honor to express to me in the name of the Board of Trade the desire of learning whether, since the publication of the note of M. Reynaud, Director of the Light-house Service, he had not arrived at some new facts in support of the observations it contained on the respective merits of the Doty and Fresnel lamps.

"Your Excellency adds that the Board of Trade would equally appreciate the reception from M. Reynaud of the fullest information relative to the merits and price of the Fresnel lamp manufactured by M. Henry Lepaute.

"The Minister of Public Works calls to mind that the note of the Director of the Light-house Service indicates that two kinds of burners were in use, giving the same results, viz, the Doty burner and the modified Fresnel burner. Now, it appears from recent information furnished by M. Reynaud that, in consequence of divers changes in details which the experiments made at the *Dépôt des Phares* had induced the engineers to adopt in each of the systems, it may now be considered that the two kinds of burners differ but little more than in name. The light-house service includes them in the same designation of mineral-oil burners.

"These burners consume the same quantity of oil, have the same luminous intensity, and offer the same guarantee of regular action.

"Their cost is the same to the French government, viz:
"Burners with 5 wicks, 60 francs.
"Burners with 4 wicks, 50 francs.
"Burners with 3 wicks, 40 francs.
"Burners with 2 wicks, 32 francs.

* * * * * * *

"Accept, &c.,
"BROGLIE.
"His Excellency LORD LYONS, *G. C. B.*"

Scotch mineral oil used. The mineral-oil used in the French service is the Scotch, which is found to be safe and quite uniform in character; and so anxious have the Scotch manufacturers been to supply the French lights, and thus gain a reputation for their oil, that they not only furnish a better article but at cheaper rates than the French refiners.

Lens-manufacturers. In regard to the three lens-makers of Paris, M. Reynaud stated that all were good and conscientious firms, and the *Commission des Phares* made no distinction between them in giving its orders for optical apparatus.

Designs furnished by Commission des Phares. Designs are furnished by the Commission, and a scale of prices is established by it; these established prices are liberal, the commission having in view the object of procuring good material and workmanship.

Competition in price likely to result in inferior quality of apparatus. M. Reynaud thought that competition in regard to prices would induce a reduction in regard to quality, which could not be thought of when the desire is to utilize for the benefit of mariners every ray of light with the least possible loss of the power with which it issues from the flame of the light-house lamp.

As before stated, all lenses made for the Commission are thoroughly tested at the *Dépôt des Phares* before being sent to the district engineers for placing in the light-houses, and to the interest felt by the government of France in lens-manufacture, first commenced, after the invention of Fresnel, in the city of Paris, is due the kind offer made to me by *Offer of M. Reynaud to test lenses ordered by United States.* M. Reynaud, to apply the photometric and other tests to any lenses which might thereafter be ordered from Paris by the American Government.

Lenses for electric lights. Referring to the electric lights, of which France has four, M. Reynaud remarked that the smaller lenses are better than the larger ones, and that diagonal bars should be used for the panels; as for the lantern, the glass can readily be cast in one piece and the use of sash-bars thus avoided, which is highly desirable for these lights.

Lanterns for oil-lights. In speaking of the construction of lanterns for oil-lights, he gave as his opinion that vertical sash-bars do not materially interfere with the beam, thus differing from the well-known opinions and practice of the English and Scotch engineers.

M. Reynaud called my attention to his *Mémoire sur l'Éclairage des Côtes de France* as containing much information on the subject of the administration of the French light-house service; also to his paper on the *Application de l'Huile Minérale à l'Éclairage des Phares*, (1873,) a translation of which will be found below. I am much indebted to M. Reynaud and to M. Allard for the facilities afforded me for gaining information while at the *Dépôt des Phares*, and the former was good enough to say that he would be glad at all times to furnish the Light-House Board of the United States with any information concerning the improvement of French lights, or on any other points that might be desired, and expressed a hope that our Board would keep the French Board advised of its progress in the science of light-house illumination.

[Translation.]

APPLICATION OF MINERAL-OIL TO LIGHT-HOUSE ILLUMINATION.

By M. Léonce Reynaud,

Inspector-General of Engineers (des Ponts et Chaussées) and Director of the French Light-house Service.

The question whether mineral-oil can be practically used for light-house illumination has been studied for several years and finally answered. A Ministerial decision of March 29, 1873, made in conformity to a recommendation of the Light-House Commission, ordered that the new combustible be substituted for colza-oil in all the lights of the coast, except the floating-lights, for which experiments have not yet shown it satisfactory. *Order of the French government for use of mineral-oil. Floating-lights excepted.*

The object of this article is to show how this measure came to be adopted, and to examine its merits, both in respect to maritime interests and to economy.

In 1856 the engineers of the central light-house service commenced a series of experiments with oil extracted from bituminous schists of the Department of Allier, and soon saw that the Maris lamp (so called after the name of its constructor) was better than all others then in use, the flame being more intense and requiring less attention. This lamp, still in use, has a single cylindrical wick. A metallic disk throws the interior air current on the flame; below the burner is a rather large cistern, into which the wick de- *Experiments with schist-oils. Superiority of the Maris lamp for schist-oil.*

scends, and the oil rises by capillary attraction, without the aid of any mechanism whatever. The flame is more extended than in colza-oil lamps, being broader and shorter, but with lens-apparatus this is an advantage rather than otherwise, especially when the catadioptric rings have been calculated to correspond. In reflector-apparatus, however, this lamp cannot be used, on account of the position and size of the cistern, this apparatus admitting only the use of constant-level lamps, the level a few centimeters below the crown of the burner.

Reflector-apparatus require constant-level lamp.

The intensity of small lens-apparatus lighted by a Maris lamp burning mineral-oil, is nearly double that obtained with an equal consumption of colza-oil, and as the former is much cheaper, it was evidently best to continue the trials, and, if prudent, to adopt it.

Intensity of Maris lamp burning mineral-oil.

In 1857 and 1858 mineral-oil was used for a few harbor-lights. Mariners were well satisfied with them, and the keepers themselves, who at first had shown some hesitation, came to acknowledge that the flames kept better than formerly, without requiring as much attention. But some accidents occurred, which, although not of a nature to cause an abandonment of the new system, showed that great caution was necessary; an ill-regulated flame smoked so badly that the light was extinguished shortly after lighting; a cistern exploded; at another place the chimney and the upper rings of the apparatus were broken.

Mineral-oil first used in 1857.

Accidents.

Experiments were then continued at the light-house depot, both with the different mineral-oils, and in regard to the measures to be taken for the desired security. At the same time the use was extended on our coasts, and soon every serious reason for hesitation disappeared, at least as far as regarded its use in single-wick lamps. At the end of 1864 the new combustible was used in forty-one harbor-lights, and the following year it was ordered to be exclusively used in all fourth-order apparatus, i. e., in all single-wick lamps.

Continuation of experiments.

But the numerous attempts made by engineers and manufacturers to use mineral-oil in lamps of the superior orders, (which have, as is well known, several concentric wicks) did not succeed; the combustion was incomplete, brilliancy feeble, smoke sometimes abundant.

While we were still experimenting, Captain Doty, an American, succeeded in solving the problem which, until then, had baffled all. In 1868 Mr. Doty brought forward a lamp for burning mineral-oil, having four concentric wicks, which was tested according to the method adopted at the

Doty lamp-burner.

light-house depot. This lamp, unlike the one with a single wick, did not give greater intensity than that obtained from lamps of the same order consuming colza-oil; but the consumption was less, and the combustible being cheaper, a considerable saving to the treasury would result from its use. Besides, navigation would gain somewhat, for although the intensity immediately after lighting was the same as that of the old lamps, it was kept up longer, and did not decrease as much toward morning.

On recommendation of the Light-house Commission, it was decided that a practical trial of this system should be made. The mouth of the Canche is marked by two first-order fixed lights, placed on a line parallel with the shore, about 820 feet apart; one of them was lighted with mineral-oil from December 12, 1868, to January 26, 1869. The subsequent investigation fully corroborated the conclusions derived from the Paris experiments. The two lights were apparently of the same intensity when first lighted; the consumption of the mineral-oil was about 17 per cent. less than that of colza; its flame was easier to manage and kept better. At first the odor was somewhat offensive to the keepers, but they became accustomed to it and did not feel the least inconvenience. *Trial of mineral-oil in light-houses at mouth of the Canche.*

Meanwhile Captain Doty had some two and three wick lamps made, which, without increasing the luminous intensity in as great a proportion as the single-wick lamps, showed themselves superior in this respect to similar ones burning colza-oil. These results were encouraging, but the fear of explosion still remained, and caused all the more uneasiness as a report kindly furnished by the Chairman of the Light-house Board of the United States of America, who had been consulted on this subject, stated that in that country, where mineral-oil is abundant, it had not been thought advisable to use it in light-houses on account of its inflammability, the dangers of which were but too well attested by numerous accidents. *Increase of intensity by use of three-wick lamps.*

But a new kind of mineral-oil, extracted from bog-head[*] and called Scotch paraffine, had been successfully used at the light-house depot, was found to be much less inflammable than the schist-oil used till then, and was used in one of the light-houses of the Canche. Consequently the engineers saw that, before proceeding farther, it would be necessary to test the comparative merits of the different oils in market, excluding, however, American petroleum, the composition and properties of which are too *Scotch paraffine.*

---

[*] Local name for cannel coal.—G. H. E.

variable. They ascertained the qualities as to luminous intensity and consumption, and Chief-Engineer Mangon had the kindness to undertake to determine, with his usual accuracy, the flashing-points and other properties useful to know.

The following table gives the figures obtained:

Table of comparative values of oils.

Designation of oils.	No. of sample.	Luminous intensity produced by the consumption of 40 grams.*	Consumption per carcel unit per hour.	Flashing-point.	Boiling-point.	Specific gravity at 32° Fahr.	Co-efficient of expansion from 32° to 212° Fahr.
		Carcel units.	Amt. in grams.	Fahr. °	Fahr. °		
Achaume, at Autun	1	1.56	25.7	77.0	284.0	0.827	0.084
J. Barse, Mineral Illumination Company of Allier, at Buxière la Grue	2	1.82	21.9	78.8	287.6	0.818	0.087
Graillot Brothers, at Autun	3	1.63	24.6	84.2	300.2	0.819	0.087
Rondeleux, mines of Coudamine, at Buxière la Grue	4	1.59	25.2	84.2	303.6	0.833	0.083
Hubinet of Soubise, at Autun	5	1.47	27.2	98.6	312.8	0.830	0.099
E. Gontier et Cie, at Autun	6	1.78	22.4	105.8	320.0	0.895	0.080
Civil Mining Company of L'Antunois	7	1.55	25.8	107.6	327.2	0.831	0.093
Anonymous Company of the Oils of Colombes	8	1.60	25.1	114.8	334.4	0.834	0.090
Roche et Cie, of Igonay, Saône et Loire	9	1.45	27.6	120.2	345.2	0.834	0.095
Young's Paraffine Light and Mineral Oil Company, Scotland	10	2.18	18.3	161.6	401.0	0.833	0.094

* It will be remembered that the unit of light adopted by the Light-house Commission is the Carcel burner, 0m.020 (0.79 inch) in diameter, consuming 40 grams of colza-oil per hour.

The superiority of the Scotch oil was thus very evident in essential respects, as it proved to be at the same time less inflammable and of greater illuminating power than the others, and lessened, if it did not completely obviate, the chances of accident. Still the Light-house Commission, wishing to be as prudent as the gravity of the interests confided to its care required, though not more disposed than in the past to repel improvements, confined itself to proposing at its session of June 26, 1869, to apply the new combustible to two-wick lamps of third-order lights, and also to such three-wick lamps as might be used where the coloration of the light required a greater intensity of the luminous focus, or where it might be thought best to increase the diameter of the burner, in order to obtain greater divergence. It was recommended, moreover, to proceed to new studies of instruments and receptacles having especially in view the prevention of accidents.

The contract for furnishing mineral-oil of French origin, made with the firm of Jules Barse, expired at the end of the year. It was not renewed, and another was made, to run for three years from January 1, 1870, with the company established in Scotland under the name of "Young's Paraffine Light and Mineral-Oil Company." It was stipulated in the conditions that the specific gravity should be between 0.81 and 0.82 at 59° Fahrenheit, and that no inflammable vapors should be produced at a temperature below 140° Fahrenheit. This is much lower than the 161.6° Fahrenheit found by M. Mangon; it was adopted so as not to create too many difficulties for the engineers as well as for the company, after numerous trials had shown that it might be considered sufficiently low as a limit and still sufficiently high to give entire security. The price of the oil had been fixed at 0.85 francs per kilogram, [about 50 cents per wine-gallon,] delivered at the light-houses, and is the same in the new contract made with the same company, to date from the 1st of January last, [1873,] all customs duties affecting the product to be refunded. The memoir on the illumination and buoyage of the coasts of France, published in 1864, states that colza-oil, the cost of which varies with the market, averages 1.51 francs per kilogram, [about $1 per wine-gallon,]* delivered at the light-houses. It will be seen that the difference is great. ^{Contract for Scotch oil.}

Mineral-oil is used in all third and fourth order lights established since the commencement of 1870, in conformity with a recommendation of the Light-house Commission, made mandatory by a Ministerial decision. We now use it in one hundred and sixteen light-houses, six of which are of the third order. There is even one light of the second order, viz, that of Pilier, which, as a considerable part of the arc of illumination had to be colored red, has been furnished with a five-wick lamp, as will be explained hereafter. Thirty-nine third-order lights yet burn colza-oil, and considering that the advantages of change are well established, it may be thought that its progress has been exceedingly slow. But besides the trouble caused by the misfortunes of the country, there were several reasons for caution on the part of the engineers of the central service. ^{Mineral-oil in use in 116 light-houses.}

Sensible of their responsibility, they asked themselves whether the new combustible would not show in actual practice some disadvantages not brought out by the Paris experiments, and whether it would retain the valuable qualities which caused it to be preferred to all other oils, and so recommended by the Commission. Besides, it was admitted by all, even by the contractors themselves, that special receptacles were necessary for mineral-oil, and hence there ^{Questions of expense, &c., resulting from a change of illuminants.}

---

* Taking the specific gravity of colza as 0.914.

4

would result expenditures, which the smallness of our appropriations would require to be several times recommenced.

No accidents. None of these reasons for delay now exist. During the three years that the Scotch paraffine has been used in the light-house service, not an accident has occurred, not a disad-

Satisfactory results. vantage has been found. Mariners, engineers on the coasts, even the keepers, all show themselves entirely satisfied. The article produced by the Scotch company has lost none of its qualities. Finally, experiments made at the light-house depot for the last two years have shown that the

Colza-oil tanks suitable for paraffine. large colza-oil tanks lined with tin, now used in our light-houses, are equally suitable for holding mineral-oil. It has even been proved that this oil can be preserved for three years, at least, in well-secured tin cans, without losing any of its qualities. In zinc, however, it does not do as well, although it does not become unsuitable for use.

Economy. A fact already mentioned, has again been placed beyond question, and it is of capital importance: for an equal intensity less mineral than colza oil is consumed, whatever may be the number of wicks, as will be seen by the following table, of which the figures may be considered as the maxima of both kinds of oil:

No. of wicks	Colza-oil.			Mineral-oil.		
	Consumption per hour.		Luminous intensity.	Consumption per hour.		Luminous intensity.
	In grams.	In fluid oz.*	Carcel units.	In grams.	In fluid oz.*	Carcel units.
1	60	2.25	1.6	55	2.29	2.2
2	175	6.55	5.0	160	6.65	6.4
3	500	18.73	15.0	360	14.97	14.0
4	760	28.46	23.0	630	26.20	23.0

* 128 to the wine-gallon.

In looking over this table it may appear anomalous that while the mineral-oil produces a greater intensity than colza in one and two wick lamps, and the same in four-wick lamps, it produces less in three-wick lamps, (fourteen in-

Change in diameters of wicks. stead of fifteen burners.) This results from the fact that, for the sake of uniformity, it was thought best to adopt in the new lamps, the same diameters for all the wicks of the same rank, which had not been done in the successive establishment of the different types of colza-oil lamps. Wick No. 3, which fixes the diameter of three-wick lamps, is 2.71 inches in diameter in colza-oil burners, and only 2.55 inches in those for mineral-oil. It may also be remarked that the saving of oil is comparatively greater in

Question of change in numbers of wicks. lamps of this order, and by somewhat enlarging the wicks both consumption and luminous intensity would be increased.

These facts established, it seemed that the proper moment had come to order the substitution of mineral for colza oil in all the light-houses; and at the same time the question occurred whether it would be advisable to allow (as had been done before in the case of third-order lights) lamps with more than the regulation number of wicks to be used in first and second order light-houses, where greater intensity appeared desirable.

This measure, as before stated, had already been adopted in the exceptional case of the Pilier light. As the Light-house Commission proposed to color this light red, in a certain angular space, to warn mariners from the shoal of Les Bœufs, and did not wish to lessen its range, it was necessary to increase the intensity of the focus; so it was thought best to use mineral-oil with five instead of three concentric wicks. This experiment was crowned with entire success.

Coloration of Pilier light.

The engineers of the central service hesitated much before deciding in favor of a change, and their hesitation increased when they perceived the consequences of innovation. While inclined to propose an increase in the number of wicks, they were prevented by a consideration most powerful with them, the profound respect which they entertain for the memory of their illustrious predecessor. Augustin Fresnel, after many experiments and mature consideration, established the relations, till now observed not only in France, but by all maritime powers, between the diameters of the burners and the dimensions of the lens-apparatus, and it will be easily understood that imperious necessity alone would excuse any alteration. It was, however, soon remembered that there was at the light-house depot a few years ago a gas-burner originating with Augustin Fresnel, which had five concentric rings of small apertures. Unfortunately this burner is lost. The brother of the illustrious inventor borrowed it to make an engraving; it was destroyed by fire, and the engraving (annexed to Fresnel's complete works) does not give the dimensions. Still the scale on which it is shown is sufficiently large (0.25) to justify some confidence in measurements taken by dividers, and thus it was found that the diameter of the exterior ring must have been about 4.72 inches, while the same diameter (taken in the middle of the thickness of the wick) is but 3.34 inches in first-order colza-oil lamps and 4.13 inches in 5-wick mineral-oil lamps. It is to be observed, moreover, that the text (vol. 3, page 514) clearly shows that the aim of the invention was to increase the duration of flashes in eclipse-lights, and that this result

Gas-burner of Augustin Fresnel.

Desired increase of duration of flashes in eclipse-lights.

could be obtained, as regards the burner, only by an increase in diameter. Thus we have the proof that this was one of the inventor's efforts to prolong the duration of the luminous apparitions in eclipse-lights; the only objection to the new lens-apparatus being that it was inferior in this respect to the old reflectors; and it is known that in the two first-order eclipse-lights constructed during the life-time of Fresnel (Cordouan and Planier) the luminous rays passing above the drum are used entirely to prolong the flashes produced by the principal lenses; the inventor thus sacrificing intensity to duration.

Fresnel lamp.

It is easily understood why he did not exceed four wicks and the resulting diameter when the combustible was colza-oil. A considerable increase of expense would have been the result, and thus an arm might have been furnished to the opponents of the new invention. There was also danger of meeting with practical difficulties, since there was no certainty that such vigilance as a four-wick lamp requires could be obtained from light-keepers; and, indeed, it was some years before we could succeed in this respect.

Thus it may be inferred that Augustin Fresnel considered the four-wick burner the highest limit which could be obtained with colza-oil, and that he sought for a new combustible which might allow him to proceed further. It is, therefore, only carrying out his ideas, to use mineral-oil in order to increase to a certain extent the dimensions of burners. It is also to be remarked that the height of the flame is not increased, so that multiplying the wicks mostly affects the horizontal divergence, which is all used for sea-illumination.

A final question was considered, whether instead of confining the advantages of the measure to a certain number of lights, it would not be better, although more expensive, to give each apparatus of the same order the same number of wicks. As this would be a more regular system, and give more security to mariners, the Light-house Commission did not hesitate to answer it in the affirmative. They proposed to make five orders of lights, establishing for each a constant proportion, more decided, however, than before, between the diameter of the burner and that of the lens-apparatus. The following comparative table of the old and new systems shows the advantage of the latter, both to mariners and the public treasury:

Proposal to regulate the number of wicks and order of light.

Comparative table of old and new systems.

EUROPEAN LIGHT-HOUSE SYSTEMS.

Comparative table of old and new systems.

	Number of wicks.	Mean diameter of outer wick.	Diameter of the burner.	Diameter of the lens-apparatus.		Intensity of the lamp.	Intensity with a fixed light apparatus, without reflector.	Consumption per burner per hour.		Consumption per burner per year.		Annual expense for oil at the average price of $1 per gal. for colza and 50 cts. for mineral oil.	Number of light-houses Jan. 1, 1873.	Annual expense for oil for each order of light.	Total annual expense.
		Inches.	*Inches.*	*Ft.*	*In.*	*Carcel units.*	*Carcel units.*	*Grams.*	*Fluid ounces.*	*Kilog's.*	*Wine gals.*				
Colza-oil:															
First order............	4	3.34	3.51	3	6.07	21	630	760	25.40	3,040	861.04	$865 94	*42	$37,209 78	
Second order.........	3	2.71	2.91	3	4.04	13	335	500	16.73	2,000	562.50	562 80	6	3,407 16	$64,537 33 coin.
Third order, (large)...	2	1.53	1.73	3	3.37	5	90	175	6.59	780	204.00	204 00	30	6,129 00	
Third order, (small)...	2	1.38	1.43	1	7.94		30	110	4.12	440	124.22	128 22	15	1,923 43	
Fourth order..........	1	0.94	1.14	1	2.74	1.6	13	60	2.23	210	60.31	69 94	230	16,086 93	
Mineral-oil:															
First order...........	5	4.13	4.32	3	6.61	30	820	900	37.42	3,607	1,181.16	$590 58	*42	24,804 36	
Second order.........	4	3.31	3.51	3	4.91	23	510	620	26.30	2,530	836.40	413 40	6	2,480 40	$44,244 47 coin.
Third order..........	3	2.35	2.15	3	3.37	14	250	350	14.97	1,440	472.46	230 23	30	7,696 90	
Fourth order.........	2	1.77	1.90	1	7.63	6.4	61	160	6.63	640	200.96	104 00	15	1,571 88	
Fifth order...........	1	0.98	1.17	1	2.71	2.2	18	55	2.29	220	72.18	36 00	230	8,300 93	

OBSERVATIONS.—No allowance is made for the less amount of oil used in tidal lights now furnished with colza-oil lamps, consuming but fifty instead of sixty grams per hour, as this is offset by the oil consumed in the supplementary apparatus placed in a few light-houses, and not included above. As to the luminous intensities, it has been considered unnecessary to mention, among the lights of the last order, either the lens-apparatus of less than 0m.375 (1 foot 2.71 inches) diameter, or the parabolic reflectors of various dimensions. To enter into these details would have uselessly complicated our statement, as the resulting difference would be quite unimportant. As before stated, floating lights are not included in the above tables.

* A deduction being made for three light-houses illuminated by electricity.   † 128 fluid ounces to the wine-gallon.

Results produced in eclipse-lights.   The preceding table, which clearly shows that the new system notably increases the intensity of fixed lights, shows nothing as to eclipse-lights, for the reason that with the latter the advantages gained are divided between increase of intensity and increase of duration of the luminous apparitions, and the figures showing the effect on each of these two phenomena vary greatly for each order of lights, according to the number of divisions of the lenticular drum, the rapidity of rotation, and the arrangements for prolonging the flashes. Therefore, without unnecessary details, we shall only give the proportional values, constant in each order, of the increase of intensity and of duration which the new lamps give to the flashes of eclipse-lights.

Table of increase in eclipse-lights.

Order of light.	Proportional increase.	
	Intensity.	Duration of flashes.
	Per cent.	Per cent.
First	7	22
Second	26	22
Third	76	50
Fourth	35	58
Fifth *		

* There are no fifth-order eclipse-lights on the coasts of France.

Besides, whether the light is fixed or eclipse, the quantity of light emanating from the apparatus is constant for each order, and it may be concluded from the above figures that, as our sea-coast lights now are, the total intensity of the *Increase of intensity by use of mineral-oil.* luminous beams sent to the horizon will be increased nearly 45 per cent. This, however, does not take into account the very considerable advantage resulting from the fact that flames fed by mineral-oil preserve their brilliancy longer than the others. The percentage would be much higher if the comparison, instead of being made when the lights are in their first condition, were made after they have been burning for a few hours. The annual saving in oil will amount to 106,676.80 francs, ($20,268.59,) about 32 per cent.

Recapitulation of advantages of the change.   Thus the advantages of the change are that 45 per cent. more light is sent to the horizon, and 32 per cent. is saved in the expense of oil.

There is, however, one objection to the new mode of illumination: it depends on the qualities of a foreign article. This may be adulterated, its price increased more or less when its merits, better known, cause a greater demand, or our supplies might be cut off in case of a maritime war, a sad calamity which we must at the present day certainly take into account. It is to be observed, however, that the formation of bog-head is very active, and aside from their own integrity the Scotch company will be more interested to maintain the reputation of their article the more their customers increase; the French manufacturers also will probably be induced by foreign competition, and perhaps by governmental regulations, (which would certainly be justifiable,) to improve the article they manufacture, if not as to luminous intensity, at least as to safety; wars, moreover, have ceased to be of long duration; finally, the burners of our mechanical lamps are so arranged that in order to go back to colza, we have only to close the tube regulating the level of mineral-oil. All our mechanical lamps will be retained and kept in use; the burners alone will be altered.

There are at present two kinds of burners in use, both of which give the same results; the Doty burner, manufactured in Paris by Messrs. Barbier & Fenestre, and another made by Messrs. Henry-Lepaute, called the modified Fresnel burner. *Kinds of burners in use*

To complete the change now commenced of colza-oil lamps to mineral-oil lamps, and to purchase the various receptacles and implements needed, will not require more than 50,000 francs ($9,500) more. This amount will be saved in less than a year by the new system.

THE LENS-MAKERS OF PARIS.

During my limited sojourn at Paris I visited the manufactories of MM. Henry-Lepaute, MM. Sautter, Lemonnier & Co., and MM. Barbier & Fenestre. These three firms were, until the establishment of the works of Chance, Brothers & Co., in England, the exclusive manufacturers of dioptric apparatus, and supplied all countries. They all have extensive establishments, and are contractors, not only for lenses, but for light-houses, (particularly small ones of iron,) for different European countries. *Visit to lens-manufactories.*

The method of grinding and polishing the prisms and lenses is almost identical at the different establishments, all of which keep on hand a large stock of prisms, adapted to different sizes of apparatus, so that orders can be filled by them with but little delay. Lenses, lamps, and lanterns were in different stages of manufacture under orders from different countries.

At M. Lepaute's, in the faubourg Saint Germain, I was shown among others a beautiful third-order lens, a duplicate of which I afterward saw at the Industrial Exhibition at Vienna.

Establishment of M. Henry-Lepaute.

Lens shown.

It was for a "fixed light varied by flashes," 180° being provided with fixed-light apparatus, and the other half divided into eight flash-panels; the characteristic of the light would therefore be, that between comparatively long periods of fixed light, there would be observed eight short consecutive flashes; quite a new characteristic and a most distinctive one for a light to be placed within short distances from other varieties of lights.

Characteristics.

M. Lepaute showed me his designs of a tower and dioptric apparatus for a light-ship the construction of which he was about to commence for Sweden. This design will be readily understood by an inspection of Plate XXXI.

Swedish light-ship.

I was much interested in modifications of apparatus for floating lights, and am indebted to him for a paper on the subject, from which the following extracts are translated. After describing the ordinary catoptric fixed light, he says:

Description of apparatus for floating lights.

"The arrangement of revolving floating lights is similar to those of fixed floating lights, but the reflectors are 36 centimeters in diameter, and but eight in number, the same as the faces of the lantern. The entire system of lamps and reflectors is supported by a chariot on rollers which has a wheel toothed on the inner side, to which a rotary movement is given by a pinion which communicates by means of a stem fixed by collars along the mast, with the clock-work placed between the decks. A special mechanism has been added to the floating light at Dunkerque, so that when the sea is too boisterous it may be hoisted only half-way up the mast and yet be made to revolve.

"For several years attempts have been made to construct floating dioptric lights. One method which has been employed, is to suspend on gimbals to an armature sliding along the mast, three harbor-lights (*feux de port*) lighting 235° each with its lantern.

Construction of floating dioptric lights.

"There has also been used a system of eight or ten small dioptric apparatus for signal-lights, 15 centimeters in diameter, suspended around a mast in a single lantern. In 1848 our father projected a system of two fixed catadioptric apparatus, 60 centimeters in diameter, suspended on gimbals in a single lantern, and each lighted by a two-wick constant-level lamp. The power of this apparatus is very great, and its service very simple.

Invention of M. Henry-Lepaute, Sr.

"In 1869 we recommenced our father's studies of an ap-

PLATE XXXI.

SWEDISH LIGHTSHIP.

SCALE FOR DETAILS

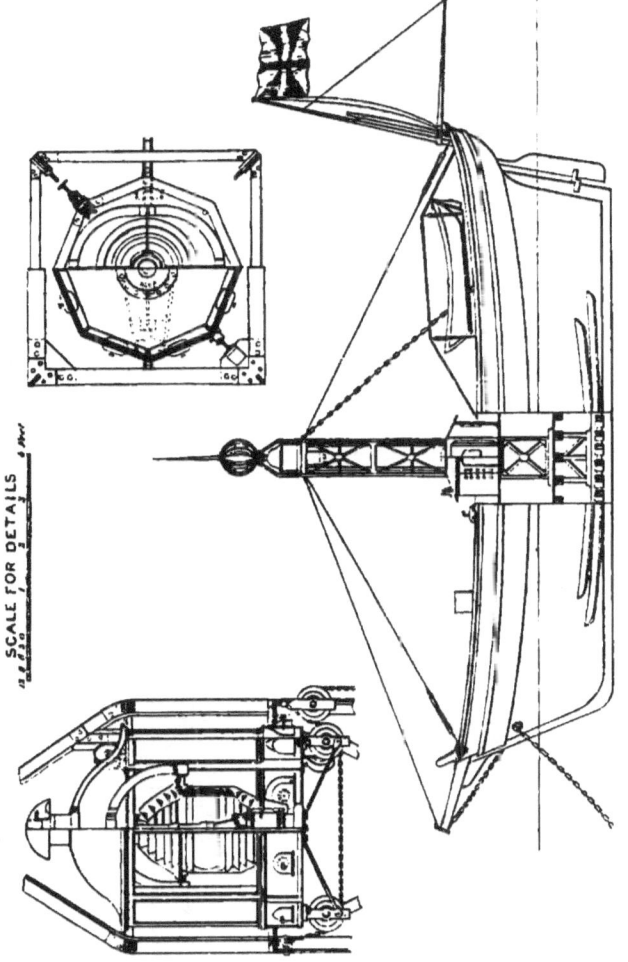

paratus for a floating catadioptric light, illuminating the whole horizon, and lighted by a single two-wick lamp. Our project was appreciated, and in 1873 we were charged with the construction for Sweden of an apparatus of this kind, placed in the center of an open iron frame-work, which takes the place of the mast of the vessel. (See Plate XXXI.) Swedish light-ship.

"The engineers, both French and foreign, who have examined this apparatus in course of construction, have expressed to us their good opinion of the result which can be obtained by this new arrangement.

"For revolving lights we have proposed to place before the reflectors, annular lenses of 15 centimeters focal distance, which will considerably augment the range of such an apparatus. Floating revolving lights.

"Plate XXXII shows a floating eclipse-light, which is entirely catadioptric, composed of two annular half-apparatus suspended on gimbals at the two opposite extremities of the same armature, and each lighted by a two-wick burner. These two apparatus are placed in the same lantern. The objection that the intervals between the flashes are unequal on account of the distance of the two half-apparatus is of but little importance, as there is not more than a half-second difference during a rotation of four minutes, which, with three lenses in each half-apparatus, produces flashes of twelve seconds' duration. Floating eclipse-light.

"We have just finished a study of a floating catadioptric light with a single burner placed in a single lantern and arranged similarly to the floating catadioptric fixed light. Floating catadioptric light.

"By making floating lights with a single burner we have the advantage of being able to make apparatus of sizes varying from the sixth order, 30 centimeters interior diameter, to the three and a half order, 70 centimeters interior diameter.

"The power of these apparatus is much greater than that of those hitherto used, and the advantage of having but one lamp makes the service much easier."

M. Lepaute has also on exhibition at his manufactory a lamp suitable for burning mineral-oil, which was designed by M. Lepaute, Sr., in 1845. Mineral-oil lamp of 1845.

The papers in support of his claim to the original invention of a lamp suitable for burning mineral-oil in lighthouses, as well as the reasons for such claim, are fully set forth in an interesting paper on the subject of light-house burners, written at my request by M. Henry-Lepaute *fils*, and which he has kindly sent me since my return to America. A translation of it will be found on page 208.

Lamp-valves. In the construction of mechanical lamps at this establishment, the valves are made of calf-bladder, and it was stated that for this use, this material is superior to any other.

Establishment of Sautter, Lemonnier & Co. Sautter, Lemonnier & Co. have a very extensive establishment near the Champ de Mars, and their shops are largely devoted to the manufacture of metal-work of light-houses.

Parapets of light-houses. The parapets of light-houses and the floors of the galleries or "decks," as we call them, are in Europe, ordinarily of stone. I saw, however, one making for Russia, of cast iron, which we have found to be the best material for this purpose, in consequence of the leaks at the joints of stone, caused by contraction and expansion due to our extremes of temperature.

M. Sautter, who speaks English with fluency, was attached to the Royal Commission which reported on the condition of the light-houses of Great Britain in 1861. He has had **Test of apparatus by M. Sautter.** much experience in the specialty of testing of prisms and lenses, and I was much interested and instructed in the account of the means and care which he takes to insure that every dioptric apparatus supplied from his establishment shall be of the highest standard of efficiency. It should be stated in this connection that the other lens-makers also exercise great care in this particular.

I observed at this establishment a handsome lantern for an electric light, designed for the Industrial Exhibition at Vienna, (represented by Fig. 24,) in which the diagonal sash-bars are reduced to a minimum thickness, viz, one-half inch.

Besides this lantern there was an iron tower for a harbor-**Farquhar burner.** light, of good design and workmanship. I was shown the "Farquhar" burner, of which Sautter, Lemonnier & Co. are the sole proprietors, and which they claim is superior to all others in that it gives a whiter and higher flame, of greater intensity, caused by the fact that the tubes for the supply of air to the concentric flames are of such capacity as to produce equal currents to the interior and exterior, so that the draught is the same to all parts. The flame is of different shape, not being pointed like those of ordinary lamps.

Establishment of Barbier & Fenestre. At the establishment of Barbier & Fenestre, I was shown a second-order lens ordered by us for Cape Elizabeth, a second-order lens for Scotland, a third-order lantern (of iron) for Uruguay, to cost 11,500 francs, ($2,300,) and fifth-order lanterns (of bronze) for Venice and for France, costing each 3,200 francs, ($640.)

Oil-cans. This firm were also making a large number of oil cans or butts for mineral-oil destined for use in the French service.

REVOLVING CATADIOPTRIC APPARATUS FOR LIGHT-SHIPS.

PLATE XXXII.

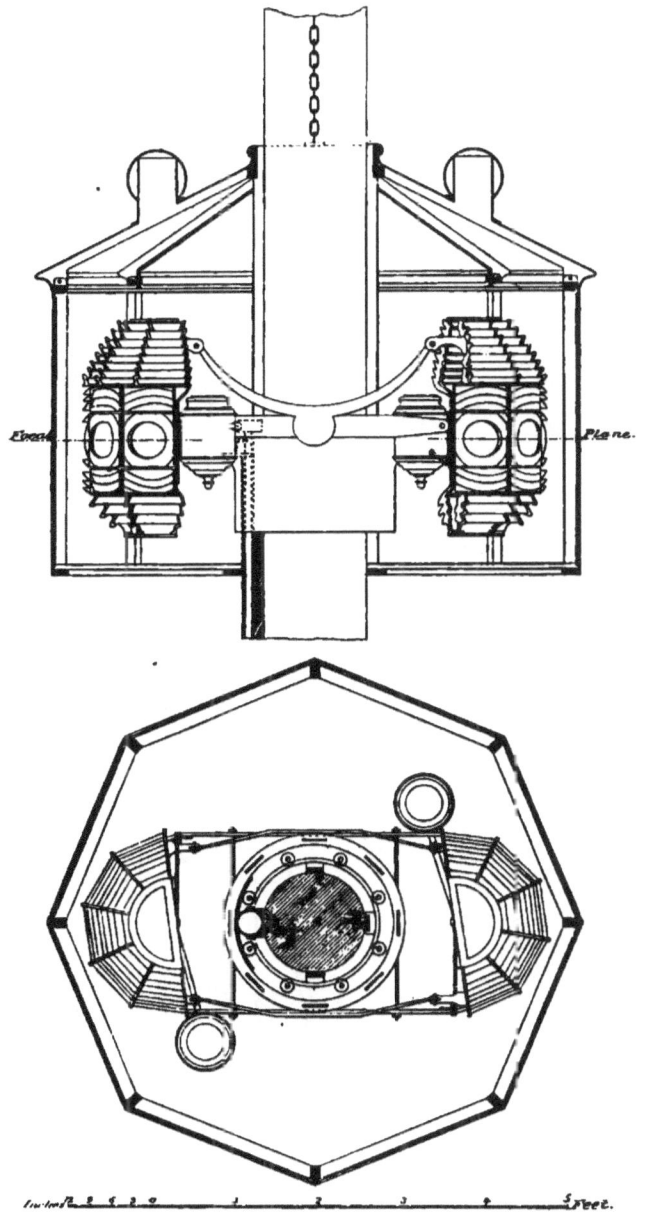

Each of these cans, which are of sheet-iron, contains seventy-five liters, (about 20 gallons,) and costs 30 francs, ($6.) They are hung in iron stands on pivots placed slightly below the middle, and the cocks are at the tops of the cans, the advantage claimed being that, as the cocks, except while oil is being drawn, are always in the air, there is no danger of leakage of this subtile fluid.

Fig. 24.

Lantern for electric light.

Barbier & Fenestre have purchased from Captain Doty his patent for mineral-oil burners, (see Fig. 2, Plate XXXVI,) except as regards Great Britain and the United States, where he reserves the right, and the French government, with which he has made special arrangement. They state that they are making mineral-oil lamps for light-houses in all parts of the world, and that mineral-oil has been adopted in most countries of Europe, as well as in South America and Canada. All the lens-manufacturers were busy in fitting the colza-oil lamps of the French light-houses for the use of mineral-oil.

Doty burner.

A variation of one millimeter ($\frac{1}{25}$ of an inch) in the height of the overflow in winter and summer is provided for.

The tops of all burners made by French lens-makers are of copper. I suggested to one of them that we had found iron to be much more durable, and he concurred in the opin-

Copper burners.

ion that iron would not submit as readily as copper to the destructive action of the flame.

Lanterns and lenses are differently constructed for different countries, some preferring the diagonal bars, on account of the less obstruction to the light, and others preferring the vertical ones, on account of the greater economy. The difference in cost between the two kinds of lenses was stated to be for the first order about 1,000 francs, ($200.)

Lenses suited to desired height above the sea-level.

Lenses are made to suit the height above the sea-level at which the different orders are usually placed, and any excess of such height requires special calculations and adjustments. Unless otherwise stated, the height at which a first-order lens is to be placed, is supposed by the makers to be between 150 and 200 feet above the sea. The photometer used by all the lens-makers in Paris is the same as I have described as in use at the *Dépôt des Phares*. I was informed that most countries use the "mechanical lamp" for the first and second orders of lights, the "moderator" for third and fourth orders, and common (capillary attraction) lamps for the fifth and sixth orders, but that the sixth-order lamp is little used.

Photometer used.

Lamps used for different orders of lights.

I found the members of the several firms all to be courteous and intelligent gentlemen, and my experience with them at their offices and manufactories confirmed the high opinion I had previously entertained, and which were derived from my correspondence with them and from the excellent quality of the apparatus furnished to us.

I should state, before concluding my notes in regard to Paris, that Mr. Washburne, the American Minister, was absent at the time of my visit. I am, however, under obligations to Colonel Hoffman, the secretary of legation, for his valuable assistance.

[Translation.]

REMARKS ON BURNERS EMPLOYED FOR LIGHT-HOUSE ILLUMINATION.

By HENRY-LEPAUTE BROTHERS.

In all countries in the world burners with concentric wicks are used in light-house apparatus, the number of wicks varying from one to four, according to the order of the lens. These burners were invented in 1821, on the principle of those of Argand, by MM. Augustin Fresnel and Arago. The concentric wicks are separated by air-spaces for supplying the oxygen necessary for good combustion.

Invention of burners with concentric wicks.

PLATE XXXIII.

BURNERS FOR LIGHT-HOUSE LAMPS, (FRENCH).

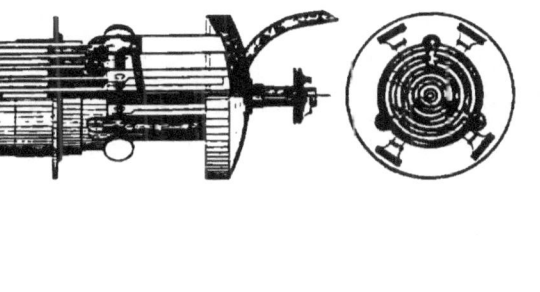
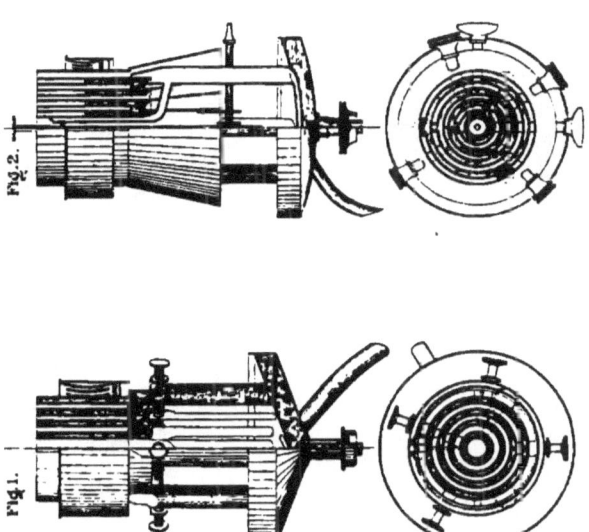

Fig. 1.   Fig. 2.   Fig. 3.

FRESNEL BURNER (1823).   CYLINDRO-CONICAL BURNER (1868).   MODIFIED FRESNEL BURNER (1879).
1st Order (4 wicks).   2nd Order (4 wicks).   2nd Order (4 wicks).

The Graphic Co. Photo-Lith. 39 & 41 Park Place, N.Y.

The conditions fulfilled by the ordinary Fresnel burners are the following:

<small>Table of ordinary Fresnel burners.</small>

Order of lens.	Number of wicks.	Dimensions of fully developed flames.		Luminous intensity in Carcel units.
		Maximum diameter.	Height taken from the burner.	
		Millimeters.	*Millimeters.*	
First	4	90	100	23
Second	3	75	80	15
Third	2	45	70	5
Fourth	2	38	65	3
Fifth	1	30	45	1.6
Sixth	1	27	37	1.3

In examining this table it will be noticed that third-order apparatus, interior diameter one meter, is lighted by two-wick burners, differing but little from those employed in fourth-order apparatus, the interior diameter of whose optical part is but half a meter.

In 1845 M. Henry-Lepaute, Sr., who had been a colaborer in the experiments and construction of the apparatus designed and calculated by M. Augustin Fresnel, sought to apply to third-order lenses more powerful burners than those then in use. After his own drawings he then had made, by M. Blazy-Jallifier, a two-wick burner of special construction. The two wicks, respectively 54 and 28 millimeters in exterior diameter, were separated by a double air-space, and an air-tube was placed around the larger wick. These air-tubes were elongated and enlarged below, and the chimney-holder slipped over the outside. This burner gave an intensity of from four to five Carcel units, and consumed 200 grams of colza-oil per hour. Some experiments were made in 1845 in the workshops of M. Henry-Lepaute, Sr., and M. Leonor Fresnel, engineer of the corps *des ponts et chaussées* and director of the French light-house service, verified the principal results by one of his assistants. These new burners were adopted by the kingdom of the Netherlands for the light-houses of Brouwershaven, (1845–1847,) Scheveningen, (1851,) Renesse, and Schiermonikoog. The French government, while recognizing the merits of these burners, did not adopt them, not wishing to alter the standard burners invented by Messrs. Fresnel and Arago. For this reason M. Henry-Lepaute continued to make them only when specially ordered.

<small>Efforts of M. Henry-Lepaute, sr., to increase the power of burners.</small>

<small>Burner of 1845.</small>

<small>Intensity of burner of 1845.</small>

<small>Adoption of burner of 1845 by kingdom of the Netherlands.</small>

In 1868 the question of the use of mineral-oil in light-houses was brought up. Previous experiments had not given very good results. An American, Mr. Doty, proposed a form of burner which succeeded quite well for

<small>Question of the use of mineral-oil in light-houses.</small>

<small>Doty burner.</small>

Experiments with burners. schist-oils, but its arrangement and dimensions differed from the standard which the light-house administration had not, up to that time, consented to modify. About that period we were making numerous experiments relating to the burning of mineral-oil, and after many attempts we succeeded in constructing a burner with air-tubes enlarged below, and with a tube exterior to the large wick. The arrangement was almost identical with that of the 1845 burner invented by our father, except that the air-space between the wicks was not double. We then experimented with the 1845 burner, feeding it with mineral-oil, and found the combustion very satisfactory.

Invention of truncated-cone burner, 1869. In 1869 we invented the truncated-cone burner, preserving exactly the dimensions of the Fresnel burner for lights of all orders. Specimens of these burners were sent in 1869 to the light-house administrations of Sweden, Norway, Denmark, the United States, and Brazil. The French administration tried our truncated-cone burners, comparing them with the Doty, and found them always equal, and sometimes superior, especially the one-wick burner. Still they charged us to try to modify the Fresnel burner so that with the same general arrangement, it might, by means of some additions, be used for mineral-oils.

Comparison with the Doty burner.

Modification of the Fresnel burner, 1872. The experiments undertaken in 1872 succeeded, and we fixed upon the arrangement of the new modified Fresnel burners of 1873, which are now being substituted for the ordinary Fresnel burners, so that schist-oil may be used in all light-houses of France. The success of these burners is due in a great measure to the addition of an exterior air-tube, as in the 1845 burner; an arrangement also used by Mr. Doty in his burners.

Exterior air-tube.

In fine, all schist burners now in use are reproductions of the burner invented in 1845 by M. Henry-Lepaute, Sr., still in actual use in a large number of light-houses in Holland. The conditions fixed by the French administration for the dimensions and intensity of the new modified Fresnel burners are the following:

Table of modified burners.

Table of modified burners.

Order of lens.	Number of wicks.	Dimensions of the fully developed flame.		Luminous intensity in Carcel units.
		Maximum diameter.	Height taken from the burner.	
		Millimeters.	*Millimeters.*	
First	5	105	110	30
Second	4	85	90	23
Third	3	65	80	14
Fourth	2	45	70	6.4
Fifth	1	25	45	2.2
Sixth	1	25	45	2.2

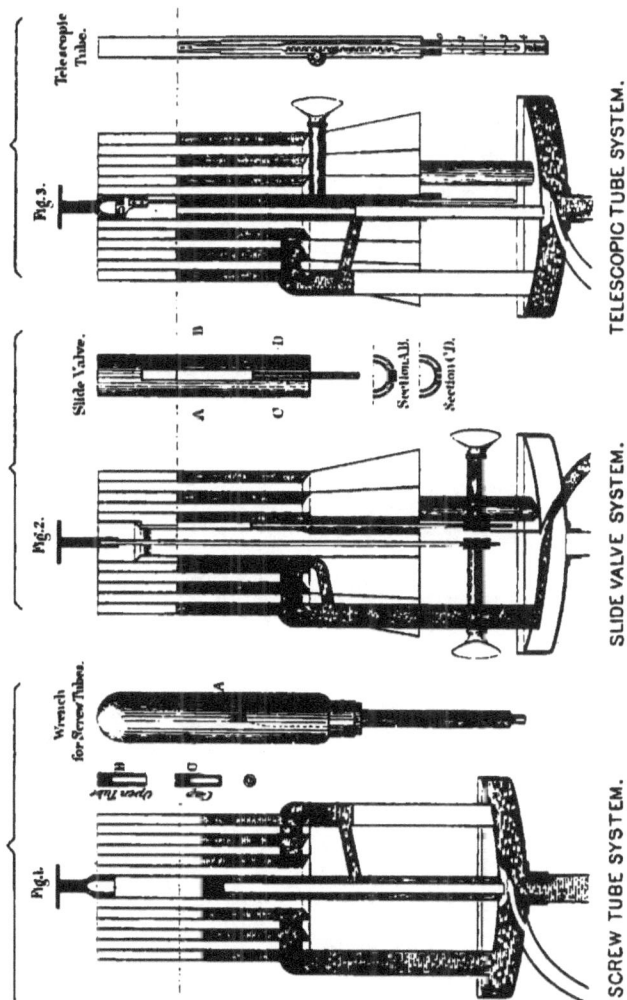

It will be seen that above the fifth order the number of wicks has been increased. There is thus obtained a greater intensity and more horizontal divergence, and as schist is cheaper than other oil the expense of illumination is not increased. *(Increase in number of wicks.)*

The following table clearly shows the differences between the old and new systems: *(Table of differences.)*

Comparative table of the dimensions, consumption, and intensities of the burners employed in the French light-house service.

Order of light.	Number of wicks.	Outer diameter of wicks in inches.					Consumption per hour.		Intensity in Carcel units.
		No. 1.	No. 2.	No. 3.	No. 4.	No. 5.	In grams.	In fluid oz.*	
ORDINARY COLZA-OIL BURNERS.									
First order	4	0.98	1.85	2.68	3.54		760	28.46	23
Second order	3	1.10	1.97	2.91			500	18.73	15
Third order, (large size)	2	0.94	1.73				175	6.55	5
Fourth order, (third order small size.)	2	0.83	1.50				110	4.12	3
Fifth order, (fourth order large size.)	1	1.10					60	2.25	1.6
Sixth order, (fourth order small size.)	1	0.94					50	1.87	1.3
MODIFIED FRESNEL BURNER FOR MINERAL-OILS.									
First order	5	1.18	1.97	2.75	3.54	4.33	900	37.42	30
Second order	4	1.18	1.97	2.75	3.54		630	26.20	23
Third order	3	1.18	1.97	2.75			360	14.97	14
Fourth order	2	1.18	1.97				160	6.65	6.4
Fifth order	1	1.18					55	2.29	2.2
Sixth order	1	1.18					55	2.29	2.2

* 128 fluid ounces to the wine-gallon.

NOTE.—The intensities obtained by burning colza-oil in the modified Fresnel burners are as nearly as possible the same as those given by mineral-oils.

Fig. 1 of Plate XXXIII shows the arrangement of the air-tubes and wicks in the ordinary Fresnel burners. The ascent and overflow of the oil is also shown.

French burner of 1821.

Fig. 1, Plate XXXVI, shows the arrangement of the two-wick burner invented in 1845 by M. Henry-Lepaute, Sr. It will be seen that the air-tubes are longer and enlarged below; it will also be remarked that the tube between the wicks is double. In all these burners the wicks are lowered and raised by racks and pinions.

Henry-Lepaute burner, 1845.

Fig. 2, Plate XXXIII, shows the arrangement adopted in our 1869 burners. All the air-tubes are elongated below, increasing in diameter. There is, as in the 1845 burner, an exterior tube, and a disk is placed in the center of the burner when mineral-oils are burned.

Cylindro-conical burner of Henry-Lepaute, 1869.

Fig. 3, Plate XXXIII, shows the form proposed by us and adopted by the French administration for all French lights. The air-tubes are arranged as in the ordinary Fresnel burner. There is also an exterior tube shorter than the others. The chimney-holder is shorter than in the ordinary burner, and there is a disk moved by a rack and pinion, which is necessary when burning mineral-oil. All the new burners are made of copper.

Modified Fresnel burner, 1873.

In burners fed with mineral-oil the oil should rise to from 35 to 60 millimeters below the crown of the burner, on account of the volatility of these oils. It is the same when a constant-level lamp is used. Furthermore, it is necessary that this level should be raised or lowered in the same burner, according to the nature of the oil and the surrounding temperature.

Height of mineral-oil in burners.

Two kinds of methods are employed for varying the level of the oil.

1ST. LEVEL VARIED BY APPLIANCES IN THE INTERIOR OF THE BURNER.

Whether the Henry-Lepaute, the clock-work, or the moderator lamp is used, an excess of oil cannot be entirely suppressed, and there should be a means of varying the level in the burner itself.

Level varied by an interior apparatus.

In the Doty burner, (see Fig. 2, Plate XXXVI,) an appendage for this purpose is attached to the outside of the cistern.* This has the disadvantage of being fragile, and

Level varied in the Doty burner.

---

* By the following communication, which I have recently received from M. Lepaute, it would seem that the French government has decided to use this appendage:

"A modification has been made in mineral-oil lamp-burners for the purpose of regulating the supply of oil. This modification consists of

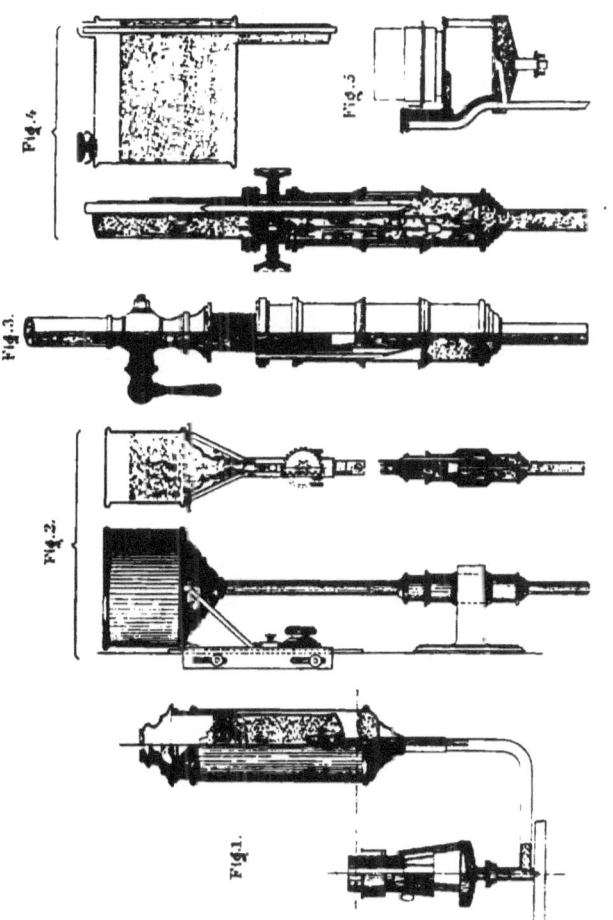

PLATE XXXV.

EXTERIOR ADJUSTMENTS OF LEVEL IN MINERAL OIL LIGHTHOUSE LAMPS.

obscures the light in one direction. All the systems proposed by us are for the interior of the burner, and have been used with success since 1869.

In the modified Fresnel burners, (Fig. 1, Plate XXXIV,) the oil overflows into a central tube, which can be lengthened by means of short sections (B) screwed on. These are of determinate lengths, and are fixed in place by a wrench, (A, Fig. 1.) *Screw-tube system.*

When it is required to burn vegetable or animal oil with an overflow, the tube is stopped by a cap, (C.) A friction-stem carries the disk. These various operations cannot be performed when the burner is lighted, and, besides this disadvantage, the level of the oil can be varied only by a considerable amount, which must be determined in advance. In France, where the temperature is moderate, and only paraffine is used, these disadvantages are not great, but they would be serious if the burners were used where the temperature varied greatly and for all kinds of mineral-oils.

We use two other methods by which the level of the oil and that of the disk can be varied while the burner is lighted.

In the arrangement shown in Fig. 2, Plate XXXIV, a double oil-tight envelope is placed inside the central wick-tube, and the tube is pierced with a longitudinal slit through which the oil flows. This slit is closed by a slide-valve raised and lowered by a rack and pinion, the stem of which passes through a packed chamber. When the slide is raised entirely it closes the slit, and vegetable or animal oil can then be burned with an overflow. The disk is carried by a guide-stem placed in the central air-tube and moved by a rack and pinion. *Slide-valve system.*

In the arrangement shown by Fig. 3, Plate XXXIV, the overflow of oil is managed the same as in the modified Fresnel burners except that the overflow-tube is lengthened by an interior sliding tube worked by means of a rack. The disk is moved by a rack and pinion passing into the overflow-tube and through a packed chamber which prevents the leakage of oil. *Telescopic-tube system.*

---

an appendage added to these burners, and which constitutes a constant-level." (It is represented in Fig. 5, Plate XXXV.)

"The two solid caps which accompany each burner are intended to be used in case it is necessary to have recourse to colza-oil; the cap with the wire-cloth (*toile métallique*) is used to cover the three tubes when mineral-oil is used.

"With burners thus modified the overflow need not be limited to a few drops, as heretofore required."

This seems to me an excellent arrangement, and not subject to the objections mentioned by M. Lepaute.—[G. H. E.]

2D. LEVEL VARIED BY EXTERIOR MEANS.

System for raising or lowering the cistern.

These various systems of our invention can be applied to all the orders. The oldest and most simple, Fig. 1, Plate XXXV, is applicable to apparatus of the sixth, fifth, and fourth orders, in which the lamp with its cistern is attached to the interior of the lens. The bottom of the cistern regulates the level of the oil in the burner, and a graduated scale shows the amount of variation. The valve-stem is long enough to always raise the valve from the mouth of the cistern. Where the capacity of the lamp is greater, the cistern is carried by a sliding armature (Fig. 2, Plate XXXV) moved by a rack fixed to one of the uprights of the apparatus or the lantern, and the level is established in an open vessel fixed at the height of the focal plane, and is regulated by the mouth of the neck of the cistern.

In these two systems the cisterns are hermetically closed, and they must be raised and inverted to be filled unless they are furnished with a movable plug at their upper part. To obviate this inconvenience, which is quite serious when the lamp is of great capacity, we employ various means.

System with screw extension-pieces.

In the system shown by Fig. 3, Plate XXXV, the level is established in an enlarged part of the conduit in which loosely enters the end of the oil-tube, on which a screw-thread is cut. On this an extension-piece slips, the lowering or raising of which varies the level as indicated by a graduated scale; a cock above shuts off the oil when the screw-plug in the upper part of the cistern is opened for filling.

System with an independent air-tube and sliding extension-piece moved by a rack.

In the system shown by Fig. 4, Plate XXXV, the air-tube is prolonged to the top of the reservoir, and is lengthened at will at the bottom by a sliding tube moved by an exterior rack. Another rack moves the valve down into its socket, and thus shuts off the descent of the oil when the reservoir is being filled. These last two systems take but little space, and are very simple and easy to repair. All these systems have been in use in Sweden, Norway, and France for several years.

Recapitulation.

To recapitulate, the requirements for burning mineral-oil are: special burners with powerful currents of air, means of varying the level of the oil in the burner according to the surrounding temperature and the nature of the oil, and means of raising and lowering the central disk, which sends air into the interior of the central flame.

Smoke-funnels.

Our experiments have induced us to modify the form of smoke-funnels (*fumivores*) and dampers, and to adopt the

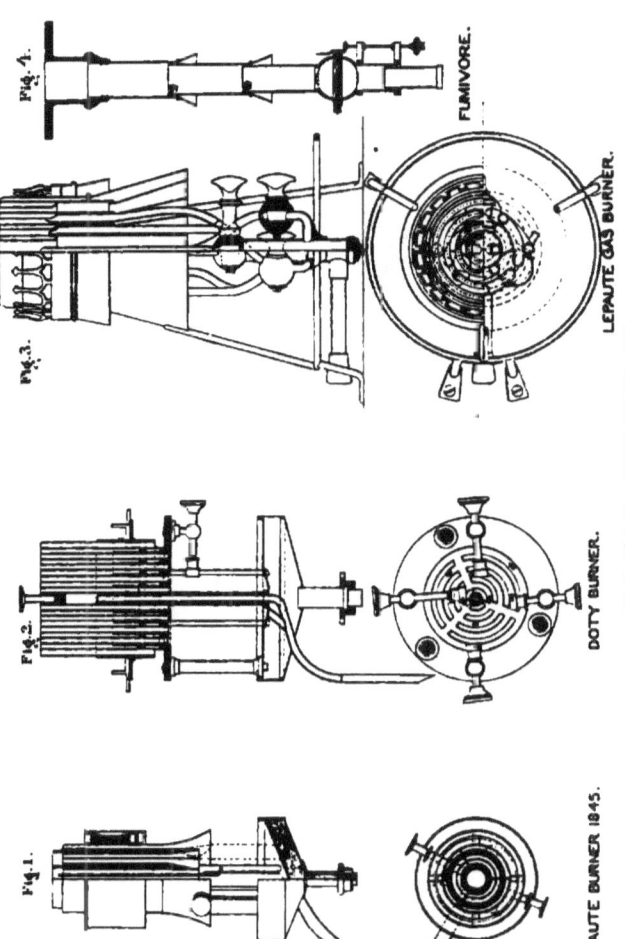

forms represented by Fig. 4, Plate XXXVI. The glass chimney is surmounted by a copper continuation of the same diameter, which sets into the damper. The latter is cylindrical, with a spherical part midway of its length, in which the register moves. A series of cylindrical smoke-funnels of increasing diameter, terminated below by truncated cones, completes the chimney. The annular spaces between the various funnels introduce air, and by cooling the smoke render the combustion more complete and the draught more regular. The enlargement of that part of the damper where the register is placed prevents the contraction of the orifice by which the smoke escapes. This arrangement is the best for burners with five, four, and three wicks, but it is not indispensable for burners of two wicks and one wick.

The principle of the cylindro-conical burners with outer air-tube has permitted us to construct gas-burners with concentric crowns giving very satisfactory combustion. Fig. 3, Plate XXXVI, shows the arrangement of a four-crown burner. It will be remarked that each of the crowns is fed at the two extremities of the same diameter, and that a special cock regulates the flow of gas to each crown. These burners have worked very well, and do not heat to any great degree.

Concentric gas burners.

REMARK.—Fig. 3, Plate XXXIII, shows that the chimney of the modified Fresnel burners is supported on a plate by means of three ears. This system, adopted by the French administration, has the disadvantage of great instability; the chimney cannot be exactly centered on the burner; besides, the chimneys not being exactly set below, there are no means of getting them quite perpendicular. We prefer the use of the old chimney-holders with movable grates on the inside, by means of which the chimney can be exactly centered and solidly held.

LIGHT-HOUSES AT THE MOUTH OF THE SEINE.

When at the *Dépôt des Phares*, M. Allard kindly gave me a letter to M. Arnoux, the engineer (*des ponts et chaussées*,) who is in charge of the administration of all public works (including the light-houses) on the left bank of the Lower Seine, and I proceeded by rail through Normandy to Honfleur, stopping at Rouen to see that ancient city, and especially the interesting antiquities in its celebrated museum.

M. Arnoux received me with great politeness, and I had an excellent opportunity of inspecting the lights of l'Hôpital and Fatouville, and the pier-light at Honfleur.

The light-keepers in the French service are known as masters (*maîtres de phare*) and keepers, (*gardiens*.)

Notes on the French light-house service.

They are appointed by the prefect or chief civil officer of the department on the nomination of the engineer of the district, who is charged with the river and harbor works, including the light-houses. Men who have served as soldiers and sailors are given the preference.

Appointment of keepers.

For appointment the following requisites are necessary:

1st. They must be French, and between twenty-one and forty years of age.

2d. They must be free from all infirmities which would prevent an active daily life.

3d. They must present a certificate of good moral character.

4th. They must know how to read and write, and have an elementary knowledge of arithmetic.

Salaries.

The annual salary of masters of lights is fixed at 1,000 francs, ($200.) The ordinary keepers are divided into six classes, salaried as follows: First class, 850 francs, ($170;) second class, 775 francs, ($155;) third class, 700 francs, ($140;) fourth class, 625 francs, ($125;) fifth class, 550 francs, ($110;) sixth class, 475 francs, ($95.) There is also allowed to each master and keeper a certain quantity of wood or coal for heating purposes, and the master and keepers of lights isolated by the sea receive indemnities for sea-rations.

Fuel and rations.

The salaries and indemnities are paid monthly. Fuel is furnished in kind and according to the decision of the engineers. Salaries of masters and keepers are subject to a deduction of 5 per cent., and these employés are entitled to retiring pensions derived from this fund. An oath must be taken by masters and keepers immediately after their appointment in order that they may be held responsible, if any dereliction is committed in the establishment to which they are attached.

Pensions.

Number of keepers at the different orders of lights.

The number and classes of keepers attached to each light is fixed by ministerial decision on the recommendation of the engineer, approved by the prefect and the director of the light-house service. The number is never less than three for first-order lights, and two for those of the second and third orders.

Maîtres des phares.

The masters are charged with the supervision of the service of several lights or beacons. The title (*maître de phare*) can also be granted to those of the principal keepers (*chefs gardiens*) who have merited it by exceptional service. At the lights served by several keepers and where there is no master one of the former takes the title of principal, (*chef.*) In case of his absence the second keeper takes his place.

The masters and principal keepers are particularly responsible for the entire service of the lights and the receipt of supplies. They are charged with the keeping of the registers and the correspondence.

The other keepers owe obedience to the master in everything that concerns the service, but have the right of appeal to the engineer.

The principal keepers assist in cleaning the apparatus, and take their watch the same as the others.

The masters are not held to this service, but they are required to visit the light at least twice each night, and, by decision of the engineer, they can be ordered to temporarily perform the duties of principal keeper when circumstances render it necessary.

Every year, on recommendation of the engineer-in-chief, a bonus not exceeding a month's salary may be allowed by the prefect to the most meritorious keepers, the number receiving such bonus not to exceed one-fifth of the total number of keepers in the department. Masters and keepers may be punished or removed by the prefect, on the report of the engineer-in-chief. *Bonus paid to deserving keepers.*

The service of beacons of secondary importance may be confided to persons who are not regularly in the light-house service. *Service of inferior beacons.*

PHARE DE L'HÔPITAL.

This light, so named from its proximity to the ancient hospital at Honfleur, is of the third order, fixed. It is a handsome structure of granite, and on both the exterior and interior no expense has been spared in the way of architectural effect and fine finish. *Description.*

The entrance is very imposing, and bears above it the inscription, "Stella Maris." The interior of the structure is lined with granite; although there is no air-space in the interior of the walls, they appeared to be perfectly dry.

Directly below the watch-room is a bed-room, furnished with a neat bed with hangings and other furniture, for the occupation of the keeper not on watch, for it is a rule of the French service that there must be always two keepers in the tower during the exhibition of the light. In cases where there are three keepers at a station, one of them can remain with his family at the dwelling, but when there are but two keepers, neither can absent himself from the tower at night. *Sleeping-room in tower.*

Watch-room. The watch-room at this station was neatly furnished with a table and easy chairs, and was nicely paved and lined with a pretty imitation of variegated marble.

Keepers dwelling. The dwelling, which is at some distance from the tower, is also a handsome building, with neatly painted walls and ceilings. Each keeper is allowed a kitchen and two bedrooms, besides certain standing furniture, as in the English service; and a list is framed and hung in each room showing what furniture therein belongs to the government and the principal keeper is responsible for. This furniture, made of hard wood, is strong and durable.

The family is not recognized in the supply of furniture, the kitchen and one bed-room only for each keeper being furnished by the government.

Rooms fitted up for district officers. Conducteur, duties of. In the dwelling are also fitted up rooms for the engineer of the district and one for the *conducteur*, (assistant engineer,) who is a subordinate of the engineer, and whose especial business it is to attend to the lights of the district, taking care that they are properly exhibited and the buildings and lenticular apparatus kept at the highest state of order and efficiency, while the engineer exercises a general supervision.

Number of keepers. There are at this light three keepers; one of them has the title of *maître de phare* and the other two are *chefs-gardiens*, all of higher rank and rate of pay than ordinary keepers. These ranks they had attained by long and faithful service.

It should be noted that in the French light-houses the keepers are not promoted and changed from station to station, as in the English service, but have a chance for promotion only at the stations to which they are first appointed.

Mineral-oil to be used. At the date of my visit the illuminant at the *Phare de l'Hôpital* was colza, but mineral-oil and a lamp for its use had been received at the station, and orders to make the change of lamps and illuminants were daily expected.

The mineral-oil was to be kept in the store-room at the dwelling, in wooden reservoirs lined with zinc, and I found that the French use much less precaution in storing it than do the English, who are building detached vaulted magazines of masonry for this purpose. The English also use sheet-iron cans or butts for reservoirs as a substitute for tin, which is not considered as good for the preservation of the oil.

FEU DE PORT AT HONFLEUR.

This is a small pier light-house of granite, with the focal plane 20 feet above the base, and shows a red light from a lens of the sixth order, when there are two meters of water upon the bar. The illuminant was mineral-oil, and the lamp, as in all the smaller orders of light-houses in France, one of the "Maris" pattern, the oil being drawn up by capillary attraction. The watch-room is neatly furnished with three linen-covered easy chairs, a table, and curtained bed. Illuminant and lamp used.

The floor of the main gallery was of stone, as was the parapet; a convenient width was given to the former by projecting the iron railing about six inches beyond the cornice of the tower, a simple mode of obtaining space around the lanterns of small towers. The keeper's dwelling is a neat stone structure standing near by on the pier. The keeper, there being but one at this station, takes care of the light and also attends to the tide-signals. Stone floor to the gallery, and extension to the same.

I found at the different light-stations which I visited, in Great Britain as well as in France, that the keepers readily accepted small gifts of coin in consideration of their services in showing their light-houses to visitors, and I asked the *conducteur* at Honfleur whether any regulation on the subject existed. He informed me that the acceptance of small sums from visitors is not forbidden, their service differing in that respect from our own; and it may be well to question whether our regulation in this respect should not be abolished, as it is constantly violated and is productive of no good.

PHARE DE FATOUVILLE.

This light, which is of the first order, is situated on a high hill on the left bank of the Seine, three leagues above Honfleur, and forms, with the *Phare de l'Hôpital* a range or "lead" to guide clear of the "Ratier" Shoals, ten miles distant from Honfleur. It is under the general superintendence of the engineer at Rouen. The tower is of stone, octagonal in plan, with keepers' dwellings on each side connected with it by passages at each story. The entrance to the tower is quite imposing, the visitor entering a large furnished polygonal hall from which a clear view of the interior of the tower to the lantern is afforded. Description of tower.

There are three keepers, (one *maître de phare* and two *gardiens* of second class,) two of whom are on duty at night, the other having the day-watch. Keepers.

The night-keepers alternate in the service of the light, one being on watch in the tower while the other sleeps in the room below.

The *gardien* who conducted me about the establishment

stated that he had been more than twenty years in the service, and much longer terms are not infrequent.

Illuminant and lamp. The lamp in use at the time of my visit was a mechanical or pump lamp, burning colza-oil. The burner had an adjustable chimney to regulate the access of the outer current of air to the flame, but it had not the "button" or the exterior "deflector" as have the English lamps. Mineral-oil was shortly to be introduced into this light-house.

Meteorological observations. In the watch-room I observed a barometer, and outside the watch-room window were wet and dry bulb thermometers. The principal keeper received 200 francs per annum for his meteorological observations. A small detached establishment contained a variety of meteorological instruments, consisting of different kinds of rain-gauges and self-registering wind-gauges. The extensive grounds about the light-house are ornamented with handsome flower-gardens and shrubbery, and are inclosed with live hedges.

Books kept at light-stations. At each light-house are kept a Visitors' Book, in which are recorded the names of all visitors; a Service-Book, in which the details of the daily service of the light (such as the consumption of oil, the visibility of other lights within range, &c.) are kept; and an Engineer's Book, in which are recorded the notes of inspection of the superintending or district engineer.

The service-lamps, three of which are always on hand, are required to be changed every two weeks, the one that has just been in use being taken entirely apart and thoroughly cleaned before it is again used.

PHARES DE LA HÈVE.

From Honfleur, I crossed the mouth of the Seine in a steamer to Havre. When I left Paris I did not intend to go farther than Honfleur, and my letters from M. Allard were only to the engineers of the *Corps des Ponts et Chaussées* at that place and at Rouen, but at Honfleur M. Arnoux kindly gave me a letter to M. Quinette de Rochemont, the engineer charged with the administration of the public works, including the lights, on the right bank of the Seine, whose office is at Havre.

Electric lights, when established. Unfortunately I did not find M. de Rochemont at home, and I took a carriage and proceeded to Sainte Adresse, near which place, and on Cape la Hève are the celebrated double electric lights, established respectively in 1863 and 1865, being the first of that kind in the world.

I arrived at the lights after dark, and requested the *maître de phare* to show me the establishment, saying that I had endeavored to find M. de Rochemont, but without

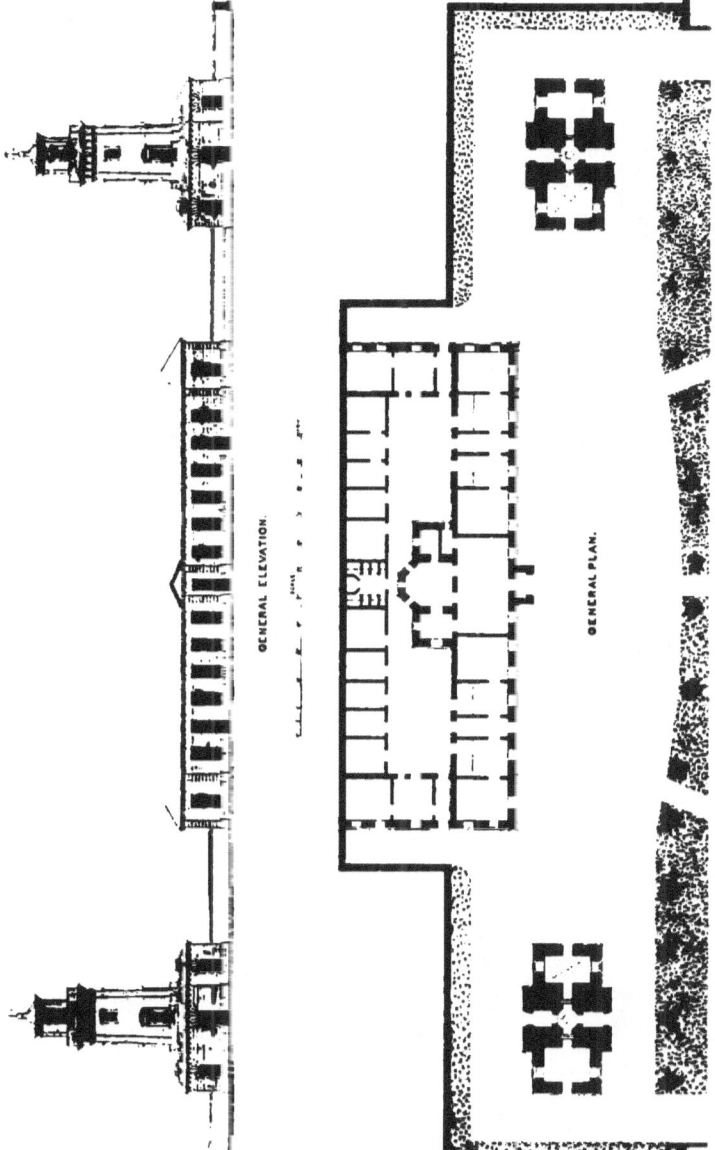

success, showing him at the same time the letter addressed to that gentleman by M. Arnoux. He would not, however, read the letter, and politely stated that he must refuse to show me the lights, as he had positive instructions not to admit any visitors within the station after lighting. He was inexorable to all my arguments, and I was at length obliged to yield, and, while praising him for his strict obedience to orders, could not but regret my unsuccessful journey, and drive of two leagues back to Havre in the rain at midnight, without seeing the interesting objects of my visit. I mention this circumstance as indicative of the character of the French light-house keepers as far as I saw them. *Refusal of keeper to admit visitors after dark.*

At all the stations I visited I found them to be bright, intelligent, and fond of their profession.

The following morning I again proceeded to Cape la Hève, and had the gratification of inspecting what is probably the most extensive light-house station in the world. The towers are very handsome, and are 65 feet high, their focal plane being 397 feet above the sea. *Focal planes.*

The dwellings, engine and machine rooms, &c., occupy the intervening space, about 300 feet, between the towers.

In the engine-room, which is kept with the utmost neatness, are two boilers, which are a combination of the upright and horizontal, and of about eight horse-power. *Boilers.*

In the machine-room are four magneto-electric machines made by the *Compagnie l'Alliance* of Paris. One machine, running with a velocity of four hundred revolutions a minute, supplies each light; and in case of fog or thick weather the other machines are added, so that the uncondensed beam, which in the former case is equal to 200 Carcel-burners, (2,000 candles,) is increased 400 Carcel-burners, or 4,000 candles. *Magneto-electric machines.* *Power of uncondensed beam.*

The lanterns which surmounted the towers when oil was used as the illuminant have been removed, and the magneto-electric lights are exhibited from what were the watch-rooms. Small cylindrical lanterns, about 2½ feet in diameter, formed of glass, cast specially for this purpose, without any sash-bars, either vertical or inclined, being projected from the square seaward angles, and illuminating about 275° of the horizon. *Magneto-electric light exhibited from the watch-rooms.* *Lanterns.*

Figs. 25 and 26* represent in plan and elevation the electric light-room, which is in two stories, each containing an entire set of apparatus.

---

* Figures 25, 26, and 27 have been taken from M. Reynaud's *Mémoire sur l'Éclairage des Côtes de France.*

In Fig. 25 A is the lower room; B, the stairway of the tower; C, steps leading to the upper chamber; D, door leading to the outside platform; K K, iron rails for the regulators; L, the lantern; O, the illuminating apparatus; R, the spare regulator; S, the luminous beam of rays emanating from a small lens placed in rear of the focus, and which throws upon the wall an image of the light. The

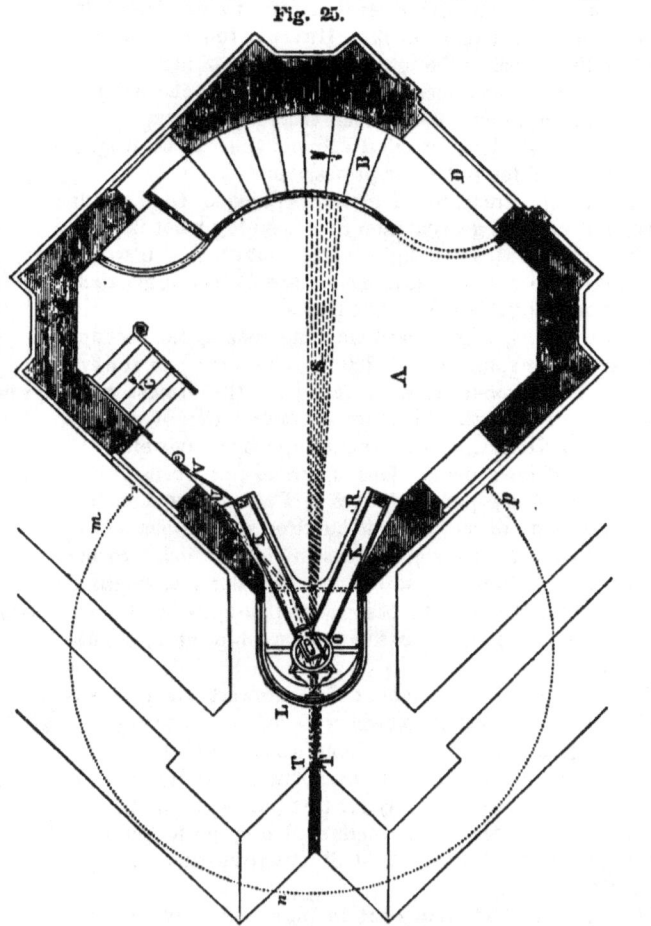

Fig. 25.

Plan of lantern and watch-room, La Hève.

position of this image, with reference to a fixed mark, indicates to the keeper whether the light is in the focus of the lens, and is of the greatest assistance to him, as it is impossible to look at the light itself without injury to the eye, on account of its dazzling intensity. T T are the conducting wires; U, the switch for changing the direction of the current; V, an india-rubber speaking-tube. The arc of visibility of the light is represented by the arc m, n, p.

In Fig. 26 A A represents the illuminating apparatus; B B, rails for the lamps or regulators; C C, the conducting wires; D D, the switches, and E E, india-rubber speaking-tube.

Fig. 26.

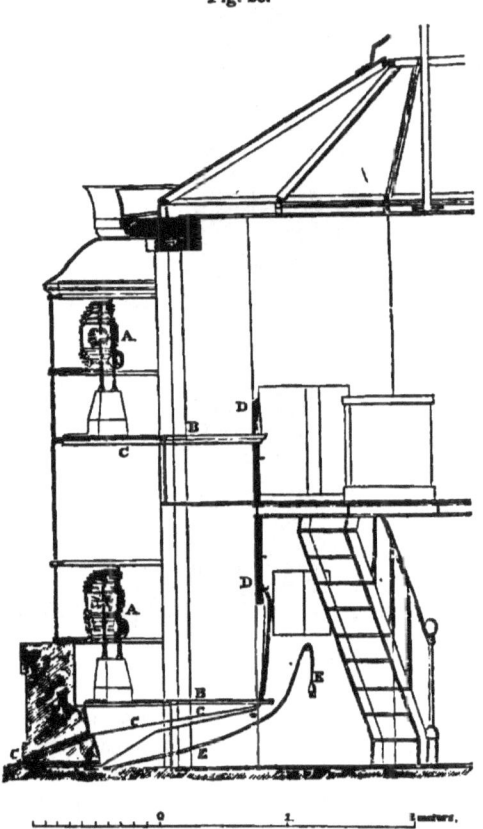

Section of lantern and watch-room, La Hève.

In each of the lanterns, two at each tower, is placed a fixed lens three-tenths of a meter in diameter, the size of the sixth order, shown in Fig. 27, and for each lens there is a duplicate electric lamp, so that, in effect, there are three reserves in case of accident.

It is found by experience, however, as I was informed by the *maître de phare*, that a second lamp is only required when changing the charcoal pencils, so that so many reserves of lamps are more than are actually necessary.

Fig. 27.

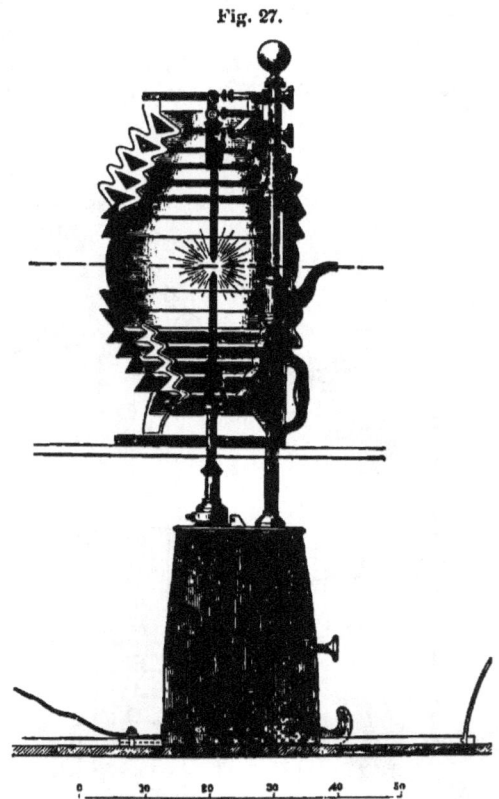

Optical apparatus, La Hève.

Apparatus in duplicate. All the other parts of the apparatus, including the engines, boilers, and machines, are in duplicate, excepting the cables or wires connecting the magneto-electric machines with the lamps. These, M. Reynaud afterward informed me at Paris, it was not thought necessary to duplicate, but I should infer that it would be well to provide in all cases a second wire, from the fact that in observing the lights from the steamer, while crossing the Seine from Honfleur to Havre, I saw that one of them waxed and waned very perceptibly, and on my questioning the *maître de phare* in regard to it, he stated that the wire leading to that light was in an imperfect condition.

In regard to the carbons, I was told that those used for Carbons used in fog. fog are much larger than those used in clear weather, in order to provide for the rapid consumption when two magneto-electric machines supply each of the lights; also that the carbons now obtained in France are much superior to French carbons superior to the English. the English, (which were used at La Hève during the war with Germany,) and that with the former the impurities (which are said to be the only source of danger to this light) are never so great as to cause its extinguishment, and they are also not subject to the production of that fine black dust which I observed at South Foreland and Souter Point as in a slight degree interfering with the full power of the dioptric apparatus.

It will be observed that the only provision at La Hève, for Provision for accidents. the exhibition of an oil-light in case of accident from any cause to the supply of the electric current to the lamp, is an oil-lamp in the small lens used for the electric light, producing an illumination of but little value for sea-coast lights of their importance.

I was informed that at Grisnez, on the Straits of Dover, Light at Cape Grisnez. where the French government has established another electric light, the old lantern and lens for the use of an oil-lamp have been retained, the electric light is exhibited from the watch-room, as at La Hève, and an oil-lamp is always ready for lighting in the first-order lens in the main lantern. A casualty, such as I have mentioned, requiring the substi- Loss of light by use of oil-lamp. tution of an oil-lamp for the electric lamp at Grisnez would reduce the intensity of the light to 6,300 candles, while at La Hève it would be reduced from 50,000 to 260 candles in fair weather, (and from 100,000 to 260 candles in fog,) and probably even less, as the lens used for electric light is not suited in any case for the exhibition of an oil-light.

Drawings illustrating the arrangement at Grisnez were kindly furnished me by Chief Engineer Allard, and are reproduced in Plate XLVI.

Neither the electric light at La Hève, nor those of Souter Point or South Foreland, which I saw in England, are sufficiently high to make necessary the plan which I am informed by General Sherman has been adopted at the new electric Electric light at Port Said, elevator in tower. light-house at Port Said, at the Mediterranean entrance to the Suez Canal, that of providing for the tower an elevator or "lift" which is operated by the steam-power used to drive the magneto-electric machines.

Up to the present time there have been established nine electric light-houses, viz:

In France, two fixed lights at La Hève, and a revolving light at Grisnez;

In England, a fixed light at Dungeness, two fixed lights at South Foreland, and a revolving light at Souter Point;

In Egypt, a revolving light at Port Said;

In Russia, a fixed light at Odessa.

The following excellent paper, which I have found in the *Annales des Ponts et Chaussées*, gives a clear exposition of the electric sea-coast lights at La Hève.

[Translation.]

DESCRIPTION OF THE ELECTRIC LIGHT-HOUSES AT LA HÈVE

BY M. QUINETTE DE ROCHEMONT,

Engineer des Ponts et Chaussées.

Dates of constructing and lighting. — The light-houses of La Hève, constructed during the latter part of last century, were lighted for the first time in 1774; wood-fires were then used. In 1810, lamps with reflectors were introduced, which, in 1845, were replaced by dioptric apparatus. A final change has been recently effected by the introduction of the electric light.

Use of electric piles. — The application of the electric light to light-houses had already been an object of investigation for a long time. Currents produced by electric piles were first tried, but their intensity very rapidly decreased when the apparatus had been in operation for some time; the expense was considerable, and, besides, it appeared rather hazardous to confide to ordinary keepers the care of keeping and regulating the piles. The system based on induction currents gave, on the contrary, very good results, in experiments made at the Central Light-House Workshops at Paris; so that in 1863 the *Test of electric light to be made at La Hève.* — Minister of Public Works decided that one of the light-houses of La Hève should be illuminated provisionally by electric light as a test. As the experiment confirmed the anticipations, electric illumination was definitively applied to both light-houses toward the end of 1865.

Description of the apparatus. — The currents are produced by magneto-electric machines worked by steam-engines, and are carried by conducting-cables to the regulators, or electric lamps, used to regulate the separation of the carbon points, between which the light is produced. The magneto-electric machines and the engines are placed in the center of the keepers' dwelling, in two rooms fitted up for that purpose. (See Plates XXXVII, XXXVIII, and XXXIX.)

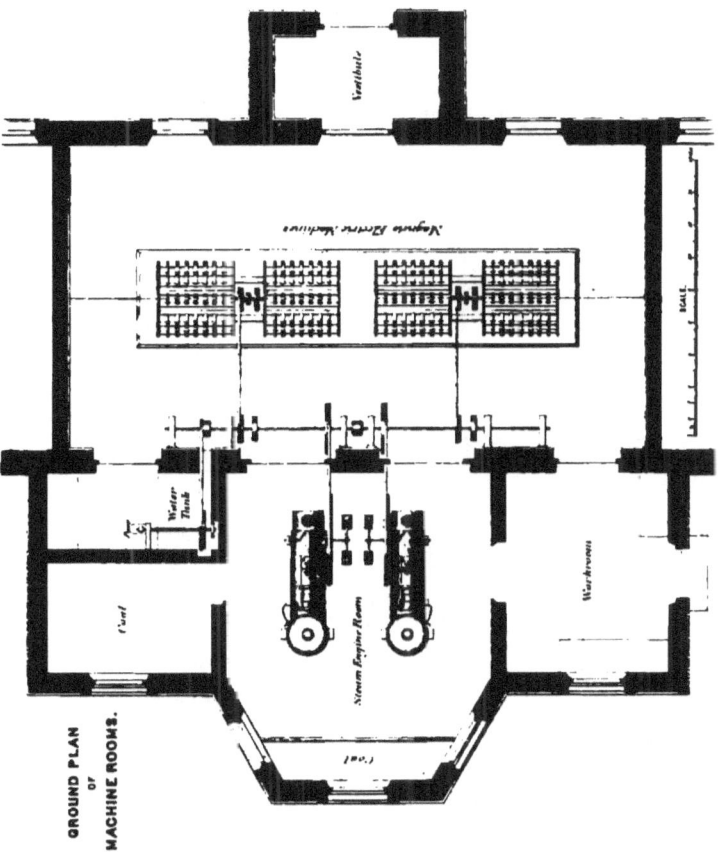

The steam-engines are two in number, and present no peculiarity worthy of notice, as they belong to the common stationary class with locomotive-boiler of eight horse-power. They are certified as capable of resisting a pressure of 70 pounds to the square inch. But one engine is generally used at a time, as it is sufficient to work two magneto-electric machines. At first it was thought necessary to keep the second engine with banked fires in case of accident to the first, but practice has shown that this precaution was needless. The motion is conveyed from the steam-engines to the machines by belts and an intermediate shaft. (Plates XXXVIII and XXXIX.) *Steam engines.*

The magneto-electric machines were furnished by the Alliance Company. They are composed (Plates XL, XLI, and XLII) of a cast-iron frame on which are placed mahogany cross-pieces which serve as supports to seven parallel series of compound magnets, all of which converge toward the central axis of the frame. The magnets of the two outer series are formed of three superposed plates, curved horse-shoe shape, the others of six plates. They are so arranged that the poles nearest each other, both horizontally and vertically, are always of opposite signs. The six-plate magnets are of a power of 145 to 155 pounds, the three-plate, about 75 pounds. Between the seven rows of magnets, there revolve six bronze disks, (Plate XLIII, Figs. 1 and 2,) mounted on an axis supported by the frame. On each of these disks, sixteen induction-spools are fixed by bronze clamps and screws. (Plate XLIII, Figs. 10 and 15.) Each spool (Plate XLIII, Figs. 10–14) is a tube of soft iron about one-third of an inch thick, $1\frac{1}{2}$ inches in exterior diameter, and $3\frac{3}{4}$ inches long, slit radially so as to more quickly lose its magnetism. Each tube is wound with eight copper wires one-twenty-fifth of an inch in diameter and about 50 feet long, so that there are about 400 feet of wire wound around the spool. These wires are covered with cotton and insulated by asphalt dissolved in spirits of turpentine; they are wound in the same direction on all the spools. *Magneto-electric machines.*

The best method of placing the spools has been determined by experiment and trial; a certain number of extremities of wires of the same sign or denomination are brought together, and an equal number of extremities of the opposite sign. All the extremities of one sign communicate with the central axis of the machine, all of the other sign with a metallic sleeve fixed on the axis but insulated from it by a plate of India rubber placed concentrically and joined to two other perpendicular plates. (Plate XLIII, Figs. 5–8.) *Method of placing the spools.*

The shaft and sleeve revolve in journal-boxes (Plate XLIII, Figs. 5, 6, 9) insulated from the rest of the frame by plates of vulcanized rubber. From these boxes and the axis start the wires which transmit the currents generated by the machine.

Induction-currents. Induction-currents are produced whenever the spools either approach or leave the poles of the magnets. Thus there are sixteen changes of direction of the current to each revolution of the cylinder; and the wires are therefore alternately traversed by currents of opposite direction and a series of discontinuous sparks is obtained. The eye, however, perceives no interval, for the number of sparks exceeds 100 per second, since the magneto-electric machines make from 390 to 400 revolutions per minute.

Grouping of the machines. Four magneto-electric machines (Plates XXXVIII, XXXIX, XL, XLI, XLII) have been set up. They are grouped two and two, each group connected with one of the lights. Under ordinary circumstances one steam-engine works two machines, one of each group, but during a fog, when it is wished to increase the intensity of the lights, each group of machines is worked by an engine. For this purpose the shaft for transmitting the motion is of two parts, which are connected or disconnected at pleasure, as either ordinary or double light is wished, (Plate XLII, and Plate XLIII, Figs. 3, 4, and 16.) In the latter case the two machines of the same group are connected.

Switches. According as it is wished to use one, the other, or both of the machines connected with one light-house, it is necessary to change the points of attachment of the conducting-wires. In order to avoid the mistakes which might result, M. Joseph Van Malderen, superintending engineer of the Alliance Company, has contrived a most ingenious switch, placed in the machine-room, (Plate XLIV, Figs. 1 to 4.) From this switch alone proceeds the cable conducting the electric currents. This cable is composed of three wires, one of which, β, communicates directly with the wire b, uniting the axes of one of the groups of machines, while the two others, a and γ, proceed from buttons at the lower part of the switch. The wires a and c, coming from the journal-boxes of the machines, are brought to two buttons at the upper part of the switch.

Communication between the wires. The communication between the wires a or c (Plate XLIV, Fig. 3) and the wire a, which, when a light of ordinary intensity is required, is always the one which conducts the current to the lights, is through a forked piece of metal, one end of which is attached by a pin to the upper button which corresponds to the machine then in operation. On the contrary, when a light of double intensity is wished, the currents arriving at the upper buttons pass to the wires a and γ through two straight rods, (Plate XLIV, Fig. 1,) and do not meet until they reach the interior of the lamp. Thus no displacement of the wires need be feared, as all are permanently attached to the machines or switches, the switch-key being all that it is necessary to touch. Each cable is carried underground to the light-house, thence through the stairway to the lantern.

In order to avoid extinctions which might result from accidents to the regulators, it has been thought necessary to have in each light-house two optical apparatus, one above the other, and for each two lamps. These lamps slide on rails fastened to a cast-iron plate, and can thus be moved into exact position. Duplicate optical apparatus.

As it was also wished to avoid displacing the conducting wires in the light-house lanterns, a second switch is used, by which the light can be doubled on either stage at will without moving the wires. (Plate XLIV, Figs. 5 and 6.) Second switch.

The wire β, coming from the axes of the two machines of the same group, connects directly with the lower cast-iron table, and the communication between the two tables is through one of the uprights of the lantern, also of cast iron. The current passes directly from the table into the lamp to reach the upper carbon point, as will be seen hereafter. The wire a communicates with a large copper bolt, A, (Plate XLIV, Figs. 5 and 6) by the metallic plate on which the bolt slides. The wire γ communicates in the same way with another bolt, B, sliding vertically like the large one, from which, moreover, it is magnetically insulated by an ivory handle.

When the bolts are pushed down, the currents arrive by the wires a and γ, pass into the wires a' and γ', and go to the lower stage; on the contrary, if the bolts are pushed up, the currents pass into the wires a'' and γ'', and thus reach the upper stage. If the light is of ordinary intensity, there is no current in the wire γ, and consequently none in the wires γ' and γ''. The currents from the wires a' or a'' reach the lamp through a flat metallic spring placed above the cast-iron plate on which the lamp rests. In case of double light the currents from the wires γ' or γ'' reach the interior of the lamp by a second spring placed at the side of the first.

In this way, when double light is used, only one of the stages of the lantern is lighted. Experience has shown that it is better to give double light on one stage than ordinary Use of single stage when double light is needed.

light on both, though at little distance the two lights (separated about 7 feet) appear as one.

Lamps, or electric regulators. Serrin's regulators, the only ones used at La Hève, (Plate XLV) are composed of two point-holders, each attached to a vertical rod which slides in a guide-tube. The rods are so arranged that when the upper one descends by its weight the other rises. To effect this, the upper rod in descending pulls the chain A, one end of which is attached to the lower end of the rod and the other to a pulley, C. This pulley, in turning, winds up another chain, B, attached to the lower end of the lower point-holder. Thus both rods move together; but, as it has been found that the lower carbon point consumes somewhat more rapidly than the other, the pulley C has two barrels of slightly different diameter; experience has shown that they should be to each other as 100 to 108 to keep the luminous point always at the same height. The motion of the pulley is transmitted by clock-work to the fly-wheel D, used to moderate the movement of the point-holders when they approach each other, impelled by the weight of the upper rod.

The lower guide-tube is carried by an oscillating parallelogram, MNOP, affected by two forces acting in opposite directions; one, the tension of the spiral spring R, which raises it; the other, the action of the electro-magnets S, on the armature Q attached to the parallelogram, which lowers it. When the lamp is in use and the points are at a proper distance from each other, the armature Q is attracted by the electro-magnets; the parallelogram descends, the pawl F, carried thereon, engages the ratchet-wheel E, mounted on the same axis as the fly-wheel D, and the points can no longer approach each other. As the points consume, their separation increases, and, consequently, the intensity of the current traversing the electro-magnets decreases; the armature is attracted with less force, and the parallelogram re-ascends, the pawl F escapes from the ratchet-wheel, and the two points re-approach, impelled by the upper rod. They continue to approach until the current has resumed its normal intensity, when the parallelogram redescends, attracted by the magnets. The points approach about one-twenty-fifth of an inch every time a tooth of the wheel escapes from the pawl.

By a screw, T, acting on a lever, the tension of the spring R can be increased or diminished so as to regulate the action of the lamp. This tension should, indeed, vary according to the intensity of the current passing about the magnets S, as these two forces should be in equilibrium when

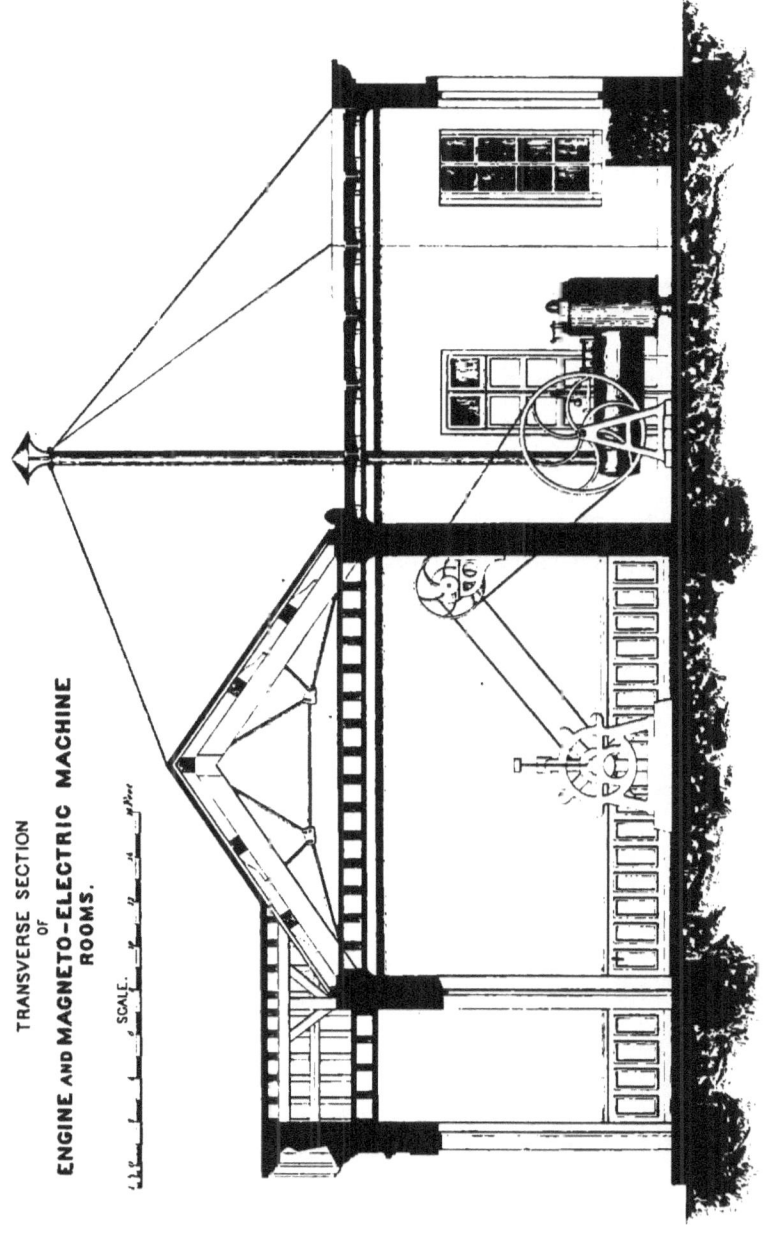

the points are a proper distance apart; the voltaic arc can thus be kept of a constant length.

As we have already stated, the currents enter the regulator by the base-plate and through either one or two springs placed beneath, as either ordinary or double light is required. In the former case the current passes from one of the springs into one of two pieces, shown together at Y; thence around the two magnets S, through the piece XX, and reaches the oscillating tube and the lower point-holder. In the second case, the other current of the same denomination, conducted by the wires γ' or γ'', passes from the second spring into the second piece at Y, and attains the piece V, where it meets the first current, without having passed about the electromagnets. The current of the contrary sign passes from the base-plate through all the other parts of the lamp to reach the upper point-holder. The parts traversed by currents of contrary signs are insulated from each other by vulcanized rubber and ivory. To assure good working of the lamp, M. Serrin has contrived some devices which we will point out.

Course of the currents.

Devices of M. Serrin.

When the upper rod is raised, the pulley C (on which are wound the chains connecting the ends of the rod) revolves; but, to prevent forcing the clock-work, a ratchet-wheel, G, is placed on the same axis as the first cog-wheel.

The two screws H and I, which meet two pieces on the side NO of the parallelogram, limit its motion.

By the screw K (Plate XLV, Fig. 1) the upper rod can be raised or lowered to alter the position of the luminous point. In his last lamps M. Serrin has adopted a much better device, by which the luminous point can be placed in exact position by moving both carbons at once, thus preventing any decrease of light. For this purpose the chains A and B are run over a little auxiliary pulley which can be moved up or down by a screw and lever so as to raise or lower both points the same distance at the same time. This device is applied to four of the lamps in use at La Hève.

A small gauge, Z, movable around the upper rod, indicates the exact height at which to place the gap between the carbons; that is to say, the luminous point. By the two screws L and L′ the carbon-points can be adjusted exactly opposite each other by moving the upper point in a direction either parallel or perpendicular to the plane of the two rods.

Finally, to prevent the point-holders burning when the carbon is consumed they stop at a distance of about $2\frac{1}{4}$ inches from each other, as then the upper rod strikes the base of the lamp and can go no farther.

Foucault's regulator. Another regulator, invented by M. Foucault and constructed by M. Dubosq, has also been tried. It was, however, not well adapted to be placed on a cast-iron table, and was twice broken within a short time, in consequence of a too great separation of the carbon points.* Its use was then abandoned.

Carbon points. The carbon points used for electric illumination are manufactured from the residue contained in gas-retorts. They are about 10 inches long and from one-third to one-half an inch thick, according as they are used for ordinary or double light.

Optical apparatus. The optical apparatus, Fig. 27, are about one foot in diameter; the catadioptric rings are symmetrical, both above and below, on account of the form of the points and the luminous center. The luminous rays are sent from the rings tangentially to the surface of the sea. The joints of the rings are placed in a direction parallel to that taken by the rays after their refraction. The luminous center being of very small dimensions, (about two-fifths of an inch by two-fifths to three-fifths of an inch,) the lantern can have no sash-bars, as the occultation of a part of the horizon which they would produce must be avoided.

Divergence of rays and position of luminous point. The divergence of the luminous rays is about 6°; about the same as that of a first-order oil-light. It is indispensable that the luminous point should remain in exactly the proper position in the optical apparatus, as a vertical displacement of one-fifth of an inch would raise or lower the luminous beam 2°. To assure a correct position for the luminous center, there has been attached to the edge of the optical apparatus a small lens, which throws the image of the points on a screen placed at the other end of the service-room. This image should be in such a position that the gap between the points appears on a line previously traced on the screen. If it does not, the position of the luminous center should be adjusted as has been indicated above. The image is magnified 22 diameters.

The lights of La Hève illuminate three-fourths of the horizon. The lanterns and service-rooms are at the unlighted angle.

Other aids. Oil-lamps with large burners can be placed in the center of the optical apparatus if there should occur an accident preventing the production of electric light; in such case one of these burners is placed on each stage of the lantern.

---

* M. Dubosq has since placed a stop-piece to prevent the too great separation of the points; the lamp works with more regularity, but we have not again used it.

LIGHTHOUSES OF LA HÈVE.
ELECTRIC LIGHTS.
MAGNETO-ELECTRIC MACHINES
FRONT ELEVATION.

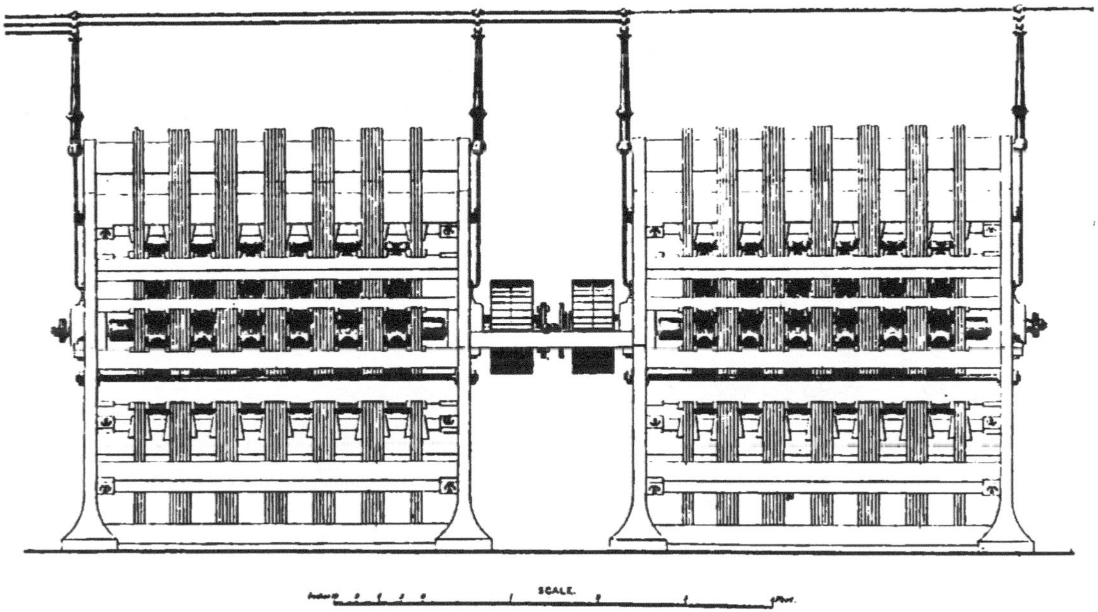

PLATE XL

Each luminous point has about the intensity of a fourth-order light; at a short distance the two lights blend and appear as one.

<small>Call-bells.</small> Call-bells with dials connect the engine-room and the light-house lanterns, so that the engineers and keepers can communicate with each other. Other bells are placed in the dwellings of the principal and other keepers.

<small>Water.</small> Water for the steam-engines and for domestic use is kept in cisterns of a total capacity of about 46,000 gallons. It is rain-water collected from the roofs of the buildings and from courts paved with asphalt, a total surface of about 2,000 square yards. It is pumped by the engines into a tank near the engine-room, (Plate XXXVIII.) There is a small workshop, so that the engineers can make all current repairs which do not require special artisans or implements.

<small>Organization of the service.</small> In charge of the lights there is a principal keeper, (*maître de phare*,) who has under his orders six assistants, (*gardiens*,) two of whom are engineers, whose special duty is to attend to the steam-engines; the others attend the lamps. The engineers have the title and rank of keepers (*gardiens*) of the first class, but as their service is much more arduous than that of their comrades, an extra compensation is allowed them at the end of the year if they have given satisfaction. With a view to subordination and harmony of relations at the lights, it seemed better to give these employés a relative rank rather than to exclude them from the service by giving them the title of engineers.

<small>Accidents, extinctions.</small> Since commencing the use of electricity (November 1, 1865) several accidents and a few extinctions have occurred. We give below a complete list.

March 5, 1866.—The suspension-chain of the upper rod of one of the electric regulators breaks.

April 8.—Light out five to six minutes, in consequence of the slipping off of several of the belts.

July 12.—The heel-screw of one of the magnets of machine No. 22 becoming loose, rubs against the spools of one series and injures them so that several have to be replaced. As this accident happened at the moment of setting the machine in motion, no extinction resulted.

July 27, September 5, September 8, September 12.—The suspension-chains of the upper rod of four regulators break; the damaged lamp being immediately replaced by another, the light was out but a few seconds.

October 2.—Four mahogany cross-pieces of machine No. 19 are broken; four spools are injured. The reserve machine being immediately set in motion, the light was out but a short time.

January 11, 1867.—The suspension-chain of the lower rod of a regulator breaks.

March 9.—Light out 3 minutes, the belt of the engine having slipped off.

March 17.—Light out 3 minutes, the fire-grates of the engine having fallen.

March 25.—Southern light out 7 minutes and 3 minutes.

March 29 to 30.—At 10.35 p. m. the southern light out. Being unable to relight it, oil-lamps are placed in the lens-apparatus. At 2.35 a. m., on again trying the electric light, it works. Electric light out about 4 hours.

May 25.—At 12.5 a. m. the water-pipe of the engine bursts, the pressure falls; reflectors are lighted at 12.20. The damage having been repaired, the electric light is re-established at 1.5.

November 20.—Southern light out 5 minutes, caused by the slipping off of the belt of the magneto-electric machine.

January 5, 1868.—Southern light out 9 minutes.

March 3.—Light out 7 minutes, caused by the belt of the steam-engine having twice slipped off.

March 19.—Light out 2 minutes. The pin of the fly-wheel becoming loose, it was necessary to tighten it.

May 4.—Southern light out 3 minutes, the belt of the magneto-electric machine falling off at the time of firing up the southern engine.

October 26.—At the moment of lighting, the joint of the steam-gauge of the southern engine bursts. The fires of the northern engine are lighted, but before steam is up the accident is repaired, so that the hour of lighting is not delayed.

October 30, 1869.—Northern light out 15 minutes, the keeper having fallen asleep.

Unavoidable nightly extinctions. In addition to the extinctions mentioned above, which were the result of accidents, there occur every night a few others of short duration, which cannot be avoided. When the lamps are changed, for instance, the light is out a few seconds; also, when double light is to be produced after ordinary light, or *vice versa*, it is necessary to throw in or out of gear the two parts of the shaft connecting the two magneto-electric machines of the same group, so the light is out 2 to 3 minutes.

Recapitulation. To recapitulate, we find that in four years there have occurred—

LIGHTHOUSES OF LA HÈVE.
ELECTRIC LIGHTS.

PLATE XLI

MAGNETO-ELECTRIC MACHINE.
SIDE ELEVATION.

SCALE.

Three extinctions of 2 minutes, 3 minutes, and 1 hour duration, caused by accidents to the steam-engine.

Five extinctions of from 3 to 7 minutes, caused by the slipping off of the belts. These extinctions were the result of the negligence of the keepers. Measures have been taken to prevent similar occurrences.

Two accidents to the magneto-electric machines, from which no extinctions of consequence resulted, as the machine in reserve was immediately set in operation.

Six lamps were sent back to Paris in consequence of the breaking of the suspension-chain of one of the carbon holders, but the extinctions caused by these accidents are not worthy of mention.

Four extinctions of 3, 7, 9 minutes, and 4 hours' duration, the causes of which we have not been able to determine.

One extinction of 15 minutes, resulting from the sleep of a keeper.

The extinctions originating from the steam-engines, or the slipping off of belts, offer nothing particularly worthy of notice; one only exceeded 7 minutes, and was the result of the breaking of the water-pipe; the others are mainly attributable to the negligence of the machinists.

Causes of accidents to magneto-electric machines. The accidents which occurred to the magneto-electric machines rendered it necessary for M. Joseph Van-Malderen, superintending engineer of the Alliance Company, to make a journey to Havre, in order to attend to their reparation. The second of these accidents was probably caused by the falling out of one of the wedges used for keeping the magnets in position, and the magnet having nothing to hold it, hit the wheel carrying the spools.

Cause of breakage of suspension-chain of the carbons. The breaking of the suspension-chain of one of the carbon-holders (an accident which occurred six times) was caused by these chains passing too near other parts from which they should have been separated, and thus becoming heated. M. Serrin has easily succeeded in remedying these disadvantages, and they will not again occur.

Extinctions by unknown causes. The extinctions of 7, 3, and 9 minutes, occurring to the southern light, are due to causes which we have been unable to ascertain; it is probable that they were the consequences of a lack of vigilance on the part of the keepers. In regard to the extinction of four hours' duration, we are unable to attribute its cause to anything but malice on the part of some one, for the light went out at 10.35 p. m., and it was impossible to relight it, as the currents no longer reached the lantern. Subsequently, however, when we arrived, about 2 o'clock a. m., the light was immediately restored without anything being done to the machines. The two conducting-wires had probably been connected at some point before reaching the lantern, and the circuit being thus closed, the currents no longer reached the lamp. It is proper, however, to state that an open investigation on this subject gave no result.

Decrease in the number of accidents. The list given above will show that accidents happen less and less frequently. Such a result might have been foreseen to a certain extent; still it is well to prove it. As the keepers become better and better acquainted with the management of the apparatus of which they have charge, they become less surprised at incidents which may occur, and can immediately apply the proper remedies to a state of things which might become grave if allowed to continue. Thus, one of the engineers, who has been in the light-houses since the end of 1863, has never had any accidents on his watch, while others serving during the same period have had only too many.

Substitution of new steam-engines. In addition to the accidents mentioned above, it was found necessary, at the beginning of 1868, to substitute new steam-engines for those which had been in the service since 1865, and likewise to replace some of the journal boxes of the magneto-electric machines.

Old engines insufficient. The old steam-engines were furnished by M. Rouffet: they were of five horse-power and certified for a pressure of six atmospheres. This was found insufficient for the labor they had to perform. From experiments made on their greatest power, in March, 1866, the results, indeed, showed that they were unable to develop continuously more than six horse-power; that, under these conditions, they gave one hundred revolutions per minute, and the mean pressure in the boiler was five and a half, without ever reaching five and three-quarters atmospheres. Now, to produce a light of normal intensity, these engines are required to make one hundred revolutions, and the pressure was five and a quarter atmospheres. The engines were thus necessarily driven to the limit of their capacity, which could not be otherwise than very injurious, resulting in rapid waste. Already, in 1867, a part of the plates of the fire-box, and the pipes of the southern engine, had to be replaced at an expense of $385.25. Other repairs made on these engines to prevent them from heating cost $234.60.

Journal-boxes. The magneto-electric machines are furnished with the self-lubricating journal-boxes of Avisse, but in the insulated parts the arbors had the same diameter beyond the journal-boxes; the oil was drawn along by centrifugal force, and the bearing was not uniformly lubricated. On this account the journals did not wear evenly, which was remedied by grooving the arbor beyond the box. The worn journals had to be re-turned, and the cushions of the journal-box renewed.

Quite a number of different apparatus are combined to produce the electric light; if any are not in proper condition they affect the light. <small>Causes of the irregularity of the light.</small>

If there is a lack of pressure in the boiler, the magneto-electric machines run slowly and the light scintillates markedly; instead of a steady light, the eye perceives successive flashes.

The bearings of the central shaft of the machines must be well insulated, as otherwise the light loses in intensity.

In the regulators the separation of the carbon points sometimes varies; the light decreases until they return to a proper position. This disadvantage has been almost entirely obviated by recent modifications made by M. Serrin. By suitably regulating the tension of the spiral spring the variations of intensity of the light, as far as depends on the regulator, may be almost entirely prevented.

The principal causes of the irregularity of light originate in a want of homogeneity of the carbon points and the displacement of the voltaic arc. <small>Want of homogeneity in the carbon points.</small>

The points, as we have already stated, are made of the carbon deposits of gas-retorts. This is sufficient to explain their want of homogeneity. They should be hard, well pressed, and give a very dry sound when broken; by their external appearance alone a just estimate of their quality can be formed. The breaking of the points when in use, although rare, will sometimes occur; this necessitates a change of lamp, and therefore an extinction of a few seconds; it is not practicable, in fact, to wait until the carbon points re-approach, as the luminous point would then be displaced. The want of homogeneity of the carbon also causes a displacement of the luminous point. In order to remedy this, M. Serrin contrived the mechanism for simultaneously lowering and raising the points. The metallic or siliceous grains found in the substance of the carbons also affect the regularity of the light by acting on the voltaic arc.

It is proper to state, however, that these oscillations of the light are much stronger and more apparent when seen near at hand than when observed at a distance, and that they are no serious disadvantage; under no circumstances could they be confounded with those of eclipse or scintil- <small>Oscillation of the light not perceptible at a distance.</small>

lating lights. These oscillations, moreover, have greatly diminished, and electric lights may henceforth be considered as quite steady.

Luminous intensity. Experiments made at the central light-house workshop show that the mean intensity of the light produced by a six-disk magneto-electric machine is two hundred burners. This intensity varies within certain limits, and it was found necessary to make a great number of experiments in order to estimate its mean. The luminous center, placed in an optical apparatus of one foot diameter, gives an intensity of about five thousand burners.

Expenses of the light. More than four years having elapsed since the definitive establishment of electric lights at La Hève, we are enabled to furnish an exact statement of the annual expenses.

The following table shows the expenses, including salaries. The items are taken from the light-house journal and from the accounts of the conductor:

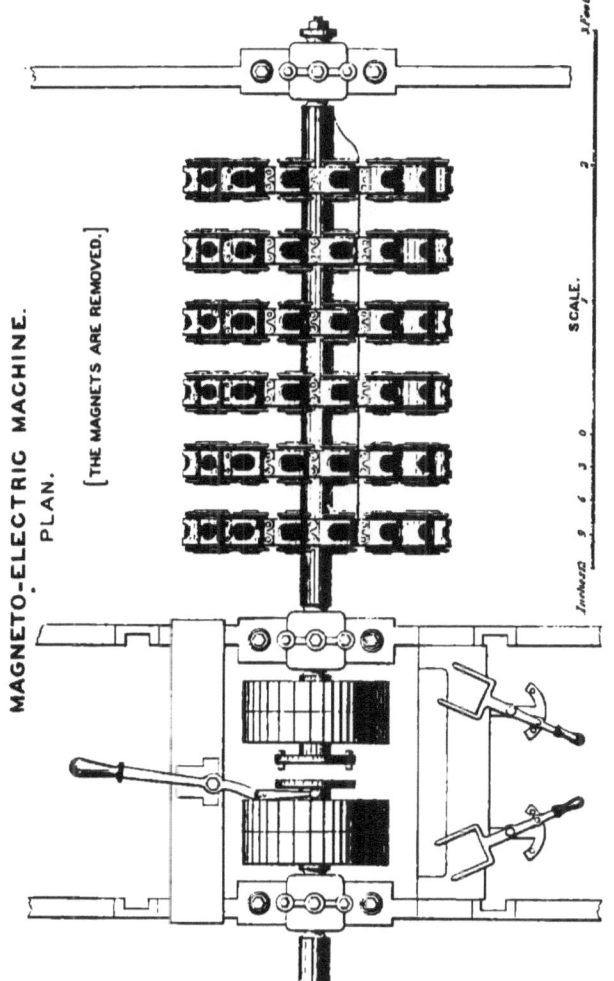

Nature of expense.	Prices.	1866.		1867.		1868.		1869.		Average year.	
		Quantities.	Cost.	Quantities.	Cost.	Quantities.	Cost.	Quantities.	Cost.	Quantities.	Cost.
Salaries			$1,148 35		$1,148 35		$1,169 83		$1,102 83		$1,153 00
Allowance for fuel			98 50		102 03		135 10		108 08		112 00
Bonus to engineers			57 50		48 25		57 00		57 90		58 00
Water	$0 30 to $6 90						04 47				
Coal, per ton of 2,240 lbs	6 83	153,350 gals	51 43	138,061 gals	70 07	156,471 gals	90 58	149,094 gals	73 15	150,000 gals	75 00
Coke, per ton of 2,240 lbs	14	14½ tons	90 30	12½ tons	818 77	104 tons	732 00	11½ tons	700 40	12 tons	730 00
Large carbon points	14	80 gyms	706 40	92½ tons	7 30	83 tons	84	90 tons	38	90 tons	7 50
Small carbon points, per lb	14	40 feet	0 00	34 feet	285 00	6 feet	280 00	11 feet	270 30	24 feet	291 00
Cotton-waste, per lb	09 to 14	2,207 feet	308 98	2,040 feet	43 08	2,000 feet	51 42	1,020 feet	51 75	2,160 feet	43 00
Tow, do	14 to 13	494 pounds	42 55	450 pounds	2 84	259 pounds	4 37	537 pounds	2 07	440 pounds	3 55
Polishing-rouge, do	08	3 pounds	43	19 pounds	2 43	35 pounds	4 39	14 pounds	1 53	22 pounds	2 00
Whiting, do	09	8 pounds	72	27 pounds	2 70	46 pounds	4 41	17 pounds	2 43	22 pounds	3 13
Grease, do	11 to 11½	12 pounds	1 17	30 pounds	13 00	49 pounds	13 39	27 pounds	9 70	23 pounds	3 12
Lubricating-oil, per gal	2½ to 1 50	86 pounds	9 48	114 pounds	71 73	135 pounds	152 03	89 pounds	129 69	110 pounds	13 00
Illuminating-oil, do	87 to 62	58 gallons	20 64	55 gallons	198 00	139 gallons	127 78	130 gallons	149 30	60 gallons	108 00
Emery-cloth, per yard	30 to 40	106 gallons	99 36	148 gallons	5 75	147 gallons	0 94	173 gallons	0 12	135 gallons	122 00
Bodu, per lb	04 to 08	16 yards	6 40	10 yards	40	17 yards	30	16 yards	35	16 yards	6 00
Spirits of wine, per quart	48 to 58	7 pounds	42	10 pounds	4 00	7 pounds	1 92	0 pounds	41	11 pounds	6 00
Towels, each	21	4 quarts	2 08	8 quarts	4 64	4 quarts	2 17	3 quarts	24	6 quarts	3 10
Linen wiping-cloths, do	11		1 94	10	5 07	7	2 73	50	4 30	10	5 25
Cotton wiping-cloths, do	48	36	7 64	97	4 18	13	2 74	32	3 12	52	4 40
Chamois-skins, do	62½	38	6 36	30	44	34	1 44	4	1 62	41	1 61
Feather brushes, do	53	2	03	1	9 51		1 36			2	1 25
Hand-dusters, do	10½	2	46	4	1 69	2	53	1		3	1 46
Silver-plate brushes, do		4	43	3	21	1	42	21	22	1	42
Various repairs and supplies to illuminating-apparatus									11		
Total			2,061 53		617 78		478 08		334 30		434 00
					3,397 74		3,360 60		3,091 90		3,215 34

NOTE.—The expenses of 1867 do not include the cost of the journey of M. Joseph Van Muldeven, which was paid at the central light-house office. The figures in the last two columns are not mathematical means of those in the preceding columns; they are based on the known results of preceding years, and show the probable consumption and expense for a single year.

Expense for water for the engines.

Rain-water is used for steam-engines, but in very dry years it has failed; it is then brought from Sainte-Adresse at an expense of about 40 cents per hundred gallons. This will explain the item of expense for water in 1860 and 1868.

Before increasing the water-collecting surface, or constructing another cistern, we wish to ascertain if the saving will justify the expense of construction.

The old steam-engines consumed about thirty-four gallons of water per hour; the new average about forty.

Expense of repairs to the machinery.

The repairs to the machinery in 1867 and 1868 were considerable on account of the inefficiency of the steam-engines; there is every reason to hope that this expense will be diminished with the new engines, which are much more powerful; we have consequently reduced this item in the last column. Other expenses for supplies and repairs include stowage of coal and cleaning of cisterns and of the water-collecting surfaces, washing of linen, and other items of little consequence. In order to obtain an exact statement of the annual expense incurred for the production of electric light, it would be necessary to add a certain amount to the figures mentioned in the preceding table for deterioration of the apparatus. It is proper, however, to state that there is but little wear to most of the machinery. The regulators, which have been in use for six years, work as well as on the first day, the only wear being to the journal-

Item of wear to be taken into account.

boxes of the magneto-electric machines. Thus, the wear of the steam-engines is almost the only item to take into account. These diminish quite rapidly in value.

Table showing number of hours of illumination and duration of the working of the engines.

	1866.		1867.		1868.		1869.		Average year.
	H.	M.	H.	M.	H.	M.	H.	M.	Hrs.
Engines working	4,288	55	4,518	33	4,509	47	4,496	21	4,540
Magneto-electric machines working	3,943	25	4,152	22	4,203	47	4,188	06	4,200
Total hours illumination	3,872	10	4,087	51	4,142	47	4,127	16	4,135
Ordinary light used	3,789	55	3,989	04	4,129	01	4,087	22	4,055
Double light used	82	15	98	47	13	46	39	54	80

Time of lighting and extinguishing.

Since May 1, 1867, the time of lighting is a quarter of an hour after sunset, and the lights are extinguished a quarter of an hour before sunrise; while formerly they were lighted a quarter of an hour later, and put out a quarter of an hour earlier; so the total annual illumination is increased $182\frac{1}{2}$ hours. This increase is taken into consideration in the last column of the preceding table. It occurred in 1868, 1869, and a part of 1867.

LIGHTHOUSES OF LA HÈVE.
ELECTRIC LIGHTS.

PLATE XLIII.

MAGNETO-ELECTRIC MACHINES.
DETAILS.

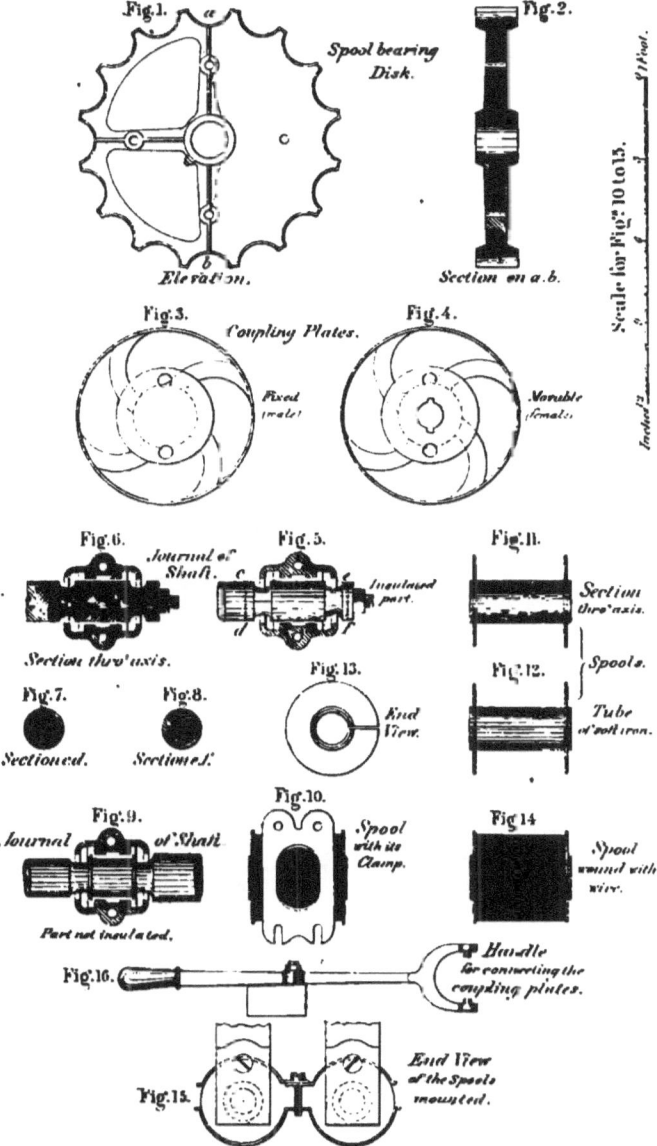

The steam-engines are fired up about an hour before the time of lighting. It takes nearly three-quarters of an hour to get up steam.

The magneto-electric machines are started about ten minutes before the hour of lighting, so that the currents may be well established.

Double light. Double light is produced whenever the fog is so dense that keepers at La Hève cannot see the beacon-lights on the north pier of Havre.

Expense of changing the system of lighting. Electric light was twice introduced at La Hève; in 1863 and 1865. At the first period the superstructure of the southern light-house was finally modified by substituting for the glass lantern, mounted on a sub-base, a structure of stone-work with the lantern in the angle.

Temporary buildings were made for the engines, but these were demolished in 1865, at which time the lantern of the northern light-house was also altered, and the engines were placed in the house formerly used as a dwelling by the principal keeper, a small building being added in the rear. Three new dwellings for the keepers were constructed, as well as a cistern, and the water-collecting surface was enlarged.

Expense of buildings. Two sums, one of $6,266.58, the other of $11,676.50, were expended for these works; but these do not include the price of the steam-engines, magneto-electric machines, regulators, lanterns, cupolas, &c., which were sent from the central light-house depot. As the details of these expenses have no general interest, we shall not give them.

In altering other light-houses the expenses, taking into account the difference of location, would probably be quite different.

Cost of special apparatus. We shall confine ourselves to giving some details of prices of special apparatus, as these prices are almost entirely independent of the situation of light-houses:

The two steam-engines and accessories cost....		$2,493 42
That is to say,		
For two engines................	$2,161 60	
Two Foucault regulators.......	77 20	
Two chimneys.................	41 56	
Two bed-plates for fly-wheel....	56 82	
Feed-pipes and mounting of the machinery..................	156 24	
The feed-pump and water-tank cost............		338 31
Shaft		378 76
Six belts, about............................		48 25
Six magneto-electric machines................		9,339 86

Namely, for—

Four machines	$9,264 00	
Columns and copper rods for conductors	56 56	
Two counters	19 30	
Four switches, with their accessories		$77 07
Two conducting-cables, about		271 20
Eight regulators, with their accessories		2,242 66
Namely, for—		
Eight regulators	1,544 00	
Alteration of first regulator in 1865	119 66	
Four cast-iron plates, with the springs	308 80	
Expenses for experiments allowed to M. Serrin*	270 20	
Four optical apparatus		1,389 60
Electric bells		262 50

Comparison of electric and oil lights. From December 25, 1863, when the southern light-house was lighted for the first time by electricity, to August 31, 1865, the date when the northern light-house was still lighted with oil, it was easy to compare the two modes of producing light. At that time the magneto-electric machines had but four disks; the intensity of the luminous point was only 125 burners, and the intensity of the beam sent to the horizon not more than 3,500 burners when ordinary light *Intensity of electric luminous point and oil-light.* was used. The intensity of the oil-lights was 630 burners.

From information obtained in 1865, it seems that the electric light always was seen before the other, even in clear weather.† The light of Barfleur was often seen at the same *Range of electric light.* time as the electric light of La Hève. According to statements of commanders of vessels, this frequently occurs now, while with the oil-lights it was very rarely observed. The electric light appears to have had a considerably greater range. We shall refrain from giving too great emphasis to a note sent us by a captain of a steamer, who asserts that in one particular instance, under favorable circumstances, he did not lose sight of the electric light until at a distance of forty miles from Havre, after having seen the light of

---

* Two hundred and twelve dollars and thirty cents was allowed in 1863, and $57.90 in 1865, to M. Serrin as compensation for experiments and models. We mention these items in order to show the exact price of the regulators; besides, M. Serrin now sells his apparatus at $289.50, instead of $193.

† We have more especially consulted the captains of steamers running to Dunkerque, Morlaix, Bordeaux, and the ports of Spain, as they have more opportunities for observing the lights of La Hève.

LIGHTHOUSES OF LA HÈVE.
ELECTRIC LIGHTS.

PLATE XLIV.

SWITCHES.

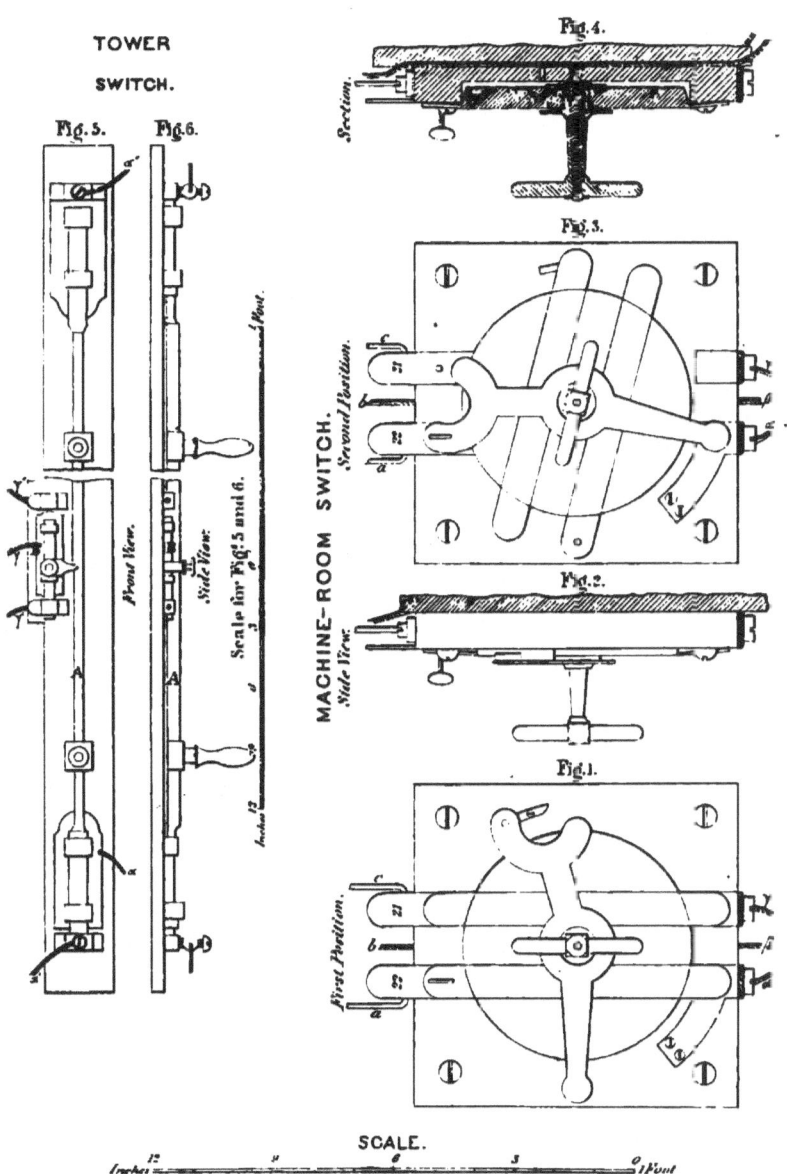

Barfleur for an hour and a half; but we shall call attention to the following statements, selected from a great number of others, which confirm the same results:

The electric light has generally been visible at a greater distance than the other by: *Statements of captains and others.*

    6 to 7 miles according to Captain Fautrel.
    4 to 5 miles according to Captain Morisse.
    2 —— miles according to Captain Lemonnier.
    3 —— miles according to Captain Rebour.
    4 to 6 miles according to captain of the Ville du Havre.
    4 —— miles according to Captain Duval.
    4 to 5 miles according to the pilot, Lecoq.
    5 to 6 miles according to the pilot, Guerrier.
    4 —— miles according to the pilot, Mazeras.

Some captains corroborate these statements by declaring that, at great distances, when the lights were first seen, the electric light seemed a little higher than the other. At some distance there was a notable difference in their brilliancy; the electric light appearing white and brilliant, the other red and smoky. The effect produced was well described by the picturesque comparison made by one of the pilots of Havre, (Pilot Savalle:) "There is as much difference between the lights as between a candle and a gaslight." This effect was moreover distinctly visible from the north pier of Havre.

The statements agreed still better in regard to the light during foggy weather. All the captains and pilots consulted by us declared that during fog there was a great difference in the range, and they all spoke highly of the services rendered them by the electric light; they assured us emphatically that a large number of vessels had been able to enter the harbor without difficulty, perceiving the southern light, while formerly they had been obliged to remain outside, the lights not being visible. Even before perceiving the electric light, its presence was marked by the illumination of the atmosphere surrounding it, and its range was thus increased, especially in foggy weather. This is an important advantage of the electric light, and may be of great practical utility, as is shown by a letter from Captain Delbeke of the steamer La Flandre, which we give below, as it well states the facts: *Light during fog.*

"In the nights of the 26th and 27th of February, 1865, I was doubling Cape Antifer at 1 a. m., three miles out, in eighteen fathoms of water, the point of the cape bearing S. E. ¼ E. The weather was very foggy on shore, but an experienced seaman could make out the point well enough to determine his position. I was then eleven miles and a half *Statement of Captain Delbeke.*

from the Fécamp light, which was not visible for the fog, and twelve miles from La Hève lights, toward which I could steer, as I saw, not the lights themselves, but their reflection. Approaching in order to make the channel and enter the inner roadstead of Havre, (it was then 3 a. m.,) I clearly saw the electric light and could not see the ordinary light."

Diminished difference in range during fog.

The difference of range between electric and oil lights diminishes rapidly when the fog thickens. Inspector-General Reynaud, in a report dated May 20, 1863, estimated the comparative range of a fixed light of the first order illuminated by oil or electricity as follows:

Table of comparative range.

Range of the unit of light.	Range in kilometers of a first-order fixed light.	
	Oil: intensity, 630 units.	Electricity: intensity, 3,500 units.
Kilometers.	*Kilometers.*	*Kilometers.*
0.1	0.160	0.177
0.5	0.93	1.06
1	2.08	2.40
2	4.80	5.77
3	8.45	10.2
4	12.9	15.9
5	18.7	23.4

It is seen that even in a slight fog the electric light is but little superior in range in spite of the great difference of luminous intensity, but it gains in proportion to the clearness of the atmosphere. This result is also shown by the following table, the result of observations made in 1864 at the light-houses of Honfleur, Fatouville, and Ver:*

Table of results of comparative observations.

Places of observation.	Distances.	Light observed.	Proportion of visibility in 100 observations.	Proportional value of electricity.
	Miles.			
Honfleur	9.32	Oil / Electricity	88 / 92	1.045
Fatouville	13.36	Oil / Electricity	77 / 79	1.026
Ver	23.89	Oil / Electricity	33 / 41	1.24

Light obstructed by vertical sash-bars.

The observations made at Fatouville seem anomalous; but this is easily explained, for one of the uprights of the

---

* This table and the remarks following are taken from the report of M. Reynaud, inspector-general, and dated March 31, 1866.

lantern of the northern light-house was placed in the direction of the light of Honfleur, thus masking a considerable part of the light emanating from the apparatus. The range of the electric light is, moreover, diminished in foggy weather, on account of its inferior power of penetration.

This arises from the different composition of the two lights, and is the more marked as the fog thickens. Experiments made at the central depot have shown, however, that if the electric light has an intensity two and a half times greater than a light of colza-oil, it will penetrate fog as well. As the intensity of the electric lights at La Hève, compared with an oil-light, far surpasses this proportion, we may be assured that the electric lights will always be superior in range.

Since 1865 six-disk machines have been used. The intensity has thus been considerably increased, and the relative range of the lights has always surpassed that shown in the preceding tables. *Use of six-disk machines.*

In order to complete the comparison of the two modes of producing light, there are yet a few words to be said as to the expense of each.

At La Hève, as we have already shown, the electric lights cost $3,215.29 annually, deterioration of machinery not included. For the same period the oil-lights before the alteration cost $2,828.88, distributed as follows: * *Comparison of costs.*

Salaries	$320 25
Fuel	77 20
Oil	1,300 58
Sundry supplies	84 53
Keeping apparatus in order	46 32

The expense is thus greater by $386.41, or about one-seventh; but this increase of cost is largely repaid by the increase of intensity of the lights.

The electric lights of La Hève have this economical advantage, that, while two in number, the expenses are far from being double what they would be for a single light. A certain number of expenses are, in fact, common to the two, or approximately so, especially salaries, the extra compensation and allowance for fuel to the engineers, the coal for the engines, &c. *Economy of double lights.*

This is not the case when oil is used, for then two lights will cost very nearly double what a single one would in the same circumstances.

---

* See M. Reynaud's *Mémoire sur l'Eclairage des Côtes de France*.

Estimated cost of electric light.

Basing our calculations on the experience of 1863 to 1865, we estimate the cost of an electric light of 5,000-burner intensity as follows:

Nature of expense.	Price.	Quantities.	Cost.
Salaries			$834 72
Allowance for fuel			77 20
Bonus to engineers			57 90
Water		71,328.60	
Coal, per ton of 2,240 pounds	$6 18	15,428	43 23
Coke, per ton of 2,240 pounds	8 69	110,200	434 25
Carbon points	43½	393.70	156 33
Cotton-waste	23	308.56	32 43
Lubricating-oil	32¼	440.80	77 20
Illuminating-oil	27	551	67 55
Grease	25	88.16	10 04
Hemp, white-lead, &c			9 65
Towels, mops, &c			11 58
Various supplies and repairs to machinery			156 52
Total			1,968 60

Cost of first-order oil-light.

The average cost of a first-order oil-light of 630-burner intensity is $1,494.40,* viz:

Three keepers, (two first and one second class).... $477 68
Fuel 48 25
Oil 903 04
Sundry supplies 42 27
Keeping apparatus in order 23 16

The electric light would cost about $475 more than the other, and the expense would therefore be increased nearly one-third. A comparison of the cost of the unit of light for oil-consuming and electric lights may easily be made from the preceding, and the figures given by Inspector-General Reynaud are verified. At La Hève, as the expense of the electric light amounts to $3,215.34 for 4,135 hours' illumination, and a mean intensity of $\frac{4,055 \times 5,000 + 80 \times 10,000}{4,135}$

or 5,097 burners, taking into account the time double light was produced; the unit of light sent to the horizon by each

Comparative costs of units of light.

light-house costs, therefore, $\frac{\$3,215.34}{2 \times 5,097 \times 4,135} = \$0.000076$.

The cost of the unit of light produced by the oil-con-

---

* See M. Reynaud's *Mémoire sur l'Éclairage des Côtes de France*.

LIGHTHOUSES OF LA HÈVE.
ELECTRIC LIGHTS.
ELECTRIC REGULATOR.

PLATE XLV.

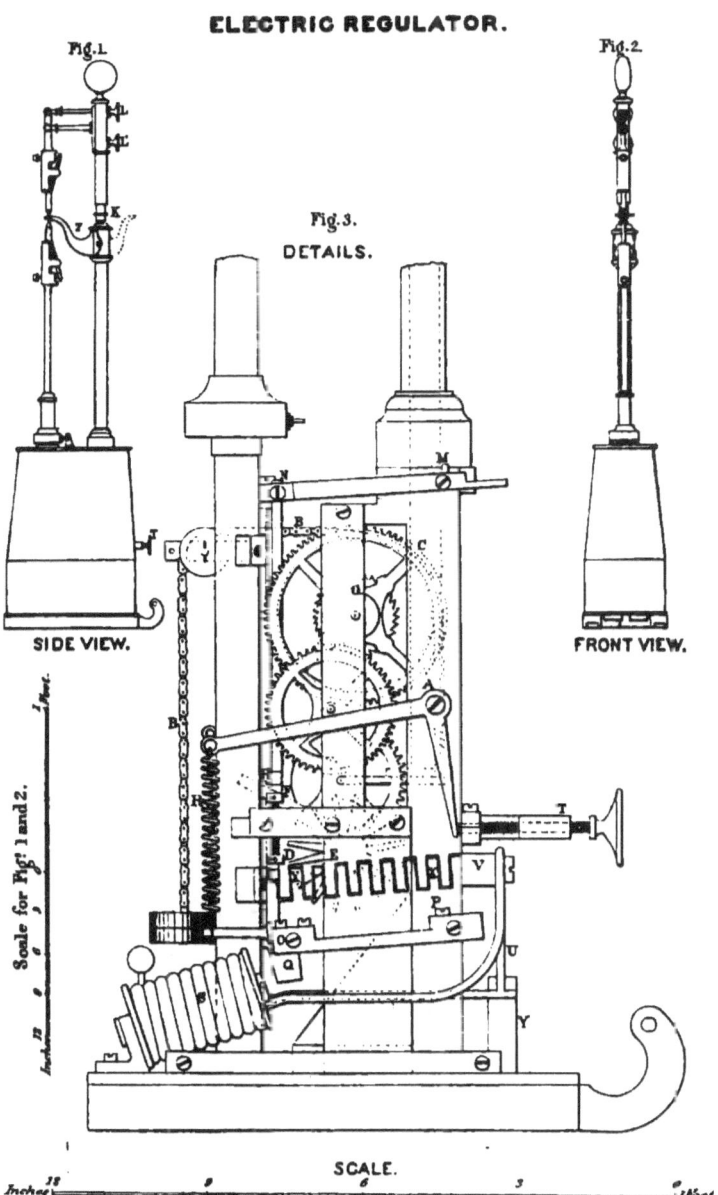

suming light-houses before their alteration amounts to $\frac{\$2,828.88}{2 \times 630 \times 3,900} = \$0.000576$,* which shows that the unit of light at the light-houses at La Hève costs about seven times less than that of oil-consuming lights. If, on the contrary, we consider a single electric and a single oil-consuming light, the unit sent to the horizon would cost for the former $\frac{\$1,968.60}{5,000 \times 4,135}$, or $\$0.0000952$, and for the latter, $\frac{\$1,494.40}{630 \times 3,900}$, or $\$0.000608$. The difference of cost is therefore somewhat more.

Conclusion.—Since the first establishment of the electric light, six years ago, sufficient time has elapsed to give an exact idea of its value for coast illumination. Recapitulation.

Navigators acknowledge with pleasure the excellent service which the electric lights render them; the advantages of the system have been keenly appreciated, the range of the lights is sensibly increased, especially during somewhat foggy weather, thus allowing a great number of vessels to proceed on their course and enter the harbor at night, which they would not have been able to do with the oil-light. Satisfaction of navigators.

The light, which at first was not as steady as could be wished, acquired a remarkable steadiness, thanks to improvements in the apparatus and the experience acquired by the keepers. The fears which were entertained, *a priori*, on account of the delicacy of some of the apparatus, have not been justified in practice. Accidents have been rare, extinctions short and few; two only of the latter during the period of six years being of notable duration; one, of one hour, was the consequence of an accident to a steam-engine; the other, of four hours, seems to have been maliciously caused. Under these circumstances there seems but little reason to be troubled about possible accidents. Steadiness of the light.

Still there are some disadvantages inherent in the system of electric illumination which necessarily limit its application. A considerable space is required for the steam-engines and the magneto-electric machinery, for storing coal, coke, oil, &c., and for collecting and preserving the water for the engines. Disadvantages.

Finally, the repairs of apparatus in use require special workmen, not usually found in the vicinity of light-houses.

---

* These calculations suppose the electric light in operation 4,135 hours, and the oil-consuming lights 3,900 hours, because under these conditions the expenses were estimated.

The keepers usually can attend only to current repairs; the more important ones have to be done in workshops better appointed than those attached to light-houses. In case of an accident of some importance, the magneto-electric apparatus can only be repaired by the Alliance Company. The lamps, under the same circumstances, have to be sent to M. Serrin at Paris. It is true that, as there are several regulators at the light-houses, one of them may be spared, so that the rare accidents which occur to these apparatus never cause but a few seconds' interruption of the light.

Proper situation for electric lights. We therefore think that in a great number of cases, especially in those of light-houses in the sea, or distant from important centers of population, or not easily accessible, the substitution of electric light for that produced by the combustion of colza-oil would be disadvantageous, or even impossible. But as this substitution offers great advantages to navigation, it seems advisable to adopt it for light-houses favorably situated; that is to say, where there is plenty of space, and they are sufficiently near to cities or easily accessible. The French light-house administration have already established a second one at Grisnez.

England, after having first tried the electric light at Dungeness, is about to establish others at South Foreland and Lowestoft. The attention of other governments is also drawn to this matter, and it appears probable that the example of England and France will be followed.

The change will cause a slight increase in the running expenses, but the difference is not so great as to cause hesitation when the increase of intensity and of range is considered.

CAPE GRIS-NEZ LIGHTHOUSE.
ELECTRIC LIGHT.

PLATE XLVI.

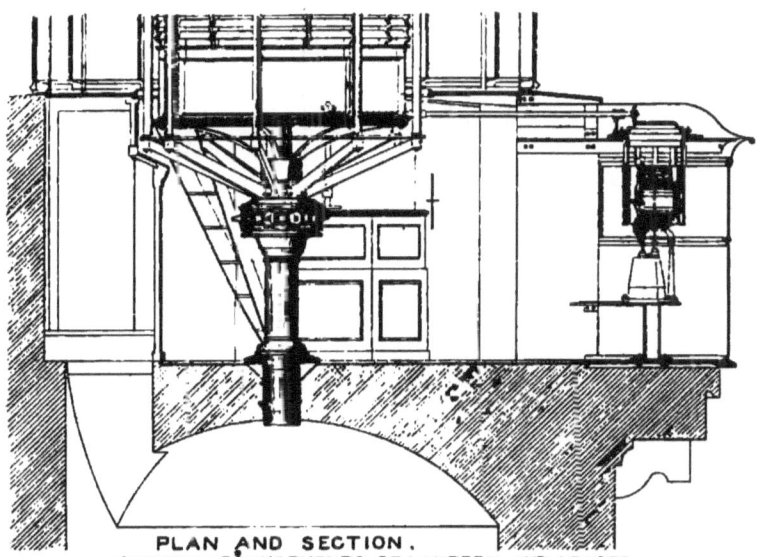

PLAN AND SECTION.
SHOWING ARRANGEMENTS OF LANTERN AND LENSES
FOR THE ELECTRIC AND OIL LIGHTS.

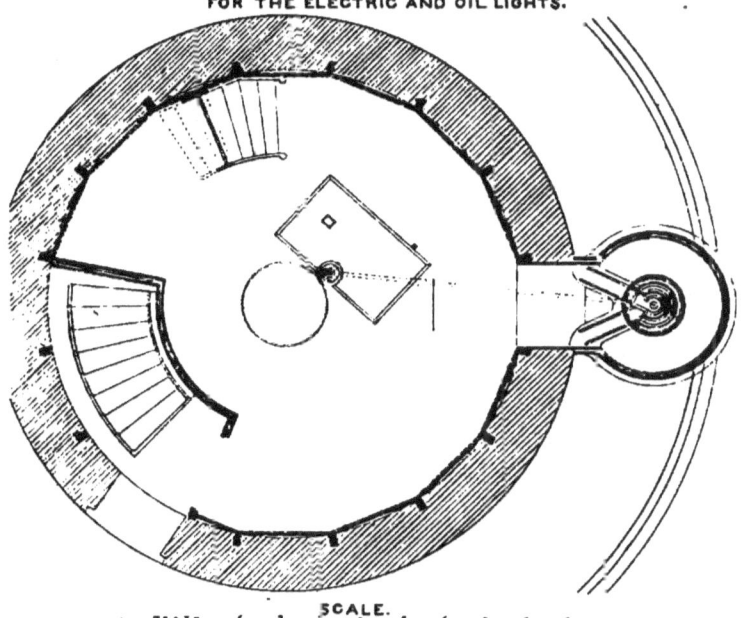

SCALE.

I had no opportunity of visiting any of the light-ships belonging to the French service. These are ten in number, one of which carries a revolving red light. Most of them are on the southwest coast of France. I have learned, however, the following particulars in regard to the regulations. Notes on the French light-ship service.

A light-ship's crew consists of a captain, mate, boatswain, (*maître d'équipage*,) and seamen. They are under the orders of the engineers or such superintendents as they may be assigned to. None can enter this service but regular sailors, who have had at least three years' service in the navy, and who know how to read and write. Crew.

Appointments are made by the prefect on nomination of the engineer-in-chief. Appointments.

The salaries are fixed by the Ministry of Public Works, and are subject to a deduction of 3 per cent., which is applied to the marine-hospital fund. Salaries.

The captain is responsible for the service of the light, and has, on board ship, all the rights of captain of the merchant-marine. He keeps the log and all the correspondence. In his absence his place is taken by the mate. Duties of the captain.

The boatswain sees that the captain's orders are executed, and he is particularly charged with the order and cleanliness of the vessel. He is not required to assist in the manual labor of cleaning the vessel, but he directs the details of the work and keeps watch like the seamen only in exceptional cases to be judged of by the engineers. In case of absence his place is supplied by a sailor selected by the captain. Of the boatswain.

The captain and mate have alternately fifteen days of service and fifteen days ashore; the sailors pass alternately a month afloat and fifteen days ashore. These leaves are, on some vessels, reduced one-half during the bad season. Services of captain and mate. Of crew.

While ashore the officers and sailors are at the disposal of the engineer-in-chief, and cannot quit their places of residence without authority. They are obliged to obey any orders that may be given them, either to return aboard, to attend to embarkations, or to any of the lights or beacons of the department. Under orders of engineer while ashore.

The captain is responsible for the provisions, and keeps the account of them. He may detail a seaman for cook, or decide that all shall take their turn.

THE INTERNATIONAL EXHIBITION AT VIENNA, 1873.

From Paris I proceeded to Vienna by way of Ulm, Augsburg, and Munich. I intended to have gone via Venice and

Trieste, but when I arrived at the foot of the Saint Gothard Pass, exaggerated accounts of the cholera prevailing at the former place, deterred me from taking that route.

On arriving at Vienna and presenting myself at the legation of the United States, I was received in the most cordial manner by the Hon. Mr. Jay, the American Minister, who proffered every assistance in his power.

Buildings. The exhibition buildings were on a grand scale, and it was said the Austrian government had already expended on this exhibition more than 20,000,000 gulden, or more than $10,000,000.

Most of the departments, particularly the department of machinery, were well filled; and the same may be said of the spaces allotted to the different countries, with a few exceptions, which include, I regret to say, that assigned to the United States. From some cause, probably our remoteness from Vienna, the American exhibition was extremely meager.

Articles on exhibition pertaining to light-houses and navigation. I was disappointed at the small number of articles of interest pertaining to light-houses and other aids to navigation, and those which were exhibited were mainly sent by the French and Austrian governments and the lens-makers of Paris.

Display by the French Department of Public Works. The "Department of Public Works" of France had a fine display of models and drawings of light-houses in connection with an extensive exhibition of views and models of other works of the *Corps des Ponts et Chaussées.* I did not observe anything particularly novel in the construction of the light-houses, but they all showed evidences of the good taste in architecture which characterize all works of the French engineers.

Light-houses represented. The following is a list of the light-houses represented:

Phare de la Palmyre, a wrought-iron range or leading light-house on a screw-pile foundation.

Phare de Royan, a masonry range or leading light built in alternate courses of stone and brick, the upper portion of the tower being of a peculiar shape, for the purpose, most likely, of serving as a day-mark.

Phare des Roches-Douvres, a wrought (plate) iron coast-light, very much like the *Phare de la Nouvelle Caledonie,* shown in M. Reynaud's *L'Éclairage des Côtes de France.*

Phare du Four, a "rock" light-house.

Phare d'Ar-men, a coast-light of stone masonry—a rock-station.

Submarine foundations. The *Corps des Ponts et Chaussées* had also models of submarine foundations for harbor light-houses built on shoals

SUBMARINE CONCRETE FOUNDATION
FOR
A HARBOR LIGHTHOUSE.

PLATE XLVII.

or bars, very much like those which I had designed for our lights in Long Island Sound, Delaware Bay, Chesapeake Bay, and other localities where light-houses are subject to injury from ice.

Two methods were represented, in one of which the iron tube or shell is left in place after it is filled with concrete; the other showed a cylindrical foundation of concrete made by means of a movable mould, which is turned around an axis, as represented by Plate XLVII.

Several novelties were shown by the lens-manufacturers of Paris.

M. Lepaute exhibited a third-order revolving lens, a duplicate of the one which I have described as exhibited to me at his manufactory at Paris, in which one-half showed a fixed light and the other half was divided into eight consecutive flash-panels. *Lens exhibited by M. Lepaute.*

This would be a striking characteristic in waters where there are already many lights, and where in placing a new one it becomes necessary to distinguish it, particularly if the flash panels were covered with red screens, which would tend to equalize the range of the fixed light and the flashes.

This lens was furnished with a mechanical or pump lamp, with a three-wick burner of the kind recently adopted by the French, which is adapted to the use of either colza or mineral oil, though the latter is designed to be used in it.

The clock-work operating the pumps was specially designed for revolving lights by M. Lepaute, and was remarkable for the small space required for it, viz, not more than six inches in diameter, a great improvement for small orders of lights. *Clock-work for revolving lens.*

Sautter, Lemonnier & Co. exhibited photographs of the flame from the Farquhar burner, (the patent-right of which is owned by this firm,) and the following lenticular apparatus:

1st. A range light or *feu de direction*, showing alternately red and green. This apparatus was composed of a dioptric and catadioptric lens for fixed light, embracing 150°, a catadioptric reflector placed in the dead angle, and two groups of vertical prisms arranged in front of the apparatus for fixed light, in the space on either side of the axis and outside the angle which it is required to light. These elements are so calculated that they concentrate the light from the fixed-light lens and distribute it as uniformly as possible throughout an angle of 45°. Between the fixed-light apparatus and the vertical prisms there is a circular screen, composed of three plates of glass, each embracing 75°. *Range-light by Sautter, Lemonnier & Co., revolving red and green.*

The two outer plates are red, the middle one green. The framework supporting this screen receives an intermittent oscillating movement, so that the light changes rapidly from one color to the other, preserving a constant color for a determinate time.

Red and green screens.

The screen passes through an arc of 75° in four seconds; then it is at rest for sixteen seconds, and then repeats the motion in an opposite direction, and again becomes motionless. This movement brings successively before each part of the apparatus first the red and then the green panes, so as to produce the characteristic required.

Pier-light by same firm.

2d. A pier-light, (*fanal de jettée.*) This light presents a very characteristic appearance, and if placed near a town it would never be confounded with ordinary street and house lights. The apparatus is composed of an ordinary fourth-order lens for fixed light, around which revolves a drum composed of vertical plano-cylindric lenses, each of which receives a luminous beam 18° wide, and concentrates it within 6°, thus diminishing the divergence and augmenting the intensity in proportion of 1 to 3. This drum reaches to just above the central lens of the apparatus, so that all the upper prisms preserve the appearance of a fixed light of sufficient intensity to be seen at least twelve miles. On account of the interposition and regular revolution of the movable lenses, the light presents a series of dilatations like equidistant pulsations, which gives it a very characteristic appearance. Experiments made to determine the best interval between two pulsations showed that it ought not to be less than a second and a half, for if the flashes were nearer together they might be confounded with the natural scintillation which lights near water have, under certain atmospheric conditions.

Manner of producing flashes.

The lens exhibited was of the fourth order, but fifth and sixth order lenses of the same kind can be easily constructed.

Scintillating lens for electric light.

3d. A very fine scintillating lens for electric light, for the following description of which I am indebted to the courtesy of Messrs. Sautter, Lemonnier & Co.:

"In most of the apparatus for electric light hitherto constructed, the optical part is composed of a cylindrical lens for fixed light 30 centimeters in interior diameter, before which, in flashing lenses, prisms of vertical elements are made to pass. We prefer to increase the diameter of the lens, and that for several reasons.

Reasons for increase to diameter of the lens.

"1st. In case there should be used a more intense electric light than the one produced by the machines now employed, (such a one, for instance, as is given by the new machines of Gramme,) it would become necessary for the preservation of the glass that it should be farther removed from the luminous focus.

"2d. Because a larger apparatus is easier to keep in order.

"3d. Because the larger the apparatus the less the inevitable variations in the position of the luminous point will affect the direction of the rays emerging from the lens.

"The apparatus (shown in Plate XLVIII) is composed of a cylindrical lens for fixed light, 75 centimeters in interior diameter, with upper and lower catadioptric zones. The metal supports of the central part are diagonal, so that in no direction does the frame completely obstruct the light. *Lens-apparatus.*

"A polygonal drum, composed of twenty-four vertical plano-cylindric lenses, envelops the apparatus from the top down to just below the central lens, and is made to revolve regularly, by means of clock-work placed in the pedestal of the apparatus. Each lens receives a luminous beam 15° wide, and concentrates it within 5°. In the apparatus which was exhibited, the drum makes a complete revolution in 120 seconds, so that the flashes succeed each other every five seconds. The duration of the flash is half the interval between the flashes, and the fixed light of less intensity is constantly visible. The intensity of the former is about eight times as great as the latter. The proportions remain the same whatever may be the intervals between the flashes; if the interval is increased the absolute duration of the flash is increased in the same proportion.

"In oil-lights a remedy is sought for the short duration of the flashes by increasing the diameter of the flame. This disadvantage will not exist in electric lights arranged as we have described, the divergence being caused not by the diameter of the luminous focus, but by the form of the lens, and existing only in the horizontal plane—that is, without loss of light.

"The electric lamp constructed by M. Serrin is placed on a revolving plate, eccentric with regard to the platform of the apparatus. This plate can receive two lamps back to back. It has two rails on which the lamps slide, and which are prolonged on a platform attached to that of the apparatus, and jutting out behind. The plate can be held by a spring-catch in two positions diametrically opposite. In the first of these positions, one of the lamps is at the focus of the apparatus, and the other in position to be attended to and its carbons changed. In the second position the reverse takes place. The current passes and the lamp is lighted of itself when placed in position.

"This apparatus should be placed in a lantern having flat steel diagonal sash-bars with an inclination corresponding to that of the lens-frame."

Iron light-house tower.

Besides the optical apparatus exhibited by Messrs. Sautter, Lemonnier & Co., they displayed an iron light-house tower of peculiar construction and excellent workmanship.

Translation of description.

The following description of it, is translated from the "*Annales Industrielles*" of September 28, 1873:

* * * * * * *

"The tower shown in elevation, (Fig. 1, Plate XLIX,) rests on a foundation formed of eight radiating iron ribs bound together by masonry, and rising about 5 meters above the terrace on which the structure rests. It is composed of a sheet-iron cylinder 12^m.50 high, with a winding stairway inside, and strengthened by eight buttresses or ribs of iron. These buttresses take the form of brackets above and support a cast-iron gallery which extends around the lantern.

Formation of tower.

"The cylinder or central tower is formed of five sheet-iron sections 0^m.006 thick, each 2^m.50 high and 1^m.80 in diameter. The interior of each is occupied by twelve steps of corrugated sheet-iron, the upper part of which forms a landing; riveted corner-pieces hold them against the outer envelope, and a cylinder 0^m.40 in diameter supports them at the center. The sections fit into each other and are held together by rivets. Each weighs 1,500 kilograms. The vertical ribs are each formed of four pieces, one above the other, and they are bound together by three bands or rigid horizontal crowns placed at equal vertical distances.

Crane.

"The tower is mounted without exterior scaffolding by means of a gallows-crane supported on the last section put in place. It is shown in detail in Fig. 4. It is composed of an upright formed of two T-irons connected by tie-pieces and held at the lower end by a cast-iron pivot. The arm, formed of two flat pieces of iron connected by braces, carries two pulley-blocks and a hook intended to receive a tackle. The fall of this tackle is conducted from the pulley-blocks to the upright axis of the crane. It descends vertically in this axis, traverses the pivot, and a final pulley carries it to a windlass used for hoisting.

Placing the first section.

"It was first necessary to place the first section on the foundation, an operation which was somewhat difficult, as the space for maneuvering was small. In order to effect it the crane was used, it being set up on the foundation and supported by guys. Raised to the required height the section was brought into place by rotating the crane. This first operation would in general practice offer no difficulty,

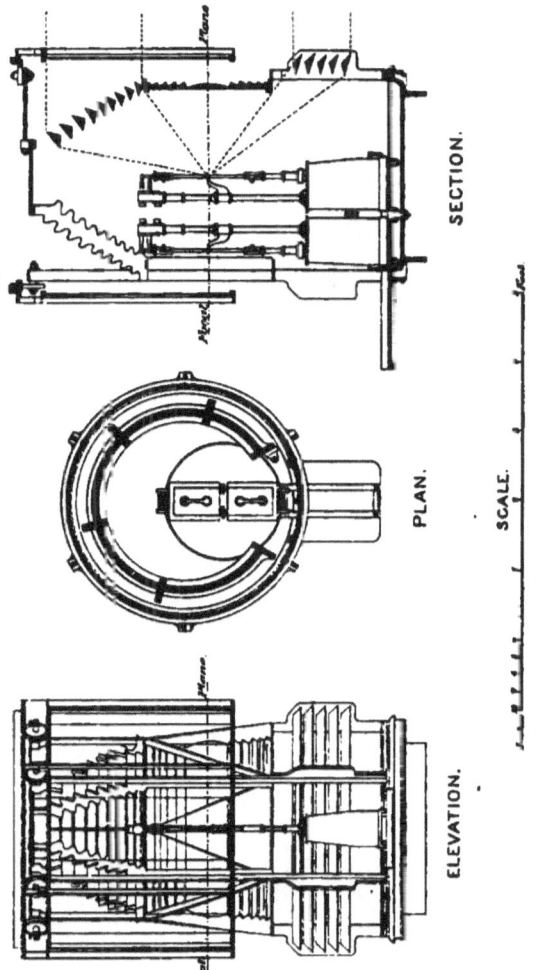

and could be done, according to circumstances, by any other means than that we have just described. But when this section is once placed and solidly bolted to the foundation, the mounting should continue and be finished without the aid of any scaffolding or support taken outside of the tower itself. The mounting comprises, first, the placing of the iron tube or the superposition of the sections of which it is composed, and which, as we have said, weigh about 1,500 kilograms each; second, the placing of the ribs, rings, brackets, galleries forming the frame-work, and the crowning of the structure with the lantern and apparatus. These latter pieces are relatively light, and can be raised to place by means of falls attached to the upper part of the central tube. If the tower is not very high the most simple mode of mounting consists in first superposing all the sections, and then fitting the ribs and other pieces of the framework. When the tower is high it is prudent to carry on at the same time the mounting of the tube and the placing of the ribs which strengthen it. {Mounting.}

"The tools employed are: {Tools.}

"The crane above described, with an extra pulley and a windlass on the ground;

"A movable scaffold shown in Fig. 5;

"Falls and cordage.

"The successive operations are: {Operations.}

"Raising and putting in place the sections;

"Bolting the sections together;

"Raising the crane after the placing of each section;

"Lowering and raising the movable scaffold for each section placed; that is to say, once before to raise the crane and put it in position, and once after to bolt the sections together.

"The placing of the ribs, the rings, the upper gallery, and finally the lantern and illuminating-apparatus.

"We have already described the crane and the way it is used to mount the first section. It is used similarly to mount the other sections, with this difference: that instead of placing it on the foundation, it is set, by means of castiron supports, on the last section placed. The bolting of the sections is done by means of the movable scaffold attached to pulleys suspended from the summit of the tube already mounted. The raising of the crane is shown in Fig. 3, and is effected as follows: Two falls are attached to the top of the section and each side of the foot of the crane, to hooks made for the purpose. The support c is unbolted, it is remounted and fixed to the top of the last section; at {Raising the crane.}

the bottom of the same section is placed the guide-support *d*, which obliges the crane to ascend vertically. Finally, when it is raised as far as desired, the support of the pivot *f* is bolted and the operation terminated.

Scaffold.

"The lowering and raising of the scaffold is done by hand, by men placed on the scaffold itself. The raising of the ribs and all the other parts of the central tube is done by hand with pulleys. The putting of them in place and the bolting or riveting of them is done by means of the movable scaffold. For all these operations six men are quite sufficient, and no greater number was employed at Vienna during the entire mounting.

"The ease and economy of mounting, is not the sole advantage of the system of constructing iron towers adopted by Messrs. Sautter, Lemonnier & Co., and applied by them to towers of all dimensions. But in this article we have only wished to call attention to the interesting fact that an edifice of great height can be rapidly, surely, and inexpensively set up by a very few men, and without the aid of scaffolding."

Lens-apparatus exhibited by Barbier & Fenestre.

Barbier & Fenestre had on exhibition the following lens-apparatus:

A third-order apparatus flashing 30″—30″, with hydraulic Funck lamps.

A fourth-order flashing apparatus, with Doty lamps and a new mechanical arrangement.

A fifth-order apparatus F. V. F., 2′—2′, 180°.

A sixth-order range-light, with two prisms so arranged as to permit of an accurate adjustment of the direction in which the light is thrown.

They had also a Doty four-wick lamp with a clock-work pump.

Models of Swedish light-houses.

The Swedish exhibition contained two models of iron light-houses resembling ours on the Florida reefs. Instead, however, of the socket-joints which we have used, the columns are connected together by means of flanges and bolts. Between the flanges are placed stout wrought-iron disks, having projections to which are fastened the horizontal, radial, and peripheral braces. The lugs or ears to which the tie-rods were made fast were on the upper or lower side of the flanges, as the strain was to be downward or upward; in other words, the flanges took upon themselves the strains upon those lugs.

These arrangements, which are shown in Figs. 28 and 29, are, I think, improvements in the modes commonly in use for similar structures.

IRON TOWER FOR 4TH ORDER LIGHT.

PLATE XLIX.

FIG. 1. ELEVATION.

FIG. 2. MOVABLE SCAFFOLD.

FIG. 3. RAISING OF THE CRANE.

FIG. 4. CRANE.

PLAN OF THE CRANE.

FIG. 5.

FIG. 6. SECTION THROUGH A, B.

FIG. 7. PLAN.

The mounting of the 5th section and placing of the ribs.

EUROPEAN LIGHT-HOUSE SYSTEMS. 257

In Fig. 28, A A are two hollow cast-iron columns fastened together by bolts through flanges; a, ear to which the tie-rods b are secured; c, large link through which the small link d is passed; e, wrought-iron plate between flanges.

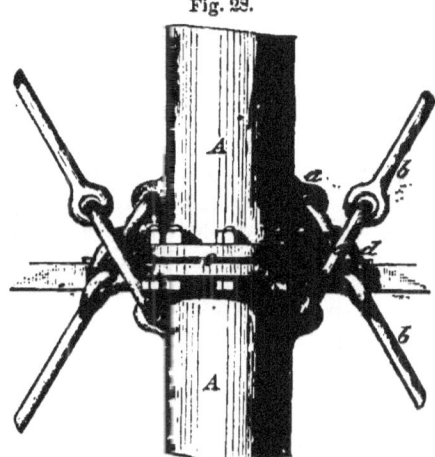

Fig. 28.

Details of Swedish light-house. Elevation.

In Fig. 29, A is the column; $a\ a$, ears for the attachment of the tie-rods; $f f f$, radial and peripheral struts of rolled iron fastened to wrought-iron plate $e\ e$, which is held between the columns as shown in the elevation, Fig. 28.

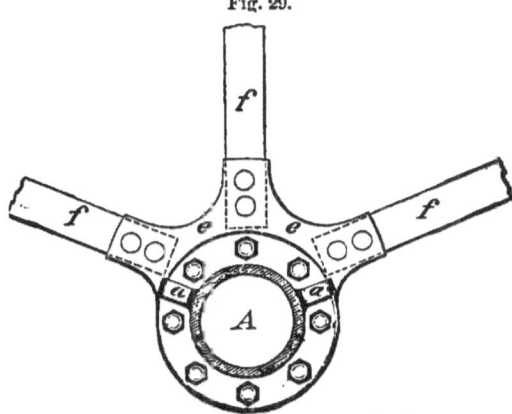

Fig. 29.

Details of Swedish light-house. Section.

In the Austrian part of the exhibition was a "Nebelhorn," Austrian foghorn.

S. Ex. 54——17

or fog-trumpet operated by steam; of this, however, I could learn but little except what was contained in the account given in the official catalogue, of which the following is a translation:

Translation of description.

"This apparatus has been constructed at Trieste, after the designs of G. Amadi, engineer.

"In foggy weather it is impossible to bring to the notice of mariners the threatening or desired proximity of land, by means of light-houses or light-signals in general. It therefore becomes necessary to carry sound far over the sea by means of vigorous acoustic apparatus, and thus to advise the mariner that he is approaching the coast. Several kinds of apparatus constructed for experimental purposes, and to which were applied the shrill notes of the whistle, could not, just on account of the high pitch of the sound, answer the purpose, and the sounds were lost at a short distance from the coast.

"In this new fog-signal there are deep notes, formed like those of an organ, by means of movable metallic reeds vibrated by steam, and they are sent out in a given direction through a trumpet or augmentor of sound. This signal has been heard, according to the experiments made, as far as sixteen nautical miles.

"In this apparatus, (a similar one is in operation at Trieste,) the notes are formed automatically by means of a steam-engine at given intervals, and the apparatus is turned at the same time on a vertical axis to reach all points of the horizon in a uniform manner."

After my return to Paris, I wrote to the Hon. Mr. Jay, and requested him to be good enough to procure, if the Austrian government would be pleased to communicate the information, drawings and descriptions of this fog-trumpet, of a reflector which I had also observed in the Austrian part of the exhibition, and of the Austrian buoys. A few days ago I had the great pleasure of receiving from Mr. Jay, through the State Department, a package containing the desired drawings and descriptions; also copies of correspondence between himself and the Austrian Minister of Foreign Affairs, in regard to my request, and of which the following are copies and translations:

Drawings sent by the Austrian government.

Letter of Hon. Mr. Jay.

"THE AMERICAN LEGATION AT VIENNA,
"*February* 23, 1874.

"DEAR MAJOR ELLIOT: In acknowledging your note of January 30, I am happy to be able to send you, as I do by this post, through the State Department, a roll which I received this morning from the Foreign Office.

" I annex a translation of the note of the Count Andrássy and a copy of my reply.

" You will see that the Ministry of Commerce is prepared to appreciate the drawings and plans which you had proposed to send, and which your note leads me to expect presently. Should there be any others that you think would be new and interesting to this government and which there is no objection to communicating, I hope you will send them. Not only the Foreign Office, but all the Ministers here, are so obliging in furnishing promptly and gracefully all information asked for, and extending facilities to our officers, that I am always glad of an opportunity of reciprocating their courtesies.

* * * * · *

" I am, dear sir, faithfully yours,

"JOHN JAY.

" Major GEORGE H. ELLIOT,
  " *Light-House Board, Washington.*"

[Translation.]

Translation of letter from Austrian Minister to Hon. Mr. Jay.

" The undersigned, Minister of the Imperial House and for Foreign Affairs, is only to-day enabled to place at the disposal of Mr. John Jay, Envoy Extraordinary and Minister Plenipotentiary of the United States of America, in compliance with his esteemed notes of 20th August and 13th December last, the drawings and description transmitted by the Imperial and Royal Ministry of Commerce.

" *a.* Of the fog-horn exhibited at Vienna in the year 1873;

" *b.* Of the parabolic reflector of Professor Osnaghi exhibited on the same occasion;

" *c.* Of the buoys in use upon the Austrian sea-coast.

" The Minister of Commerce, to whom the compliance with the wish of the Envoy has afforded especial pleasure, has at the same time requested the undersigned to plead with the Envoy, as excuse for the delay in the transmission of these drawings, the circumstance that the authorities at Trieste who exhibited at the Vienna exhibition the objects in question were able but recently to obtain a description of the fog-horn.

" While the undersigned has the honor to communicate to the Envoy the above fact, he begs further to state that the Minister of Commerce would gratefully acknowledge the courtesy should Major Elliot, in return for this collection, place the Imperial and Royal Government in possession of the promised drawings of those safety-signals which in the United States are in use and have attained such high perfection.

"The undersigned avails himself of this occasion to renew to the Envoy the assurance of his distinguished consideration.

"Vienna, February 22, 1874.

"For the Minister for Foreign Affairs,

"ORCZY."

Letter of Hon. Mr. Jay to Austrian Minister.

"The undersigned, Envoy Extraordinary and Minister Plenipotentiary of the United States of America, has the honor to acknowledge the receipt this morning of the note of his Excellency the Count Andrássy, Minister of the Imperial House and for Foreign Affairs, dated February 22, accompanied by a sealed roll of drawings and descriptions transmitted by the Imperial and Royal Minister of Commerce—

"*a.* Of the fog-horn exhibited at Vienna in the year 1873;

"*b.* Of the parabolic reflector of Professor Osnaghi, exhibited on the same occasion;

"*c.* Of the buoys in use upon the Austrian sea-coast;

"These drawings are to-day transmitted to the Department of State for Major Elliot, of the governmental Light-House Board at Washington, and will probably reach that office in time to be used in the preparation of his forthcoming report.

"The undersigned begs to add in reference to His Excellency's remark, that the Imperial and Royal Minister of Commerce would gladly acknowledge the courtesy should Major Elliot, in return for this collection, place the Imperial and Royal Government in possession of the promised drawings of the fog-signals which in the United States are in use, and which, His Excellency is pleased to say, have attained such high perfection, that the undersigned has received a note from Major Elliot, dated the 30th of January, saying that in a few days a parcel would be dispatched to this legation for the Imperial and Royal Ministry.

"The undersigned has the honor to present his thanks to His Excellency, and, through His Excellency's obliging intervention, to the Imperial and Royal Minister of Commerce, for the valuable information now afforded upon a subject so interesting and important to the commerce of the United States, and for their courtesy in furnishing the same to this legation immediately upon its transmission by the authorities at Trieste.

"The undersigned embraces this opportunity to renew to His Excellency the assurance of his distinguished consideration.

"JOHN JAY.

"FEBRUARY 23, 1874.

"His Excellency the COUNT ANDRÁSSY."

THE AUSTRIAN FOG TRUMPET. (NEBELHORN.)

Plate L, which is copied from the photograph sent me by Mr. Jay, represents the fog-trumpet as I saw it at the exhibition at Vienna, and the following is a translation from the Italian of the description which accompanied it:

"Since the introduction of acoustic signals, used in America as well as in Europe to mark dangerous points on the coast in foggy weather, it has become desirable to have a more perfect instrument, an apparatus that can be used not only at light-stations in foggy weather and snow-storms, but also on board of ships, especially on steamers, not only as an alarm but as a signal for correspondence. *Amadi's fog trumpet.*

"This object has been fully accomplished by the invention of Giovanni Amadi, of the Technical Institute of Trieste. His trumpet was exhibited at the universal exposition at Vienna, and was awarded a medal of merit.

"This apparatus, which consists of a trumpet, formerly operated by compressed air, but now directly by steam, is provided with an automatic distributing steam-valve, and with a special valve with finger-board (operating keys) so as to produce sounds at will. *Description.*

"The instrument has a most extraordinary power in proportion to its dimensions and to the pressure of steam required to produce the vibrations; it can be put up either directly over the boiler or separately, and connected with it by a pipe, and it can be turned to any part of the horizon. *Power.*

"In addition to its use as a fog-signal on shore, it may be applied on board of steamers of whatever steam-power, and is especially advantageous on board of men-of-war.

"By means of the finger-board, one is enabled to give long and short sounds at will with great accuracy, and communications may be made at night, in fog, or in snow-storms, by means of an alphabetic formula similar to that used in telegraphy. *Use of finger-board.*

"The trumpet (shown in Plate L) is operated by a steam-boiler of eight horse-power and a pressure of twenty-five pounds per square inch. *Steam-pressure required.*

"The boiler is more than sufficient to produce thirty blasts in thirty seconds, which are audible at a distance of fifteen nautical miles in clear weather. *Boiler.*

"Connected with the boiler is a small machine, which operates the automatic distributing steam-valve and can be so regulated that the different intervals in the sounds distinguish the different stations where trumpets are used. *Method of producing distinctive sounds.*

"In the Technical Institute at Trieste, where this trumpet was constructed, it was very particularly tested, and the government officers at Trieste testified that the sounds were

Sounds heard fifteen miles. plainly audible at a distance of fifteen miles, the height of the trumpet being thirty English feet above the sea; also that when operated with the finger-board, the signals, according to Morse's method, could be plainly distinguished at a distance of six nautical miles.

Trumpet placed on Point Salvore, Istria. "The Austrian government purchased this trumpet to operate it on Point Salvore, Istria, after trial of a smaller one of the same kind (audible five miles) near the light-house at Trieste.

Trumpet to be placed on light-ship at Grado. "A third trumpet, with a steam-generator of two horse-power, and audible eight miles, has also been ordered and finished, and will be put in operation on board of the light-ship anchored at Grado."

The following is a translation from the German of the description of the reflectors before referred to, and, with the sketch as reproduced in Fig. 30, illustrates clearly its principles:

Fig. 30.

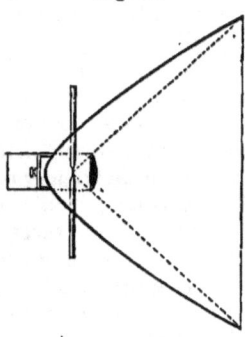

Osnaghi's reflector.

"PARABOLIC REFLECTOR FOR INTERMITTENT LIGHTS, DEVISED BY PROFESSOR FERDINAND OSNAGHI.

Object of the apparatus. "The principal object in devising this apparatus is to collect as many rays of light as possible into a beam parallel to the axis of the reflector, thus obtaining the greatest possible amount of light at the points illuminated by the beam, and by making the best use of all the light produced in one direction avoid the considerable loss which occurs in most apparatus of this kind.

Plano-convex lens. "1st. The rays are united in a beam by means of a plano-convex lens placed in front of the luminous focus.

Spherical mirror. "2d. A spherical mirror is inserted at the vertex of the parabolic surface.

"According to the laws of reflection from curved surfaces, if a luminous body be placed at the focus of a parabolic reflector, the rays will be reflected parallel to the axis, but the cone of rays tangent to the circular rim, and all rays within this cone, will radiate divergently and be scattered. To bend these rays parallel to the optical axis the plano-convex lens is used. *Use of plano-convex lens.*

"If the vertex of the parabolic reflector were retained the rays of light which, directly behind the lens, are rendered parallel by reflection, would, on passing through the lens, be concentrated into its focus and thence proceed divergently, so that nothing would be gained; but if we remove the parabolic vertex and substitute a spherical mirror with the luminous focus for its center, the rays will be reflected directly back through the focus, strike the lens at the same angle as if they had come directly from the source of light, and after refraction proceed parallel to the optical axis. In this way all the rays are united in a luminous beam of parallel rays of great intensity, and the loss of light is reduced to a minimum. *Effect by use of the parabolic reflector without the mirror.*

"Some loss occurs on account of the apertures in the parabolic surface, through which enter either the carbon points of the electric lamp or the burners of ordinary oil or petroleum lamp, and furthermore from the absorption which always occurs when light is reflected from metallic surfaces. Although there is more absorption by reflection than by refraction, it may be asserted that this apparatus would have an advantage over the dioptric Fresnel lenses now in use, for the reason that it concentrates the *entire* light into one beam, while with dioptric apparatus the rays are collected into from eight to twelve divergent beams, each of which gives only an eighth or twelfth of the total light transmitted. *Loss of light.*

"Flash-lights which appear with full intensity at certain intervals and then disappear require a rotation of the optical apparatus. In the Fresnel system where there are several flash-panels the motion may be slow, for to produce one flash per minute the revolution takes place only once in eight minutes if there are eight panels. The new apparatus, however, must make a revolution in one minute if it is required to show a flash every minute. This accelerated motion can easily be obtained, but, as the apparatus is light, finer and more accurate clock-work is required. *Use in flash-lights.*

"The principal dimensions of this apparatus when adapted for electric light are as follows: *Dimensions of the reflector for use in electric light.*

"Focal distance of the paraboloid, 30 millimeters.

"Opening distance of the paraboloid, 375 millimeters.

"Diameter of the spherical mirror, 85 millimeters.

"Diameter of the plano-convex lens, 85 millimeters.

"Focal distance plano-convex lens, 45 millimeters.

"Height of axis of paraboloid above the bottom plate, 280 millimeters.

"In the vertical plane the paraboloid can be moved around a horizontal axis passing through the focus, and in the horizontal plane it may be rotated with its standards around a vertical pivot. When ordinary lamps are used the dimensions should be increased to correspond with the size of the luminous body. Notwithstanding the small size of the apparatus, it has given very good results with a petroleum-lamp, quite surpassing a fourth-order Fresnel apparatus.

Gain of light effected. "Photometric tests showed that with a paraffine-candle at the focus of the reflector the co-efficient of concentration is 21,000, that is, 21,000 such candles would be required to produce the illumination of a plane surface that one would give at the focus of the reflector.*

Use of reflector for fixed lights. "To use this apparatus for fixed lights, the rays, already concentrated in one parallel beam, must be converted into a luminous disk by a second reflection. A considerable part of the original intensity will, however, be lost by spreading the rays around the entire horizon. To effect this the parabolic reflector is placed in a vertical position with the mouth or opening upward, and the rays from it are received by a set of totally reflecting prismatic rings arranged conically. The loss of light by reflection will be but trifling.

Cost. "Nothing can be positively stated as to the cost of the reflector, but it is certain that even a large one would not be as expensive as a dioptric apparatus."

Professor Osnaghi's combination of catoptric and dioptric agents is not new, having been previously invented by Mr. Thomas Stevenson under the name of "catadioptric holophotal reflector,"† and has been used in many cases in Great Britain.

In Fig. 31 will be found what I conceive to be an illustration of the professor's ideas in regard to a fixed-light apparatus.

In this figure a is the focus; b, the spherical mirror; c, the plano-convex lens; $d\ d'$, the parabolic reflector; $e\ e\ e$, the totally reflecting prisms; $p\ p$, the carbon pencils for the electric light. It will be observed that the major part of the light proceeding from the focus a of the parabolic reflector impinges against and is reflected by the surface of the latter; that the remainder of the rays are caught and refracted by the plano-convex lens c; that both the

---

* This co-efficient is excessive, and is, without doubt, an error.—G. H. E.
† See Light-house Illumination; Thomas Stevenson, 1871.

reflected and refracted rays are bent into a vertical beam which impinges against the conical surface of total reflec-

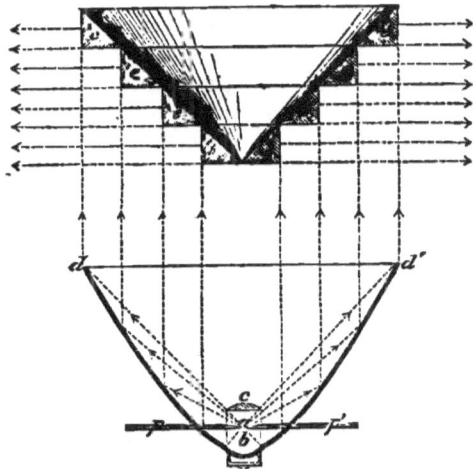

Fig. 31.

Reflector as adapted for a fixed light illuminating 360°.

tion formed by the prisms *e e e*, &c., and that from this surface the rays are uniformly distributed around the horizon. In this apparatus for fixed light a considerable loss of light would occur— *Loss of light.*

1st. In the reflection of the rays by the parabolic reflector and the refraction by the plano-convex lens;

2d. In the passage of the rays from the bottoms of the totally reflecting prisms to the surfaces of total reflection, and thence to the vertical sides of the prisms.

I should judge this apparatus for *fixed* light inferior to the dioptric apparatus of Fresnel, in which but a single agent is used.

The Austrian government had also on exhibition some bell, mooring, and other kinds of buoys, drawings of which accompanied the drawings and descriptions of the fog-trumpet and reflector received through Mr. Jay; but I observed nothing else of sufficient importance to take account of in this report.

The drawings and other information concerning our light-house establishment, referred to by Mr. Jay in his letter to me of February 23, were some weeks ago forwarded to him by the board, through the State Department, for presentation to the Ministry of Public Works of Austria. *Drawings forwarded to the Austrian government.*

I should not forget to mention that the photographs and paintings of some of our light-houses occupied a prominent place in the American part of the exhibition, and it has been announced that they were awarded a diploma of honor.

Diploma taken by American pictures of light-houses.

I cannot close my account of the exhibition at Vienna without expressing my warmest thanks to the Hon. Mr. Jay, the American minister, for his great kindness to me at Vienna and since my return, as well as for the interest he has shown in furthering the object of my visit, particularly by his successful efforts to procure for me from the Austrian government, for publication in this report, drawings and descriptions of some of the aids to navigation which that government had on exhibition.

Kindness of Mr. Jay.

RETURN VOYAGE.

I sailed from Liverpool in the Cunard steamship Cuba on the 30th of August, after an absence of four months from the United States.

Conversation with Captain Moodie.

On my return voyage to America I had several conversations with Captain Moodie, one of the oldest and most experienced commanders of the Cunard line, respecting the light-houses of Great Britain and the United States, and an interesting fact was mentioned by him in regard to one of the gas light-houses on the Irish coast, viz: that on the night in which we came out of Saint George's Channel, the weather being thick, he observed the light on Tuskar Rock at a distance, as he supposed, of six miles, judging by his former experience with this light, but when he had come up to it he found he had run more than twelve miles after he first observed it. He subsequently found (probably when we stopped at Queenstown for the mails) that since his last voyage *the light had been changed from oil to gas*, and he remarked that he was confident he saw the gas-light at least twice as far as he would have observed the oil-light in the same condition of the atmosphere.

Tuskar Rock gas-light.

I asked him his experience and opinion in regard to the low fog or "occasional" light at South Stack, on the coast of Wales, which I have described, and he stated he had found it of much value when not a ray could be seen of the upper light, which is often obscured in fog.

Fog-light at South Stack.

Captain Moodie thought our lights efficient as far as he had observed them, and spoke particularly of the great value of the revolving light at Fire Island, on the outer coast of Long Island, which is the first light ordinarily made by over-sea steamers approaching New York, but he is of the opinion that our aids for using Gedney's Channel into the

Aids to navigation near New York.

harbor, (which he says cannot now be used in the night-time, particularly in thick weather) are insufficient, and suggests that a light-ship should be moored where the fair-way buoy is, inside the bar; that the fair-way buoy outside the bar should be replaced by a bell-buoy, and that the present light-ship should be moored north-northwest from her present position and into line with the bell-buoy and the light-ship inside the bar.

Captain Moodie mentioned the difficulties of entering the harbor of New York in the winter-time when the iron buoys are removed and spar-buoys are substituted for them, stating that in harbors much farther to the northward no difficulty is found in maintaining the larger buoys in position in winter;* also the difficulty in making out the positions of the leading or range light-houses in the day-time when snow is upon the ground, and suggested that they be painted some dark color instead of white, to better serve as day-marks, a change which has been effected since my return. *(margin: Buoyage of harbor of New York)*

Captain Moodie's views as to the desired ameliorations in the system of aids to navigation at the entrance to the harbor of New York are entitled to much weight, and the investigation of the subject which I have made since my return, confirms me in the opinion that the changes suggested should be made without delay.

In my preliminary report I mentioned my obligations to Mr. Pelz, chief draughtsman of the Board, and I cannot close, without again tendering to him my sincere thanks for his zeal and interest in preparing for publication the plans and other drawings which I obtained in Europe; also to Mr. Baker, the talented financial clerk of the Board, for his excellent translations of the French papers which are found herein, and for his valuable assistance in the preparation of this report for the press.

---

* I do not think this would be possible in the harbor of New York without great loss. The ice-fields, moving with great velocity in the spring, carry off any buoys which may be placed, and spar-buoys, which are inexpensive, are, for this reason, used *in winter*.

INDEX.

ACCIDENTS from use of mineral-oil, 194, 195, 198
— to electric light at La Hève, 233
— provision against, at Grisnez, 225
———— Haisborough, 105
———— Howth Baily, 159, 162
———— La Hève, 223, 225, 232
———— Souter Point, 122
———— South Foreland, 70
———— Wicklow Head, 161, 165
——— in gas-light apparatus, 162, 164
Acid, used for removing sea-weed, 147
Administration, light-house, foreign, superior to that of the United States, 268
Aids to navigation New York Harbor, 266, 271
Air, how admitted to English lanterns, 107, 154
——— at the Wolf, 143
Air-tubes, Douglass burner, 77, 82, 86, 151
— Doty burner, 74
— Farquhar burner, 206
— Fresnel burner, 208, 212
— Lepaute's 1845 burner, 209, 212
—— truncated-cone burner, 210, 212
—— modified Fresnel burner, 210, 212
Allard, M., 185, 193
——— drawings received from, 225
Allen, Mr. Robin, 19
Allowances to keepers, English service, 98
———— French service, 216
——— Scottish service, 177
Amadi's fog-horn at Vienna, 261
American lights, opinion of Captain McCauley, 15
—————— Moodie, 266
—burners compared with Douglass's, 88
Anchors for English light-ships, 93
——— buoys, 96
Anglesea, island of, visit to, 149
Apparatus, catoptric. *See* Reflector-apparatus.
— dioptric. *See* Lenticular apparatus.
— revolving. *See* Machinery.

Apparatus for electric light, 67, 123, 221, 241
——— gas-light, 75, 104, 159, 160, 161, 162, 165, 166, 167, 170, 171
——— testing oils, 188
— designs furnished by Commission des Phares, 192
— prices fixed by Commission des Phares, 192
— at Dépôt des Phares, 186
—— Industrial Exhibition, Edinburgh, 180
—— Vienna Exposition, 251
Application de l'huile minérale, etc., by M. Reynaud, 193
Appointments of keepers, English service, 98
——— French service, 216, 249
——— Scotch service, 177
——— boatmen, Scotch service, 178
Arago, observations on sound, 34
Areas best for red and white panels in flashing lights, 118, 142, 270
Areometer, for testing oils, 188
Argyll, Duke of, cited, 55
Ar-men, phare de, model at Vienna, 250
Armstrong, Sir Wm., visit to his ordnance-works, 120
Arnoux, M., 215
Arrow, Sir Fred., 18, 74
Atmosphere, effect of, on sound, 27, 29, 33, 39, 40, 41, 43, 44, 46, 49, 50, 51, 52, 53, 55, 58, 59, 62
Austrian fog-horn, at Vienna, 257, 261
— government, drawings sent, 265
——— received from, 258

BALLYCOTTIN light, view of, 17
Barbier and Fenestre, visit to manufactory of, 206
——— agents for Doty burner, 207
——— their articles, exhibited at Vienna Exposition, 256
Basin for rain-water in lantern-floor at Haisborough, 107

S. Ex. 54——18

Beacons, day, towers painted to serve as, 17, 108
—— on Wolf Rock, 141
—— French system of coloration, 186
— illumination of, by gas, 172
— French service of, 217
Beazley, Mr. M., 147
Bell-boat off Queenstown, 17
Bell-buoy, mouth of the Mersey, 18
— Rundlestone, 144
Bell-buoys, English, 97
— French, 186
— Rundlestone, 144
Bell Rock, light-dues at, 20
—— cost and contents of, 142
Bells, signal, La Hève, 233
Bilge-keel in English light-ships, 104
Bill of Portland lights, 135
Bishop light-house, cost and contents of, 142
Blackwall depot, visit to, 76
Board, United States Light-House, extract from report of, 9
—— —— —— preliminary report to, 9
—— —— —— sent siren for English fog-signal experiments, 36
— of Trade, English, controls British light-house affairs, 21
— of Commissioners of Irish Lights, letter from, 158
—— —— —— visit to, 158
—— —— —— Scottish Lights, visit to, 175
—— —— —— establishment and organization, 178
Boat for landing, Wolf Rock, 143
Boatmen, appointment of, in Scottish service, 178
Boilers, best for fog-signals, 25
—— —— for fog-signals at Vienna, 261
—— —— La Hève, 221
—— —— Seven Stones light-ship, 146
—— —— Souter Point, 126
—— —— South Foreland, 67
Bonus paid French keepers, 217
Books, at English stations, 108, 138
—— French stations, 220
—— Scottish stations, 176, 177
Breakwater at Holyhead, 149
—— Plymouth, 137
Broglio, Duc de, letter on French burners, 191
Bronze for window-frames, rock-stations, 143
Buoyage, English system, 96
— French system, 186

Buoyage of New York Harbor, 267
Buoys, Coquet, 127
— English, 95, 96, 102
— French, 186
— designation of, 96, 271
— for strong tide-ways, 95
—— channels, 95
— Godrevy, 148
— Herbert's, 95
— models at Vienna, 8
—— —— Dépôt des Phares, 186
— moorings for, 96
— mouth of the Mersey, 18
— Queenstown, 17
— Rundlestone, 144
— water-ballasted, 95
Buoy-indicator, Yarmouth, 102
Buoy-list, English, 96
Buoy-shed, Blackwall, 97
— Yarmouth, 101
Bureau des Longitudes, experiments by, 34
Burner, Carcel, the French unit of light, 187, 196
Burners, comparative tables of old and new systems, 87, 88, 201, 202, 211
— Doty's, 74, 181, 190, 191, 194, 203, 207, 209, 212, 256
— Douglass, for colza and mineral oils, 82, 83, 84, 86, 87, 104, 150, 169, 181, 190
—— gas, 79
— Farquhar, 206, 251
— Fresnel, for colza, 190, 208, 209, 212
—— gas, 199
— Lepaute's, of 1845, 205, 209, 212
—— truncated cone, 210, 212
—— modified Fresnel, 191, 203, 210, 212, 215
—— gas, 215
—— paper on, 208
— foreign, are superior to those used in the United States, 267
— means of varying oil-level in, 207, 212-215
— received from Trinity House, 89
— results of improvements in, 20, 79, 84, 85, 86, 88
— Silber's gas, 79
— six-wick, not used by the French, 180
— statement of French Minister, 191
— table of experiments with, 81, 82, 83, 84, 86, 168, 169, 170
— Wigham's gas, 75, 104, 159, 160, 161-171
—— —— triform, 166, 167, 168, 171
— at Haisborough, 104
—— —— Holyhead, 150

INDEX.

Burners at Howth Baily, 159
—— Vienna Exposition, 251, 256
—— Wicklow Head, 160
— used in French service, 190, 191, 203, 210
Button, central, in Doty burners, 74
—— —— Douglass burners, 74, 81, 86
—— —— gas-burners, 79
—— —— Lepaute's burners, 212, 214
By-pass in gas-burners, 160, 162, 172

CABLES, conducting, at La Hève, 224, 229
— for English light-ships, 93
Cans, oil, at Blackwall, 91
—— manufactured by Barbier and Fenestre, 206
Cape Elizabeth, lens manufactured for, 206
Carbon-points used in electric lamps, 68, 123, 225, 232, 237
Carcel burner the French unit of light, 197
Certificates required from applicants, English light-keeper service, 97
—— —— French light-keeper service, 216
— given applicants, English light-keeper service, 98
Chains, English, terms of contracts for, 93
Chain-wheels, used in revolving apparatus, 150
Chance, Bros. & Co., manufactory of, 183
—— —— manufacturers of lens at Flamborough Head, 118
—— —— —— —— Holyhead, 149
—— —— —— —— Longstone, 130
—— —— —— —— Souter Point, 121
—— —— —— —— South Foreland, 68
—— —— —— —— Wolf Rock, 142
Channels, English system of buoyage, 93
Chart of Coquet, 128
—— Souter Point, 122
—— The Wolf, 141
Children of Scottish keepers, instruction, 177
Chimney-cap, Faraday's, 156
Chimneys, lamp, at Inner Farne Island, 130
—— Douglass's, 87, 151
—— for gas-light, 105, 162
—— modified Fresnel burner, 215
—— preserving from breakage, 119
—— smoke-funnels for, 214
—— test of, 92
Church-service to be read by English light-keepers, 115
Clergyman of Church of Scotland sent to stations, 177

Clock-work of revolving machinery, provision for accident to, 122
— exhibited at Vienna, 251
Coal used at Haisborough gas-light, 106
Collinson, Admiral, 99, 131
Colza-oil. See Oil.
Commission des Phares, visit to, 184
—— —— executive officers, 185
—— —— furnishes designs for apparatus, 192
—— —— fixes scale of prices for apparatus, 192
— of Irish Lights introducing gas, 104
—— —— visit to, 158
—— Scottish Lights, 175, 178
Conducteur, French service, 218
Consumption of carbon-points at electric lights, 123
—— gas, Douglass burner, 79
—— —— Wigham burner, 106, 163, 165, 169, 173
—— oil, comparative, 81, 82, 83, 86, 87, 182, 190, 196, 198, 201, 211
—— —— tested in English service, 90
—— —— Doty burners, 195, 198, 201
—— —— Douglass burners, 83, 87, 190
—— —— Lepaute's burners, 190, 191, 211
—— fuel, electric lights, 67
Coquet Island, description of light at, 127,
—— chart, 128
—— division of area around, 129
Corporation of Trinity House, 20
Corps des Ponts et Chaussées in charge of the administration of the French lighthouse service, 185
—— —— —— models exhibited by them at Vienna, 250
Cost of change to electric light at La Hève, 241
—— —— of oils in France, 203
—— colza and mineral oils, 88, 190
—— dovetailing stone at the Wolf, 141
—— English buoys, 97
—— lights, decrease in, 18
—— light-ships, 93
—— lantern at The Start, 135
—— maintenance of electric lights, 245
—— —— gas-lights, 171
—— new French burners, 191
—— repairs to machinery at La Hève, 240
—— rock light-houses, (table,) 142
—— station at Souter Point, 125
—— substitution, gas for oil, 103, 270
—— —— of electric for oil lights, 72, 270
—— —— mineral for land oil, 88

Cost of unit of electric light, 246
———— light in France and the United States compared, 269
Contents of rock light-houses, (table,) 142
Contracts, English, for oil, 89
———— ship-cables, 93
— French, for oil, 189, 197
Council chamber, at Dépôt des Phares, 185
Cowes, depot at, 132
Crane for moving buoys, Yarmouth, 101
—— setting up iron light-house tower, 254
Crew of English light-ships, pay, rations, pensions, &c., 114
—— French light-ships, 249
—— Seven Stones light-ship, 145
Cunningham, Mr. A., 176, 183
Cut-off for gas, automatic, 161.
Cylinder to produce red cut, 138
Cylindro-conical burner of Henry Lepaute, 212

DABOLL trumpet, on Newarp light-ship, 113
—— at Dungeness, 40, 46
Day-beacons erected on Wolf Rock, 141
— towers serving as, 17, 108
— French system of coloration, 186
Deflector, interior. *See* Button, central.
Deflectors, arrangement of, 77
Delbeke, Captain, cited, 243
Department of Public Works, French, models at Vienna, 250
Dépôt des Phares, visit to, 185
— at Blackwall, 76
—— Yarmouth, 101
— on Isle of Wight, 132
Depots on English coasts, 97
Derham, cited, 38, 51, 52
Derrick used at Wolf light-house, 142
Diagram, illustrating revolving intermittent gas-light, 163
Diameter of flames at Haisborough, 105
— lantern at The Start, 135
— English lantern, 155
— wicks of new French lamps, 201
Dimensions of burners fixed by French government, 201, 211
Dingeys carried by English light-house tenders, 131
Dinner at Lord Mayor's of London, 19, 74
—— Trinity House, 132, 140
Disadvantages of electric lights, 247
Divergence of light from large flames of gas-lights, 105, 170

Divergence of light from mineral-oil flames, 200
Doty, Captain, interview with, 74
—— claims of, 74, 191
—— compensation by Scottish board, 181
———— French board, 191
— burner, description, 74
—— adopted by Scottish board, 181
—— gives same results as the modified Fresnel, 191, 203
—— cost of, 192
—— first brought forward, 194, 209
—— patent purchased by Barbier & Fenestre, 207
—— variation of level in, 190, 207, 212
—— exhibited at Vienna Exposition, 256
Douglass, Mr., meeting with, 19
—— at Holyhead, 149
—— his report on mineral-oil, 80
—— details of his improvements in burners, 78, 80, 86, 150
—— finished Wolf Rock light-house, 140
—— extract from his narrative of the construction of Wolf Rock, 142
Douglass burner, for oils, description of, 77, 82, 86, 150
———— variation of level in, 82
———— results with, 83, 84, 87, 169, 190
———— compared with American light-house lamps, 88
—————— gas-light, 104, 109, 110
—————— at Flamborough Head, 113
—————— Haisborough, 104
—————— Holyhead, 150
—————— value as compared with Doty burner not settled, 181
—— for gas, 78
Dove, cited, 33
Dover, fog-signal experiments at, 22
— Castle, pharos in, 72
Drawings furnished by M. Allard, 225
—— Austrian government, 265
Duane, Gen. J. C., extracts from his report on fog-signals, 58
Duke of Argyll, cited, 55
Dungeness, fog-trumpet at, 40, 46
Dunkerque floating-light, 201
Dwellings, keepers, English, 108
—— Scottish, 177
—— at Gunfleet, 100
——— Honfleur, 219
——— La Hève, 241
——— Pharo de l'Hôpital, 218
——— South Foreland, 66, 71

INDEX.

EARTH-CLOSET used at light-stations, 71

E hoes, aerial, 32, 35, 38, 39, 44, 46, 47, 50
Eddystone, description of, 136
Edinburgh, Duke of, 74
— industrial exhibition, models at, 180
Edmundson & Co., manufacturers of gas-apparatus at Haisborongh, 184
Edwards, Mr., 19
Elder Brethren of Trinity House, 21
— — yearly inspection made by, 131
Electric light at Grisnez, 225
— — — Odessa, 226
— — — Port Saïd, 225
— — — Souter Point, 120
— — on Westminster clock-tower, 75
— — how produced, 67, 123, 221
— lights, causes of irregularity in, 68, 237
— — comparison with oil-lights, 72
— — — — gas-light, 75, 121
— — cost of maintenance, 72, 239, 246
— — — — machinery for, 241
— — — — substituting for oil, 72, 125, 241, 270
— — — — unit of, 246
— — disadvantages of, 247
— — intensity of, 70, 121, 238, 270
— — lantern proper for, 192, 232, 253
— — — for shown by Sautter & Co., 203
— — lenticular apparatus proper for, 121, 192, 252
— — of the world, 226
— — opinion of Captain McCauley, 15
— — proper situation for, 248
— — power of penetrating fog, 121, 126, 243, 244
— — range of, 242, 244
— — regulations at, English, 125
— — recommended for the United States, 260
— — at La Hève, 220, 223
— — — South Foreland, 68, 70
Elevator in light-house at Port Saïd, 225
Emerson, Mr., buoy-indicator of, 102
Engine, caloric, will work the siren, 62
Engineer of Trinity House, 19
— — Scottish light-house board, 178
— at electric lights, 125
— des Ponts et Chaussées, 185
Engine-room at La Hève, 221
— — — Souter Point, 122
— — — South Foreland, 63
— — at La Hève, 227, 236
— — — — replaced for more powerful ones, 236
— — — Souter Point, 122

Engines, steam, at South Foreland, 67
— — on Seven Stones light-ship, 146
Examination of English light-keepers, 97
— — French light-keepers, 216
— — Scottish light-keepers, 176
Exhibition, Industrial, at Edinburgh, 180
— — — Vienna, 249
Experiments with fog-signals at Dover, 25, 22
— — lights on Westminster tower, 75
— — colza and mineral oil, 80, 181, 193, 195, 196
— — new Douglass burners, 87
— — oil and gas-lights, Haisborough, 110, 104
— — — — — at Howth Baily, 167, 152
— — Doty burner, 195
Exterior deflector of burner, 77
Extinctions of electric-lights, La Hève, 233

FARADAY, Prof., cited, 151
— — wind-guard invented by, 156
Farquhar burner at Sautter, Lemonnier & Co.'s, 206
— — photographs of flames at Vienna, 251
Fastnet Rock, view of, 17
Fatouville, 219
Feu-de-port at Honfleur, 218
Fire, means of extinguishing, at South Foreland, 72
Flag-staffs at English stations, 108, 139
Flag, Trinity House, displayed, 108, 139
Flamborough Head, description of, 117
— — fog-gun station at, 118
Flames of gas-lights, photometric values of, 170
— — mineral-oil burners, broader than colza flames, 194
— — — — brilliancy obtained, 190, 195, 202
— — old and new burners, 20
— — Farquhar burner, photographs exhibited at Vienna, 251
Flashes in eclipse-lights, efforts of Fresnel to prolong, 200
Flashing-point of mineral-oil, 188
Flexibility of gas-light, 111, 237
Floating lights, paper of M. Lepaute on, 204
— — apparatus for, 204
Fog, at Haisborough, 110
— — Souter Point, 123
— effect on electric lights, 126, 243, 244
— — — sound, 27, 49, 53, 54, 57, 59, 60
— obscuring high lights, 123, 157
— when dense, impenetrable to light, 111, 121, 126

Fog-bell, machinery for, constructing at Blackwall, 93
Fog-light at South Stack, 157, 266
—— would be of advantage on Pacific coast, 127, 270
Fog-powers of Douglass burner, 151
—— electric light Souter Point, 121
———— La Hève, 221, 241
—— gas-lights, 105
Fog-signals. *See* also Sound.
— adapted for different localities, 63, 64
— best form of boilers for, 25
—— site for, 56, 63, 64
— causes of fluctuations in range, 29, 33, 38, 52, 53, 58, 59
— effective range of, 56, 57, 61, 64
— intervals between blasts, 57, 65
— sound-reflectors for, 179
— experiments near Dover, how conducted, 22
———— signals used, 22, 23, 61
———— pressure used, 22
———— questions considered, 23
———— Tyndall's report, 25
———— Sir Fred. Arrow's remarks, 69
—— by General Duane, 25, 58
— American, opinion of Capt. McCauley, 15
— gun, qualities of, 23, 25, 42, 43, 46, 48, 49, 63
— reed instruments, objections to, 62
— steam-whistle, qualities of, 23, 62
—— best form of bell for, 25
— Austrian *nebelhorn*, 257, 261
— horns used simultaneously, 26
— Daboll's trumpet, 40, 42, 46, 113
— gong, 113
— Holmes' trumpet, 40, 42, 43, 125, 145
— siren, 37, 38, 39, 41, 45, 46, 47, 48, 51, 53, 56, 62
— Wigham's gas-gun, 174
— at Flamborough Head, 118
—— Holyhead, 149
—— Howth Baily, 160
—— Newarp light-vessel, 113
—— Seven Stones light-vessel, 145
—— Souter Point, 125
—— South Stack, 156
—— St. Anthony, 137
—— St. Catherine, 133
—— The Lizard, 140
—— The Start, 135
———— Wolf, 143
—— in Scotland, 180
—— on light-ships, 113, 145, 271

Fog-signals on transatlantic steamers, 16
Foundation for harbor light-house, 250
Four, phare du, model at Vienna, 250
French light-house service, notes on, 216
Fresnel, Augustin, first lens made by, 186
—— concentric gas-burner of, 199
—— efforts to prolong flashes in eclipse-lights, 199
— burners, invention of, 208
—— dimensions and intensities of flames from, 209
—— air-tubes and overflow in, 208, 212
—— French engineers hesitate to alter, 199
—— modified by Lepaute, 210, 212
Fuel, tar used as, 106, 171
— for French light-house keepers, 216
Fund, mercantile marine, 21
Furniture, manufactured at Blackwall, 92
— allowed English keepers, 109
—— French keepers, 216

GALLERY, adjustable, in Douglass's burners, 87
—— Doty lamp, 74
— photometric, at Blackwall, 93
Gang-planks for small boats, 131
Gardiens of French service, 216
Gas, its use for light-house illumination, 75, 104, 159, 160, 169, 173
— how manufactured, 106, 159, 160
— consumption of, 79, 106, 166, 169, 171
— cost of, 171
— economy in use of, 162, 172
— means of producing uniformity of pressure, 159
— for illumination of beacons, 172
Gas-burner of Mr. Douglass, 78
—— Fresnel, 199
—— M. Lepaute, 215
—— Mr. Silber, 79
———— Wigham, 75, 104, 159–171
———— for triform light, 166, 167, 168, 171
Gas-gun for fog-signal, 174
Gas-engine at Howth Baily, 160
Gas-holder at Haisborough, 106
—— Howth Baily, 159
Gas-light apparatus, cost of, 170, 171
—— at Haisborough, 104
——— Howth Baily, 159
——— Tuskar Rock, 266
——— Wicklow Head, 160
—— on Westminster clock-tower, 75
—— Wigham's patent, 161, 162, 165, 166, 167

Gas-light, comparison with electric lights, 75, 121
— — oil-lights, 104, 109, 167
— cost of maintenance of, 132, 171, 172
— — — substituting for oil, 108, 270
— divergence of, 105, 165
— experiments with, 167
— flexibility of, 104, 111, 173
— heat produced by, 105, 160
— intensity of, 76, 166, 169, 270
— less trouble than oil-lights, 161
— opinion of Professor Tyndall, 173
— power of penetrating fog, 76, 110, 173
— at Haisborough, 104
— — Howth Baily, 159
— — Wicklow Head, 160
— on Westminster clock-tower, 75
— recommended for United States light-houses, 269
Gas-meters at Haisborough, 106
Gas-referees of London, cited, 112
Gas-regulator at Howth Baily, 159
Gas-works, Haisborough, 106
— Howth Baily, 159
— Wicklow Head, 160.
Gauge, for testing chimneys at Blackwall, 92
Gedney's Channel, New York Harbor, suggestions of improvements in the marking of, 266
Glass, cylindrical, for lanterns, 107, 151, 154
— used for observing lights, 110
— lantern, broken by sea-fowl, 107, 154
Godrevy light-house, description of, 148
— — — manner of landing at, 148
Gong, on board Newarp light-ship, 113
Grace Darling, home and tomb of, 131
Gratuities allowed Scottish light-keepers, 177
— — French light-keepers, 219
Gravity, specific, of mineral-oils, 80, 83, 80, 183, 196, 197
— — — mineral-oil required by French contracts, 188, 197
— — — — — English contracts, 89
Grisnez, electric light at, 225
Gunfleet light-house, description of, 100
Gun, fog at Flamborough Head, 118
Gun-metal, its use for window-frames, 143

HAISBOROUGH, description of, 104.
— observations of oil and gas lights at, 105

Hanois light-house, cost and contents of, 142
Hawes, Mr., inspector Irish lights, 159, 475
Hawkshaw, Mr., 149
Head of Kinsale, view of, 17
Heat produced by gas-light, 105, 160
— — — electric light, 252
— — — six-wick burners, 180
— destroys ordinary burners when mineral-oil is used, 83
Heaters used at American fog-signals, 146
Helices in magneto-electric machines, 67, 227
Herbert's Buoy, 95
Hetling, cited, 55
Hoffman, Colonel, 208
Holophone, Stevenson's, 179
Holophote, Stevenson's, 179
Holyhead, description of, 149
Honfleur, fen-de-port at, 218
Hôpital, Phare de l', 217
Hopkinson, Dr., 184
Howth Baily, description of, 159
— — experiments at, 167
— — gas-gun at, 174
Humboldt, cited, 58

ILLUMINANT, question of the best, 19, 190
Illuminants. See Electric light, Gas-light, and Oil.
Inner Farne Island light, 130
Inscription, Phare de l'Hôpital, 217
Inspection, annual, made by Elder Brethren, 131
Instructions, to English keepers, 97
— — Scottish keepers, 176
Instruments, meteorological, at Fatouville, 220
Insurance on life of English keepers, 98,
— — — — Scottish keepers, 177
Intensity, comparative, of gas and electric lights, 75, 121, 270
— — — — — oil-lights, 104, 109, 167, 169, 173
— — — oil and electric lights, 70, 72, 121, 238, 270
— — — mineral-oils, 80, 196
— — — — and colza oil-lights, 81, 82, 83, 85, 86, 87, 181, 190, 195, 198, 201, 202, 211
— — — — — lard-oil lights, 83
— — sound, 51
— — Doty burner, 181, 191, 195, 203

Intensity of Donglass burner, 78, 82, 83, 84, 87, 151
—— Farquhar burner, 203
—— Fresnel burner, 209
—— Lepaute's burners, 210, 211
—— Wigham's gas-burners, 75, 110, 166, 169
———— triform gas-burners, 75, 166, 169, 170
—— Maris lamp, 194
Irish Board, opinion as to use of gas, 172
—— visit to, 158
— lights, gas introduced at, 104
—— view of those on the southern coast, 158

JAY, Hon. Mr., 250, 266

Jets of Wigham's gas-burner, 105, 160
Junior Brethren of Trinity House, 21
————— dinner in honor of, 132

KEEPERS, English, regulations concerning, 97
—— transportation of family, 109
—— salaries of, at electric lights, 125
— Scottish, regulations concerning, 176
— French, to test oils, 189
—— regulations concerning, 216
— foreign better than those of the United States, 271
— at Dungeness, 125
—— Fatouville, 219
—— Gunfleet, 100
—— Haisborough, 103
—— Howth Baily, 160
—— Honfleur, 219
—— Longstone, 131
—— Lizard, 140
—— La Hève, 221, 233
—— Pharo de l'Hôpital, 217
—— Souter Point, 125
—— South Foreland, 71, 125
—— St. Catherine, 134
—— The Wolf, 143
Key West light-station, shadows cast by sash-bars, 153
Kinsale, Old Head of, light-house as daymark, 17

LA HÈVE electric lights, description of, 220, 226
Lamp-chimneys. See Chimneys.
Lamp-guard at Whitby, 118
Lamp-shop, Blackwall, 76

Lamp-valves, how made, 203
Lamp, Doty. See Doty burner.
— Donglass. See Donglass burner.
— Fresnel. See Fresnel burner.
— electric. See Regulator, electric.
— Maris, description of, 193
— constant-level, when used, 194
— oil, used at electric and gas lights in case of accident, 70, 105, 122, 129, 159, 162
— at Fatouville, 220
—— Holyhead, 150
—— Souter Point, 123
Lamps, used for testing oil at Blackwall, 90
——— different orders of lights, 189, 208
— keepers instructed in use of, 97, 176
— changed semi-monthly at French stations, 220
— tested at Dépôt des Phares, 157
— Landing at Godrevy, 148
—— Wolf Rock, 142
Lantern at Blackwall, 92
—— Gunfleet, 100
—— Grisnez, 225
—— Holyhead, 151
—— La Hève, 221, 232
—— Souter Point, 124
—— Spurn Point, 116
—— The Lizard, 124
— for The Start, 135
— model at Edinburgh, 180
— made by Sautter, Lemonnier & Co., 205
——— Barbier and Fenestre, 206
Lantern-gallery, floor of, 143
Lanterns, for electric lights, 71, 107, 125, 192, 206, 232, 253
—— gas-lights, 162
—— oil-lights, 151, 192
—— light-ships, 92, 205
— glass for, 107, 154, 192
— English, dimensions of, 71, 155
—— have diagonal sash-bars, 19, 125, 143, 151
—— parapets of, 19, 154, 206
—— ventilation of, 107, 143, 154, 155
—— painted by keepers, 108
Lard-oil. See Oil.
Lees, Mr., 158, 175
Lens, for verifying position of electric light, 222, 232
Lens-makers, Barbier & Fenestre, 207
— Chance Bros. & Co., 183
— M. Lepaute, 204
— Sautter, Lemonnier & Co., 205

INDEX. 277

Lenticular apparatus, for electric lights, 193, 241, 252
——— floating lights, 204, 205, 271
——— gas-lights, 162, 165, 166, 167
——— range-lights, 180, 251
—— calculated for different heights, 208
—— revolving, worked by hand, 122, 270
——— proper intervals of, 57
——————areas of red and white panels, 118, 143, 270
—— manufactory of Chance Bros., 183
—— manufactories at Paris, 203
—— M. Reynaud offers to test, 192
—— at Coquet, 128
——— Flamborough Head, 118
——— Holyhead, 149
——— Inner Farne Island, 130
——— La Hève, 223, 226, 229, 232
——— Plymouth, 138
——— Souter Point, 121, 122
——— the Eddystone, 137
———— Longstone, 130
———— Needles, 134
———— Start, 135, 184
———— Wolf, 142
———— Vienna Exposition, 251
——— Wicklow Head, 160
——— Whitby, 119
—— for Cape Elizabeth, 206
——— Longships light-house, 184
——— Swedish light-ship, 204
—— used at Howth Baily experiments,
——— on Westminster clock-tower, 75
Lepaute, M., visit to his manufactory, 204
—— burners manufactured by, 191, 203, 205, 209, 210, 212, 215
—— paper on burners, 208
———— floating lights, 204
—— articles exhibited at Vienna, 251
Light-house administration, benefits of permanence in, 183
—— English, vested in Trinity House, 20
—— French, vested in Corps des Ponts et Chaussées, 185
—— foreign better than that of the United States, 271
— illumination, requirements of, 18, 111, 121, 173
— List, English, extracts from, 117, 122, 128, 134
Light-houses, models at Dépôt des Phares, 185
———— Edinburgh, 180
———— Trinity House, 122

Light-house, models at Vienna Exposition, 250
— screw-pile, 99
— iron at Vienna Exposition, 256
Light-keepers. See Keepers.
Light-ship, needed on Rose and Crown Shoal, 15
— lantern for, constructing at Blackwall, 92
— design of one for Sweden, 204
Light-ships, apparatus for, 18, 76, 115, 181, 204
— bilge-keel for, 104
— moorings for, 93
— regulations for English service, 114, 146,
——— French service, 249
— repaired at Blackwall, 93
— of iron no longer built in England, 93
— none in Scotland, 181
— with revolving lights, 271
Letter from Commissioners Irish lights, 158
—— the Duc de Broglie to English embassador, 191
—— Mr. Douglass, 79
—— Hon. Mr. Jay, 258
—— M. Reynaud, 190
Level, oil drawn from more than one, 91
— means of varying, 212, 214
— proper for oil in burner, 81
Libraries at light-stations, 108, 177
Lightning-rods at light-stations, 108, 133
Light, fog, at South Stack, 157
— increase of, at Howth Baily, 159
— intermittent, at Wicklow Head, 160
— obstructed by vertical sash-bars, 151
— utilization of the rear or landward, 122, 271
Lights, electric. See Electric lights.
— gas. See Gas-lights.
— range. See Range-lights.
— oil. See Oil.
— American, opinion of Captain McCauley, 15
— high, obscured by fog, 127, 157, 270
— kept by English local authorities, 20
— results desired for, 112
Lime for removing sea-weed, 147
Liverpool, docks at, 18
Lizard lights, description of, 159
Longships light, description of, 146
Longstone light, description of, 130
Low light, use of, on Pacific coast, 127, 157, 270
—— at Coquet, 127

Low light at Plymouth breakwater, 138
— — — St. Anthony, 137
— — — Souter Point, 70, 122
— — — South Stack, 157
— — — The Start, 135, 184

MACHINE-SHOP at Blackwall, 92
Machine-rooms at La Hève, 221.
Machines, magneto-electric, at La Hève, 221
— — — South Foreland, 67
— — — Souter Point, 123
— polishing, at Chance Brothers', 183
— — — Paris, 203
Machinery, at La Hève, cost of repairs, 240
— for firing gas-gun, 175
— — fog-bell, constructing at Blackwall, 93
— — fog-bells in English service, 137
— — fog-signal, Holyhead, 149
— — — Seven Stones light-ship, 146
— — — Souter Point, 125
— — — South Stack, 156
— — — St. Anthony, 137
— — — The Wolf, 143
— — intermittent gas-light, 162
— revolving, flat wheels for, 150
— — at South Stack, 157
— — — Vienna, 251
McCauley, Captain, conversation with, 15
Magnets used at electric lights, 67, 123, 227
Magneto-electric machines at La Hève, 221, 227, 236
— — — South Foreland, 67
— — — Souter Point, 123
Maintenance of electric light, cost of, 72
— — gas-light, cost of, 162, 171, 172
Maitre de phare. *See* Keepers, French.
"Manacles, The," low light at St. Anthony, to mark, 137
Mangon, M., results of tests of, with paraffine, 196
Manufactory of Chance Brothers & Co., 183
— — Barbier & Fenestre, 207
— — Lepaute, 204
— — Sautter, Lemonnier & Co., 206
Manufactories near Souter Point, 120
Maplin Sands light-house, description of, 99
Maris lamp, description of, 190.
Masonry, comparative sections of, 141

Mast, steel, for lantern of light-ship, 92
Mayor of London, dinner given by, 19, 74
Measurements, Seven Stones light-ship, 146
Media, glass, for observations, 110
Medicine-chests at English light-stations, 108
Members of the Corporation of Trinity House, 21
— — — Scottish light-house board, 178
— — — French light-house board, 184
Mercantile marine fund, English, 21
Mersey, buoyage of, 18
Meteorological observations at South Foreland, 72
— — on English light-vessels, 115
— — at French light-stations, 220
Meters, gas, at Haisborough, 106
Middle ground in channels, English system of buoying, 96
Mineral-oil. *See* Oils.
Minot's Ledge light-house, 142
Mirrors on French bell-buoys, 186
Model of Stevenson's holophone, 179
Models in Dépôt des Phares, 185
— at Trinity House, 122
— — Vienna Exposition, 250
— of light-houses at Edinburgh exhibition, 180
Modified Fresnel burners, chimneys of, 215
— — — description of, 210
— — — tables of, 210, 211
— — — used in the French service, 191, 203
Moodie, Captain, conversation with, 266
Moorings of buoys, 93
— — English light-ships, 93
— — Newarp light-ship, 113
— — Rundlestone bell-buoy, 144
— — Seven Stones light-ship, 93, 146
Morton, Mr., 129
Mucking light-house, description of, 99
Museum at Dépôt des Phares, 186

NANTUCKET SHOALS, badly lighted, 15
Naval review at Spithead, 132
Nebelhorn, Austrian, description of, 257, 261
Needles, The, description of, 134
Netherlands, Kingdom of the, adoption of Lepaute's burner by, 209
Newarp light-ship, description of, 113

New Castle, ordnance-works at, 120
North Stack fog-signal station, description of, 155
Note of fog-signals most useful, 38, 39, 45, 46, 48, 113

OAK, for window-frames, 143
Observation of lights at Haisborough,- 109
— — — — Howth Baily, 167
— — light at Souter Point, 126
Observations, meteorological at Fatonville, 220
— — South Foreland, 72
— — on English light-ships, 115
Odessa, electric light at, 226
Oil-butts, St. Catherine's, 133
— South Foreland, 71
Oil-cans at Blackwall, 91
— for mineral-oil, 206
Oil-cellars, Haisborough, 107
Oil, foreign contracts, 89, 197
— how stored at Blackwall, 91
— cans for delivery of, 91
— not stored by the French, 187
— special trial of, made by Scottish keepers, 178
— Trinity House, purchases for other governments, 91
Oil, colza, color and odor of, 71
— — qualities compared with mineral-oil, 84, 85, 87, 182, 201
— — required by foreign contracts, 90, 199
— — results when used with new burners, 83
— — tests of, 90, 189
Oil, mineral, accidents from use of, 194, 195, 198
— — adopted by foreign governments, 19, 181, 190, 193, 268
— — economy in use of, 19, 85, 89, 182, 202, 235
— — experiments with, 20, 80, 131, 193, 195, 196
— — qualities of different samples, 80, 196
— — — as compared with colza-oil, 84, 85, 87, 182, 201
— — — — — lard-oil, 82, 268
— — — required by contracts, 89, 197
— — tests of, 89, 90, 187
— — cost of, 85, 89, 182, 190, 197
— — precautions taken with, 99, 178, 189
— — cans for, 206

Oil, mineral, regulation of overflow, 82, 190, 207, 212
— — Chairman United States Light-House Board on, 195
— — Mr. Douglass on, 80
— — M. Reynaud on, 193
Ordnance-works of Sir Wm. Armstrong, 120
Orfordness light, description of, 101
Osnaghi's reflector, 262
Outer Farne Island light, description of 130
Overflow of oil, adjustment of. 82, 190, 207, 212

PACIFIC COAST, suggestions for foglights on, 127, 157, 270
Painting of towers and buildings, English service, 108
Palmyre, phare de la, model at Vienna, 250
Panels, flash, for new lens at Start Point, 154
— — proper ratio of red and white, 113, 142, 270
Paraffine. See Oil, mineral.
Parapet of English lanterns, 19, 154, 203
Patent, Doty's, infringement claimed, 74
— Silber's, for gas-burner, 79
— Wigham's, for gas-lights, 169
Pelz, Mr. P. J., 271
Pensions to English light-keepers, 98
— — — light-vessel crews, 115
— — French light-house keepers, 216
— —Scottish light-house keepers, 177
Permanence desirable in light-house service, 183
Pharos in Dover Castle, 72
Photographs of American light-houses, at Vienna, 233
— — Farquhar burner, at Vienna, 251
Photometer used at Blackwall, 90
— — — Dépôt des Phares, 187
— —by French lens-makers, 203
— Hopkinson's, 184
Photometric experiments, rooms for, at Trinity House, 22
— — — — — Blackwall, 93
— — — — — Dépôt des Phares, 187
Piles, electric, formerly used at La Hève, 226.
— of Gunfleet light-house, 100
— — Maplin Sand light-house, 99
— — Swedish light-house, at Vienna, 257
Pinnace, carried by English tenders, 131

Plymouth breakwater light, description of, 137
Poe, Col. O. M., 144
Point Bonita, Cal., resembles Start Point, England, 136
Point Roche, view of, 17
Port Said, electric light and elevator at, 225
Power of burner for fixed gas-light, 169
— — English lights increased, 78, 85
— — first-order sea-coast lights, 121
— — fog-signal at Vienna, 261
— — gas-light Westminster clock-tower, 76
— — lens at Souter Point, 121
— — light at Grisnez, 225
— — — — La Hève, 221, 238, 242, 244
— — — — South Foreland, 70
— — machines at South Foreland, 67
— — — — Souter Point, 123
— — magnets at South Foreland, 67
— — Osnaghi's reflector, 264
Powers of oil and electric lights compared, 72, 121, 225
— illuminating, of oils, 80
— — gas-lights, 169
Prices of apparatus fixed by Commission des Phares, 192
— — — for electric lights, 241
— — oil, colza, 81, 197
— — — mineral, 85, 88, 89, 182, 190, 197
— — — lard, 88
— — Wigham's gas-apparatus, 170, 171
Prisms, machines for polishing, 183, 203
— removed for triform apparatus, 166
— tested by M. Sautter, 206

Q
QUEENSTOWN HARBOR, buoyage of, 17
Questions to be considered at Dover fog-signal experiments, 24

R
RAIN, its effect on sound, 26, 37, 40, 45, 46, 48, 50, 52, 59, 60
Ramsay, Captain, U. S. N., 22
Range, effective, of fog-signals, 56, 57, 61, 64
— of electric light, 112, 242, 244
— — fog-signal at Vienna, 262
Range-lights, apparatus for, 180, 251
— Bill of Portland, 135
— Fatouville and l'Hôpital, 219
— Orfordness, 101
— Whitby, 118

Range-lights on English coast, 116
Ratios for English keepers, 98, 114
— — French keepers, 216
— — Scottish keepers, 177
Ratio of areas of red and white panels in flashing-lights, 118, 142, 270
— — increase of lights by apparatus, 169, 264
Raynolds, Lieut. Col. W. F., 144
Rear light, utilization of, 122, 271
Red-cuts at Coquet, 127
— — Godrevy, 148
— — Longships, 147
— — Orfordness, 101
— — Plymouth breakwater, 138
— — Souter Point, 122
— — Spurn Point, 116
— — Start Point, (in new lens,) 184
— — The Needles, 134
— — Whitby, 118
— recommended for United States light-house service, 270
Reflector-apparatus, for light-ships, 76, 271
— still in use in English light-houses, 76
— requires constant-level lamps, 194
— made by the elder Stevenson, 179
— for harbor and ship lights, 179
— — fog-light, South Stack, 157
— — Dunkerque light-ship, 204
— at Gunfleet, 100
— — South Stack, 157
— — The Cockle, 115
— — — Eddystone, 137
— — — Lizard, 139, 140
— — — Longships, 147
— — Vienna Exposition, 262
— — Whitby, formerly used, 119
Reflector for sound, 23, 58, 179
Regulations as to care of oils, 99, 178, 189
— concerning English service, 97, 114
— — French service, 216, 249
— — Scottish service, 176
Regulator, electric, at La Hève, 230
— — — Souter Point, 123
— — — South Foreland, 68
— gas, at Howth Baily, 153
— oil, in Douglass lamp, 82
— used in fixed gas-light, 161
Repairs, English keepers taught to make, 98
— general, made by superintendents English lights, 131
— to machinery at La Hève, 240
Report of Chairman United States Light-House Board on dangers of mineral-oil, 181

INDEX.

Report of Mr. Douglass on oils and burners, 60
— — General Duane on fog-signals, 58
— — United States Light-House Board for 1873, extract, 9
— — Scottish board on lamps and oils, 181
— — Professor Tyndall on effect of sash-bar shadows, 152
— — — — — gas-lights, 163, 173
— — — — — — fog-signals, 25
— preliminary, to the Light-House Board, 9
Review, naval, at Spithead, 132
— of troops at Dover Castle, 72
Revolving apparatus. *See* Machinery.
Reynaud, M., meeting with, 199
— — offers to test apparatus for the United States, 192
— — his paper on mineral-oil, 193
Rings of jets in gas-burners, 101
Robinson, Dr., cited, 27, 51, 56
Rochemont, M. Quinette de, his paper on electric lights at La Hève, 226
— — — — cited as to shadows of sash-bars, 153
Roches Douvres, phare des, model at Vienna, 250
Rock light-houses, sections of, 141
— — examples in the United States, 142
— — cost and contents of, 142
— — rations of keepers at, 98, 216
Rooms for engineer at French stations, 218
— — photometric experiments, Trinity House, 22
— — — — Blackwall, 93
— — — — Dépôt des Phares, 187
— — officers, South Foreland, 71
Rose and Crown Shoal, needs to be marked by a light-ship, 15
Rouen, visit to, 215
Royan, phare de, model at Vienna, 250
Rundlestone bell-buoy, 144

S AILING, date of, 9, 15
Sailing-directions for Coquet, 128
— — Souter Point, 122
— — The Needles, 134
St. Anthony light, description of, 137
— Catherine light, description of, 133
Salaries of Elder Brethren of Trinity House, 21
— — English keepers, 98, 125
— — — light-ship crews, 114
— — French light-keepers, 216

Salaries of French light-ship crews, 249
— — Scottish light-keepers, 176
Sash-bars, diagonal, best for electric light, 192, 232
— — used in English lanterns, 19
— — first designed by the elder Stevenson, 130
— — at Holyhead, 151
— — — Haisborough, 107
— — — Souter Point, 125
— — — Spurn Point, 117
— — — The Stack, 135
— vertical, effect of shadows cast by, 151, .152, 153
Sautter, Lemonnier & Co., manufactory of, 206
— — — — articles exhibited at Vienna, 251
— — — — tests of prisms and apparatus by, 206
Saving to the United States by adoption of mineral-oil, 88
— — — — — — new burners, 88
Schenck, General, 22
Scottish light-house board, members of, 178
— — — report on oil and lamps, 181
— — — subordinate to Board of Trade, 21
— — — when established, 178
Screw-pile light-houses, Gunfleet, 100
— — Maplin Sand, 99
— — comparative cost of, 101
Sea-fowl break lantern-glass, 107, 154
Sea-weed, means of removing, 147
Seven-Stones light-ship, description of, 146
— — moorings of, 93, 145
Shadows cast by vertical sash-bars, 151, 152, 153
— sound. *See* Sound-shadows.
Shoals off Queenstown, how marked, 17
Signals to vessels in danger, on English light-ships, 115
Silber, Mr., his patent gas-burner, 79
Siren, principle of, 37, 45
— superiority of, 39, 41, 46, 47, 48, 51, 53, 56, 62
— echoes produced by, 38, 39
— best pitch for note, 47
— means of rotating desirable, 56
— with compressed air, 56, 62
Skerryvore light-house, 142
Smalls light-house, 142
Smeaton, John, 136
Smoke-funnels, Lepaute's, 214
Sound, causes of fluctuations in range, 29, 33, 38, 52, 53, 58, 59
— — — acoustic opacity, 53, 55

Sound, effect of changing direction of, 41, 50, 51, 64
— — — fog on, 27, 49, 53, 54, 57, 59, 60, 115
— — — height on, 26
— — — pitch on range of, 38, 39, 45, 46, 48
— — — rain on, 26, 37, 40, 45, 46, 48, 50, 52, 59, 60
— — — snow on, 52, 58, 59, 60
— — — wind on, 46, 51, 61
— intensity of, 51
— reflection, aerial, of, 31, 32, 35, 38, 39, 44, 46, 47, 50, 58, 59
— Arago, cited, 34
— Derham, cited, 38, 51, 52
— Dove, cited, 33
— Duke of Argyll, cited, 55
— Hetling, cited, 55
— Humboldt, cited, 58
— Robinson, cited, 27, 51, 56
Sound-shadows, effect of, 16, 26, 41, 42, 43, 56, 64, 133
Souter Point light, description of, 120
— — — observation of, 126
South Foreland, fog-signal experiments at, 22
— — lights, appearance from Dover pier, 70
— — — description of, 66
— — — means of extinguishing fire at, 72
— Stack, description of, 156
— — low light at, 157, 266
Speaking-tubes, South Foreland, 71
— La Hève, 223
Specific gravity. *See* Gravity, specific.
Spectacle Reef light-house, by whom built, 144
Sperm-candle, the English and American standard unit of light, 81
Spithead, naval review at, 132
Spools of magneto-electric machines, 67, 227
Sparn Point light, description of, 115
Stairs at Haisborough, 106
— — South Foreland, 71
— — Whitby, 118
Start Point light, description of, 135
— — — new lens for, 184
Steamers, need of fog-signals on, 16
Stevenson, Mr. Thomas, 179, 183
— — — his differential reflector, 179
— — — — apparatus for ship and harbor lights, 180
— — — — holophone, 179
— — — — holophote, 179
— Messrs., engineers of Scottish board, 178
— the elder, his lantern with diagonal bars, 180

Stone-courses, Wolf Rock light-house, 141
— Longships, 147
Stones, The, buoy off Godrevy, 148
Stores, Scottish keepers report quality of, 178
Store-houses, Yarmouth depot, 101
Store-rooms, South Foreland, 71
Strata, moving, at St. Catherine's light, 133
Superintendents of Trinity House make repairs, 131
— — — have charge of tenders, 97
Supernumeraries in English service, 97, 98
Swedish light-house at Vienna, 256
— light-ship made by M. Lepante, 204
Switches for changing currents at electric lights, 228, 229

TABLE showing qualities of different mineral-oils, 80, 196
— of results of experiments with Douglass burner and different oils, 81, 82, 83, 84, 86, 87
— — — — — old and new French burners and different oils, 193, 201, 209, 210, 211
— — salaries of English light-house keepers, 98
— — — and rations of English light-ship keepers, 114
— — — — — Scottish light-house keepers, 176, 177
— — cost and contents of rock light-houses, 142
— — experiments with gas-light, 168
— — illuminating powers of gas-lights, 169, 170
— — increase of intensity in eclipse-lights by use of mineral-oil, 202
— — expenses of electric light at La Hève, 239, 246
— — hours of illumination and the working of the engines, La Hève, 240
— — comparative range of electric light, 244
— — observations of oil and electric lights, 244
Tanks, mineral-oil, in Scottish light-houses, 178
— — — French light-houses, 198
Tar, use of, as fuel, 103, 171
Tenders, English, in charge of superintendents, 97
— — description of, 131
Test, for chain-cables for English light-ships, 93

Test of apparatus at Blackwall, 97
— — — — Dépôt des Phares, 187
— — — offer of M. Reynaud, 192
— — — at Chance Brothers, 184
— — — Sautter, Lemonnier & Co's, 206
— — lights, at Haisborough, 110
— — oil at Blackwall, 90
— — — — French light-houses, 189
— — — — Scottish light-houses, 178
Tips for burners in Douglass lamp, 77, 83, 86
— removable, 83, 86
Tower at Coquet, 129
— — Eddystone, 136
— — Longships, 146
— — Plymouth breakwater, 137
— — St. Anthony, 137
— — St. Catherine on moving strata, 133
— — Sonter Point, 124
— — Wolf Rock, 141
— exhibited at Vienna by Sautter, Lemonnier & Co., 254
Towers as day-marks, 17, 108
— at Blackwall for testing purposes, 97
— — Haisborough, 105, 107
— — Phare de l'Hôpital, 217
— — South Foreland, 71
— — Spurn Point, 115
— English and American, comparative merits, 106
Transportation, English keepers, 109
Triform gas-light. See Gas-lights.
Trinity House, London, 18
— — acts as agent for purchase of oils, 91
— — ceased to build iron light-ships, 93
— — corporation of, 20
— — dinner given by, 132, 149
— — engineer of, 19
— — lamps and burners received from, 88
— — members of, 21
— — powers of superintendents of, 97
— — subordinate to Board of Trade, 21
Tripoli, Cunard steamer, accident to, 17
Trumpet, Daboll's, at Dungeness, 40, 46
— — in Dover experiments, 42
— — on Newarp light-ship, 113
— Holmes, in Dover experiments, 43, 46
— — at Souter Point, 125
— — Seven Stones light-ship, 145
Tuskar Rock, view of, 17
— — gas-light on, 266
Tyndall, Professor, his opinion of gas-lights, 104, 163, 173
— — — — sash-bars obscuring light, 152
— — — report on fog-signal experiments, 25

UNIFORMS worn in English light-house service, 98, 109, 114, 139
Unit of light, photometric, English, 81
— — — — French, 187, 196
— — — — comparative values of French and English, 187
— — — cost of, in France and the United States, 269
— — — cost of, when produced by electricity and oil, 246
Utilization of landward light, 122, 271

VALVES, lamp, made by Lepaute, 206
Ventilation of English lanterns, 107, 143, 154, 155
Vestal, description of, 131
Vienna Exposition, 249

WALES, PRINCE OF, 149
Walls at Longstone, 130
— — Spurn Point, 117
Watch of keepers, 71, 126.
Watch-room at Souter Point, 122
— — Phare de l'Hôpital, 217
— — English, size of, 71
— — painted by keepers, 108
Water at South Foreland, 72
— supplied at La Hève, 233
Washing, allowance for, to Scottish keepers, 177
Webb, Captain, 18, 132, 144
Weller, Captain, 99, 131
Westminster clock-tower, lights on, 75
Wheels of revolving machinery at Holyhead, 150
Whistles, best form of bell, 25
— on steamers as fog-signals, 16
— qualities as fog-signals, 23, 62
Whitby light, description of, 118
Wicks, care taken in purchase of, 189
— concentric, separation in burners, 200
— effect of different oils on, 84
— for French mineral-oil lamps, 201
— in six-wick lamp, 151
— tested as to effect of combustion on, 90
— used in Douglass burners, 81
Wigham, Mr., description of his gas-lights, 161–167
— — — his apparatus at Haisborough, 104
— — — burner, used at Westminster clock-tower, 75
— — — gas-gun for fog-signal, 174
— — — offer to erect apparatus in United States, 171

Mr. Wigham, his plan for illuminating beacons with gas, 17
Wind, its effects on sound, 46, 51, 61
Wind-guard at North Stack, 155
Wind-vanes at English stations, 103
Window-frames at English stations, 143
— at the Wolf, 143
— of low light-room, Souter Point, 122
Wolf Rock light-house, plans of, 19
— — — description of, 140

Wood, used for sound-reflectors, 179
Wrecks, English system of buoying, 95

YARMOUTH, buoy-depot at, 101
Young's paraffine used by French government, 197

ZONE of maximum intensity in six-wick burners, 78

www.ingramcontent.com/pod-product-compliance
Lightning Source LLC
Chambersburg PA
CBHW030344230426
43664CB00007BB/528